Experimental Fluid Mechanics

C. Arcoumanis · T. Kamimoto
Editors

Flow and Combustion
in Reciprocating Engines

 Springer

Prof. C. Arcoumanis
City University London
School of Engineering &
Mathematical Sciences
Northampton Square
London EC1V 0HB
United Kingdom
c.arcoumanis@city.ac.uk

Prof. T. Kamimoto
Tokai University
Dept. of Mechanical Engineering
1117 Kitakaname
Hiratsuka-shi, Kanagawa
Japan
TakeKamimoto@aol.com

ISBN: 978-3-540-64142-1   e-ISBN: 978-3-540-68901-0

Library of Congress Control Number: 2008926729

Cover design: WMX Design GmbH, Heidelberg

Printed on acid-free paper

9 8 7 6 5 4 3 2 1

springer.com

Introduction

The internal combustion engine fuelled by mainly gasoline and diesel has been the dominant powerplant for well over 100 years. Although the continuation of its dominance worldwide has been questioned by environmentalists, who see cars threatening the planet's climate through the effect of CO_2 on global warming, and by some engineers who have been overoptimistic about the potential, and the timing of introduction of fuel cells and electric vehicles into mass production, the reciprocating engine is here to stay for the foreseeable future. The key to its success has been its continuous re-invention into more efficient and cleaner modes of combustion, the on-going development and refinement of catalytic converters and the successful partnership between engines, transmissions and electric systems. Another interesting development is the different approach of automotive engineers in the various continents; for example, Europeans have consistently put more faith into diesel engines, with over 50% of new cars in Europe being diesels, while Japan and the US have faithfully supported the relatively less efficient gasoline engines and, more recently, gasoline-hybrid vehicles which represent a reasonable compromise. There are also others who believe that the time has come for a convergence in the global markets through the introduction of the diesel hybrid vehicle but there are uncertainties about its ability to acquire 'mass production' status before fuel cell powered and electric vehicles become affordable by customers worldwide.

This volume attempts to bridge a serious gap in the existing literature between conventional textbooks such as the highly successful one authored by John Heywood in 1987 and the significant technological breakthroughs presented in worldwide conferences during the last ten years on direct-injection gasoline engines, advanced diesels and homogeneous-charge compression-ignition engines. The multi-authored volume consists of eight chapters written by world experts from industry, government laboratories and academia. Each of the chapters is self-contained and, therefore, independent from the others in that it covers its central theme in depth, and width, although prior knowledge of the fundamentals remains a prerequisite. As such it is expected that this volume will become an essential reference text of engineers involved in research and development in global automotive and consultancy companies, research engineers in government laboratories and academic researchers involved in fundamental and applied research on various aspects of the flow, mixture preparation and combustion in reciprocating engines.

Chapter 1 considers the fundamentals of spark-ignition, combustion and emissions in conventional port fuel injection engines which are in terms of numbers the preferred powerplant worldwide. Emphasis is placed on the characteristics of the ignition system, the modeling of premixed turbulent combustion, initiation and development of knock and the pollutant formation mechanisms.

Chapter 2 describes in great detail the research performed over a period of more than ten years on direct-injection, two-stroke gasoline engines which, despite their obvious promise, never reached production in the automotive market. This chapter also provides detailed information about the various optical diagnostic techniques widely used in the research laboratories of all major automotive companies and universities which, together with computational fluid dynamic (CFD) models, have become standard engineering tools in R&D. Although there are no two-stroke automotive engines driving road vehicles today, the research approach described in Chap. 2 has been instrumental in the development of the four-stroke direct injection engines which appeared in the market in the late 1990s.

Chapter 3 presents in its first part an overview of the general features of the first-generation four-stroke DI engines, their advantages as well as their limitations, while the second part focuses on more recent research on the second-generation of stratified DI engines which are entering production equipped with the latest technology of high-pressure gasoline injectors. Details are provided about both multi-hole and pintle injectors operating in the spray-guided concept, and their potential for becoming standard technology is discussed in both naturally-aspirated and supercharged/turbocharged engine configurations.

Chapter 4 describes the flowfield in direct-injection diesel engines operating in the presence of swirl, where particular emphasis is placed on the mean and turbulent flow structure around top-dead-center (TDC) of compression as characterised by both laser-based-experiments and CFD predictions. The implications of squish and swirl, as well as fuel injection on combustion are also described and discussed.

Chapter 5 provides a comprehensive analysis of the penetration of diesel fuels, sprays and jets under quiescent conditions, in order to avoid the implication of swirl and mean flow-structure on spray development, the subsequent flame initiation and lift-off, as well as the link between mixing-controlled fuel-vaporisation, and combustion under diesel-engine thermodynamic conditions.

Chapter 6 complements the previous two chapters by focusing on auto-ignition, combustion and soot and NOx emissions in conventional diesel engines. Emphasis is placed on the factors affecting auto-ignition, its modeling approach, the soot formation and oxidation mechanisms, as well as the practical means for controlling combustion in second-generation direct-injection diesel engines.

Chapter 7 presents an overview of the various types of advanced diesel combustion which, at present, are the subject of intense investigation by researchers in both industry and academia. Although HCCI (homogeneous charge compression ignition) is the most well-known form of advanced combustion, it is widely accepted that premixed, controlled-autoignition, low-temperature combustion represents the most promising concept for eliminating both NOx and soot emissions in the next generation of diesel engines.

Finally, **Chapter 8** summarises in the first part the fuel effects on combustion and emissions for petroleum-based hydrocarbon fuels (gasoline and diesel) while in the second part the emphasis is placed on alternative and renewable fuels, including synthetic fuels, that offer promise to be used in reciprocating engines.

March 2008   *Dinos Arcoumanis*
                                                                                            Take Kamimoto

Contents

Chapter 1
Spark Ignition and Combustion in Four-Stroke Gasoline Engines

Rudolf R. Maly and Rüdiger Herweg

1.1 Introduction

Today, and even more so in the future, significant and simultaneous reductions of emissions and fuel consumption are the key issues in engine combustion. Since the S.I. engine is highly developed already, common trial and error methods alone will no longer be adequate to meet future requirements. Fuel and engine properties form such a complex system of mutually interacting processes that a detailed knowledge of all its properties is required if possible improvements shall be successfully iden-tified, implemented and optimized. Therefore, a close link between practical and theoretical work is mandatory right from the start of conceiving new combustion concepts. New ideas must be complemented with adequate diagnostics to assess benefits or drawbacks and also with models, preferably predictive ones, for guiding and monitoring progress.

This chapter presents an overview on where we are today and identifies what we will need for future S.I. engine combustion requiring knowledge both from practical engine work and from fundamental combustion studies. Since there is a wealth of information available in both areas it will be impossible to cover all details explicitly, and therefore, emphasis is placed on main traits. Citations will be made preferably to review-type publications whenever possible facilitating tracing back to older work and retrieving additional in depth information if required. Recommended sources for review-type literature are: Heywood's text book [1], the journal: Progress in Energy and Combustion Science, the proceedings of the Symposia (Int.) on Combustion and the proceedings of the international symposia on Diagnostics and Modeling of Combustion in internal combustion engines (COMODIA). The paper focuses on the more fundamental aspects since a thorough understanding of the underlying physics

Rudolf R. Maly
Consultant Engines and Future Fuels, formerly Daimler AG, Research Division, 71065 Sindelfingen, Germany, e-mail: rudolfmaly@t-online.de

Rüdiger Herweg
Work carried out at Institut für Physikalische Elektronik, University of Stuttgart, Germany, e-mail: rue_be@t-online.de

C. Arcoumanis, T. Kamimoto (eds.), *Flow and Combustion in Reciprocating Engines*, DOI: 10.1007/978-3-540-68901-0_1, © Springer-Verlag Berlin Heidelberg 2009

1

and chemistry is mandatory to solve the practical problems in front of us. The outline follows the temporal sequence of processes initiated by the ignition system.

1.2 Spark Ignition

Good keys to the older literature are the books by Penner and Mullins [2] and by Lewis and von Elbe [3] as well as the papers by Müller, Rhode and Klink [4] and by Conzelmann [5]. Most of the subsequent material is based on the comprehensive and most complete review of Maly [6].

1.2.1 Electrical Characteristics

Although numerous - apparently different - ignition systems can be found in the literature, both commercial and experimental ones, their principal characteristics are all characterized by the single equivalent circuit presented in Fig. 1.1.

The current voltage characteristics of the resulting spark are determined by the actual selection of the circuit components and the driving high voltage signal which both may vary widely. Their specific selection controls formation and properties of the spark plasma and hence the ignitability of an actual ignition system. Since the Transistorized Coil Ignition (TCI) system has an unsurpassed benefit/cost relation it is still the dominating system wordwide. The electrical energy is stored in the inductance of a coil and released comparatively slowly over about 2 ms. Typical Current–Voltage (*I–V*) and Energy-Power (*E–P*)characteristics are displayed in Figs. 1.2 and 1.3.

We notice a very short (ns) first phase - *the breakdown phase* - during which the spark current rises to a first current maximum of several hundred amperes. Its peak value is determined by the ignition voltage U_0 and the impedance of the near-gap circuit:

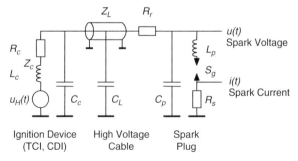

Fig. 1.1 Generalized circuit of ignition systems (non-functional high voltage distribution gap omitted). $u_H(t)$: high voltage signal, L_c: coil inductance, R_c: coil resistance, C_c: coil capacitor, Z_c: coil impedance, Z_L: high voltage cable impedance, C_L: high voltage cable capacity, R_r: radio interference damping resistor, C_p: plug capacitor, L_p: plug inductance, S_G: spark gap, R_s: current shunt (for measuring purposes only), $u(t)$: spark voltage, $i(t)$: spark current

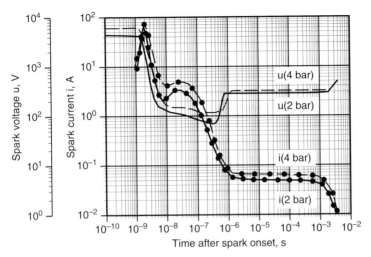

Fig. 1.2 *I–V* characteristics of a commercial TCI in air at 300 K supplying 30 mJ to a 1 mm spark gap. The pressure of 4 bar corresponds roughly to engine conditions at ignition timing. The ignition voltage U_0 is shown at $t < 10^{-9}$ s

$$\hat{I}_B = U_0/Z_p \approx 10\,\text{kV}/50\,\Omega = 200\,\text{A}$$

Within a few ns the gap voltage drops to very low values of around 100 V. This phase is uniquely determined by the capacitive ($C_p = 5 - 15\,\text{pF}$) and inductive components ($L_p + L_G \approx 5\,\text{nH}$) of spark plug and spark, respectively. This phase is followed by a second one - *the arc phase* - lasting only for about 1 μs. During this time the capacity of the high voltage cable ($C_L = 40 - 100\,\text{pF}$) and the coil

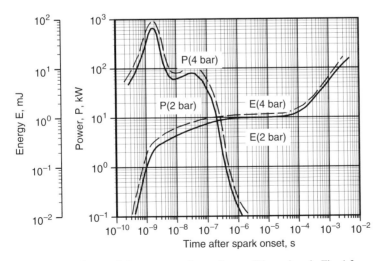

Fig. 1.3 *E–P* characteristics corresponding to the conditions given in Fig. 1.2

capacity ($C_c \approx 50\,\text{pF}$) discharge through the radio interference damping resistor ($R_r = 1 - 10\,\text{k}\Omega$) in series with the cable impedance. Hence typical values for the second current peak are:

$$\hat{I}_B = U_0/(R_r + Z_L) \approx 10\,\text{kV}/2\,\text{k}\Omega = 5\,\text{A}$$

Finally, a third phase follows - *the glow phase* - delivering the lion part of the originally stored electrical energy into the spark discharge. In a TCI this phase lasts for several *(typically 2)* milli-seconds. The peak glow current is mainly controlled by the coil impedance ($Z_c \approx 200\,\text{k}\Omega$) and decays approximately linearly with time (the logarithmic scales are deceiving on a first glance):

$$\hat{I}_G = U_0/(Z_C + Z_L + R_r) \approx U_0/Z_C \approx 10\,\text{kV}/200\,\text{k}\Omega = 5\,\text{mA}$$

If the glow current exceeds about $100\,\text{mA}$ the discharge may transit into an arc. This occurs for low impedance coils or in Capacitive Discharge Ignition systems - CDI - , where the electrical energy is stored in a primary capacitor and the coil is replaced by a pulse transformer. In these cases rapid voltage oscillations may occur due to rapid transitions between arc and glow phases.

The three discharge modes - *Breakdown, Arc and Glow* - are also uniquely characterized by their power and energy release properties. The *Breakdown* offers highest power levels of up to several megawatts at rather small energy levels ($0.3 - 1\,\text{mJ}$ in commercial ignition systems). The *Arc* ranges in between, whereas the *Glow* discharge operates at lowest power (tens of watts) and highest energy levels ($30 - 100\,\text{mJ}$). This is mainly due to its extremely long discharge interval of $1 - 2\,\text{ms}$.

Any ignition system is composed of different combinations of these three basic discharge modes and its ignitability depends therefore on the details of the layout of the electrical system. Knowing the plasma physical properties of the three discharge modes - which will be treated in the following sections - the ignition characteristics of any system can easily be deduced from the knowledge of its *I–V* characteristics.

1.2.2 Spark Plasma Characteristics

1.2.2.1 Pre-Breakdown

The initially perfectly insulating gas within the spark gap region is ionized due to the applied spark voltage ($\approx$linear slope, rise times: TCI : $10\,\text{kV/ms}$, CDI: $\approx 100\,\text{kV/ms}$). The randomly existing starting electrons ($\approx 200\,\text{cm}^{-3}$, corresponding to 1 electron at any given time in the gap volume of a conventional spark plug with 3-mm electrodes and a 0.7-mm gap), produced by hard background radiation from space and earth - are accelerated towards the anode. If the applied electrical field has reached sufficiently high levels ($E \approx 50 - 100\,\text{kV/cm}$) the accelerated elec-

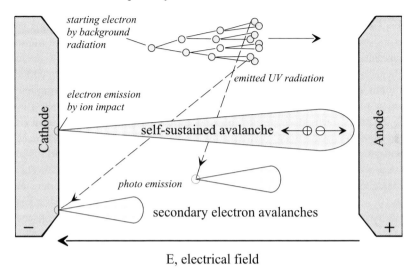

Fig. 1.4 Formation of conductive channels in the spark gap due to multiplication processes in electron avalanches at electrical fields $E \approx 50 - 100\,\mathrm{kV/cm}$

trons will electronically excite and ionize the gas molecules by collisions generating radiation and additional electrons and ions which in turn are accelerated, too. Thus the number of electrons increases rapidly very much like in an avalanche. As indicated in Fig. 1.4 the primary electron avalanche may generate secondary avalanches also in the cathode region due to its emission of UV radiation ($\lambda < 300\,\mathrm{nm}$).

During the pre-breakdown phase the discharge is dependent on the applied external field. The gas temperature remains near its initial value and the average electron density below $n_e \approx 10^{15}\,\mathrm{e/cm^3}$. However, in the electron avalanches (streamers, channels) the electron density may exceed $n_e \approx 10^{18}\,\mathrm{e/cm^3}$. This mixed ionization cloud fills the spark gap in all places where sufficiently high field strengths exist. The field depends thus both on the applied voltage as well as on the electrode shape. For very slowly rising voltages the pre-breakdown phase may last for appreciable time intervals (minutes and longer). The faster the voltage rise, however, the higher an effective over-voltage can be achieved rendering the ionization process much more efficient thus shortening the transition time to the sub-μs region.

1.2.2.2 Breakdown Phase

The pre-breakdown process becomes self-sustained as soon as enough UV radiation is emitted liberating sufficient photo electrons at the cathode surface. Now the much slower ions are created near enough to the cathode surface to liberate new starting electrons by ion impact onto the cathode surface. An over-exponential increase in discharge current results being assisted by a self-created space charge *(self-sustained avalanche, leader)*. In practical cases this transition occurs for pre-breakdown currents in excess of ≈10 mA. Since there is no current limiting mechanism in the discharge itself the spark current rises within nanoseconds to values of hundreds or thousands of Amperes until a further increase is limited by the impedance of the near-gap circuit *(i.e. the spark plug)*. The spark voltage and the electrical field drop accordingly *(breakdown)* to very low values of ≈100 V and ≈1 kV/cm, respectively. The minimum energy required for creating a complete breakdown at 1 bar and a 1-mm gap is 0.3 mJ.

During the breakdown phase the current confines itself to a region which happens to have the highest conductivity - which is the reason for the statistical location of the spark discharge in the gap region and the origin of the statistical fluctuations in the breakdown voltage - leading to extreme current densities in excess of $10^7 A/cm^2$ in this spark channel with an initial diameter of about 40 μm. The electron density rises up to 10^{19} e/cm^3 so that an efficient energy exchange via Coulomb forces takes place. In consequence electrical energy will be transferred very efficiently from the plug capacitor to the electrons and ions inside the spark channel. In consequence extremely high degrees of ionization and electronic excitation result as well as a fast heating of the gas to temperatures of up to 60,000 K. Details are presented in Figs. 1.5 and 1.6.

Due to the high degree of ionization enormous amounts of energy can be stored rapidly in the channel in form of potential energy being slowly much later released in form of thermal energy over tens of micro-seconds. This is illustrated in Figs. 1.6 and 1.7 showing composition and specific heat, resp., of spark plasmas as function

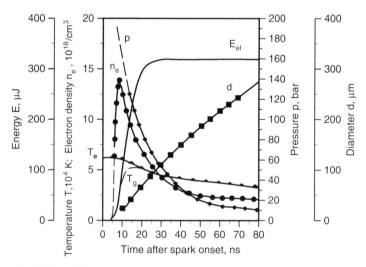

Fig. 1.5 Spark plasma parameters of a breakdown in air at 300 K, 1 bar and a 1-mm gap. Duration of the breakdown current: 10 ns. E_{el}: electrically supplied energy, T_e: electron temperature, T_g: gas temperature, n_e: electron density, p: over-pressure and d: diameter of the spark channel, respectively

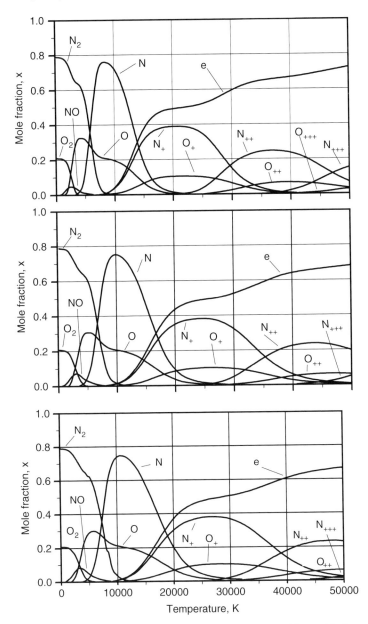

Fig. 1.6 Composition of a spark plasma in air as computed for an ideal gas in thermodynamic equilibrium for 1, 10 and 100 bar (top to bottom) [7]

of temperature and pressure [7, 8] so that the plasma properties are fully characterized in combination with the data given in Figs. 1.10 and 1.12.

The computed plasma data agree well with the measured data presented in Fig. 1.8. where the sum of the dissociation and ionization energies of all plasma

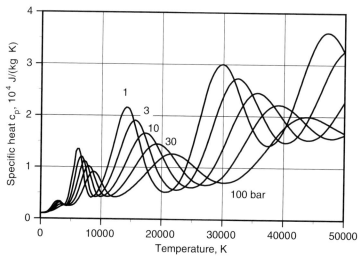

Fig. 1.7 Specific heat of a spark plasma in air as computed for an ideal gas in thermodynamic equilibrium in the range 1 – 100 bar [7]

particles is represented by the term potential energy. The peaks and valleys characterize the stepwise onset of different energy transfer processes.

The high over-pressure (> 200 bar) in the spark channel causes its supersonic expansion and the emission of a spherical shock wave. The shock energy is regained, however, prior to ignition so that a large, toroidal plasma kernel (*diameter at 50 μs : ≈2 mm*) is created with very steep gradients, a structure being most favorable for

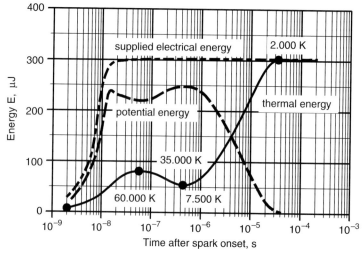

Fig. 1.8 Temporal redistribution of the electrically supplied energy in a breakdown phase from initially potential energy (dissociation, ionization) into thermal energy

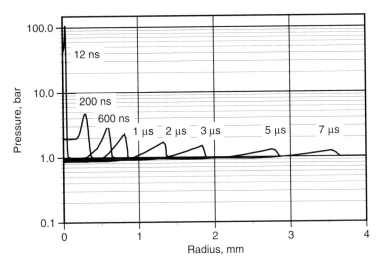

Fig. 1.9 Computed pressure profiles of a spark plasma in air at 1 *bar* and 300 K as initial conditions. Ignition system: breakdown phase: 0.275 mJ with a subsequent glow current decreasing linearly within 1 ms from 100 mA to zero. [7]

ignition (see Figs. 1.9 and 1.12). This ideal structure is caused by the strong shock wave transporting internal plasma mass to the plasma surface and by the rarefaction wave propagating into the plasma center. The model results in Fig. 1.9 corroborate

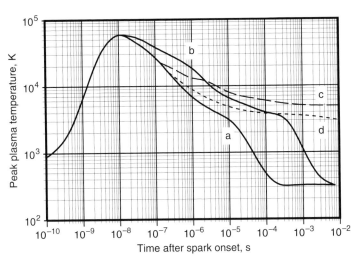

Fig. 1.10 Measured histories of the maximum gas temperature in different discharge modes (ignition systems). (**a**): *CDI, 3 mJ, 100 μs*; (**b**): *Breakdown, 30 mJ, 60 ns*; (**c**): *CDI plus superimposed constant current (2A) Arc, 30 mJ, 230 μs*; (**d**): *CDI plus superimposed constant current (60 mA) Glow discharge, 30 mJ, 770 μs*

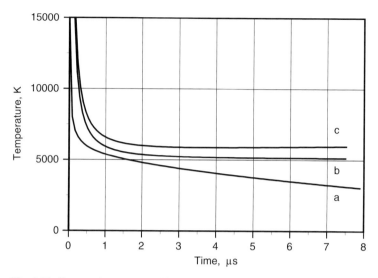

Fig. 1.11 Computed temperature histories of spark plasmas in air with 1 bar and 300 K as initial conditions. Ignition systems: **a:** *0.275* mJ *in the breakdown phase;* **b, c:** *same as* **a** *with a subsequent glow* (**b**) *and arc* (**c**) *current decreasing linearly within 1* ms *from 80* mA *and 300* mA, *resp., to zero.* [7]

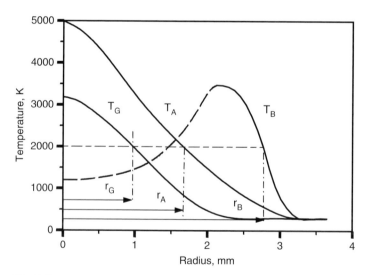

Fig. 1.12 Measured temperature profiles T_B, T_A and T_G for breakdown, arc and glow discharges, resp., in air at 300 K, 1 bar for a 1 mm gap. Due to different power levels the final profiles will be reached at different time intervals. The breakdown profile is shown at ceasing over-pressure in the spark kernel. r_B, r_A, r_G: hypothetical kernel radii for an assumed flame temperature of 2000 K [10]

Table 1.1 Energy balance for Breakdown, Arc and Glow discharge plasmas under idealized conditions using thin electrodes

	Breakdown	Arc	Glow
Radiation losses	<1%	≈5%	<1%
Heat conduction	≈5%	≈45%	≈70%
Total energy losses	≈6%	≈50%	≈70%
Total spark plasmaenergy	≈94%	≈50%	≈30%

experimental data from interferometric studies of plasma and flame kernel development in sparks [6, 9, 10].

Heat losses (heat conduction into the electrodes) and cathode losses (cathode mechanisms) are small due to the rapid channel expansion and the supply of feedback electrons at the cathode by photo and/or field emission. Table 1.1 gives an overview of the energy balance for all three discharge modes.

Prolonged duration of the extreme high current densities in a breakdown discharge (> 10 ns) leads from field emission via thermionic-field to thermionic emission from hot spots formed by over-heating and melting the microscopic surface spikes initially carrying the field emission currents. This indicates the end of the *Breakdown* and the start of the *Arc* phase.

1.2.2.3 Arc Phase

An arc must always be preceded by a breakdown phase to create the electrical conductivity in the gas and the thermionic emission from hot molten cathode spots. In practical ignition systems the threshold is characterized by currents in excess of ≈100 mA. An upper bound is set only by the external circuit impedances. The burning voltage is very low ($U_A ≈ 50 V$ *at 1* bar *in air and a 1-mm gap*) and splits into ≈15 V for the cathode fall, ≈10 V for the arc plasma and ≈25 V for the anode fall. The cathode fall is required to maintain numerous molten hot spots on the cathode surface providing the necessary feedback electrons via thermionic emission (spot diameter: $10 − 40\mu$m, T_{sp}: *up to 3,000* K, i.e. evaporation temperature of the cathode material at the relevant gas pressure).

Due to these hot spots heavy electrode erosion (material evaporation) results which increases with increasing peak ignition voltage since the fraction of energy stored in C_c and C_L rises with the square of the ignition voltage ($E = 1/2\,CV^2$). This effect can be reduced somewhat by chosing electrode materials with high evaporation temperatures, however, in principle it can not be eliminated. The only strategy to minimize erosion is to avoid the arc phase completely [11].

Although spark discharges are instationary processes, the arc (and the glow discharge) attain quasi-stationary values. The electron density and the center temperature are $n_{eA} ≈ 10^{17}$ e/cm^3 and $T_A ≈ 5,000 − 6,000$ K, resp., and vary only slightly with arc current. An arc following a breakdown does not enforce the breakdown plasma. Rather it slows down cooling of the plasma and maintains arc conditions over the duration of the arc discharge. In Fig. 1.10 measured center line temperature dependencies on time are given for all 3 discharge modes. Data computed by a

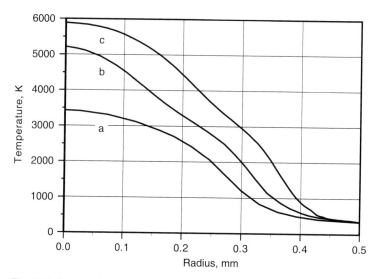

Fig. 1.13 Computed temperature profiles of spark plasmas at 7.5 μs after onset in air at 1 bar and 300. Ignition systems: **a:** *0.275 mJ in the breakdown phase;* **b, c:** *same as a with a subsequent glow* **(b)** *and arc* **(c)** *current decreasing linearly within 1 ms from 80* mA *and 300* mA, *resp., to zero.* [7]

dedicated and detailed spark model (see Figs. 1.11 and 1.13, [7]), are in very good agreement with the experimental data in Fig. 1.10.

Due to the low electron density, the energy exchange between particles is slow and a mainly thermal plasma character results. Energy is transfered from the energy input on the axis to the plasma surface by heat conduction and mass diffusion rather than by over-pressure. Hence Gaussian temperature profiles result (see Figs. 1.12 and 1.13). These processes are comparatively slow ($\approx$ms) and become the more inefficient the larger the plasma radius grows. At the center temperature of 6,000 K the degree of dissociation is very high in spite of the low ionization levels (see Fig. 1.6) providing a rich source of reactive species - albeit limited to the near axial region - for initiating combustion in addition to the temperature itself. The temporal evolution of the plasma volumes (spark kernels) initiated by the different spark modes are shown in Fig. 1.14.

1.2.2.4 Glow Phase

The high pressure *Glow* discharge is very similar in nature to the *Arc* discharge except for lower center temperatures and lower dissociation levels as well as a cold cathode. Hence the same arguments apply for the energy transport processes as were given for the arc discharge. In commercial plugs the peak glow current is limited to about 100 mA. If the electric circuit forces higher currents a transition into an arc mode occurs (see Arc Phase). The field strength in the gap is low ($E_G \approx 10^3$ V/cm *in air at 1* bar) and the anode fall has the same value as in an arc. For a gap of 1 mm this sums up to 500 V for the total glow voltage across a spark gap. The

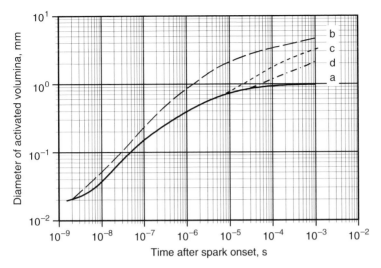

Fig. 1.14 Measured diameters of plasma volumes initiated by different discharge modes. **a:** *Breakdown, 0.3* mJ, *10* ns; **b:** *Breakdown, 30* mJ, *60* ns; **c:** *CDI plus superimposed constant current (2A) Arc, 30* mJ, *230* μs; **d:** *CDI plus superimposed constant current* (60 mA) *Glow discharge, 30* mJ, *770* μs

field strength E_G increases with pressure ($\approx$500V/cm bar *in air at 300* K) so that the ratio of energy in the plasma to the energy lost to the cathode improves with rising gas pressures. The quasi steady state values for the electron density and the kernel temperature on the axis of a high pressure glow discharge are $2 \cdot 10^{14}$cm^{-3} and *3,000* K, respectively. Characteristic glow data have been shown already in Figs. 1.10–1.15.

Its feedback electrons are obtained by the very inefficient electron emission by ion impact onto the cold cathode. Rather high cathode falls have to be maintained therefore to provide enough feedback electrons. Since the cathode fall region is very thin ($<$ 0.1mm) and is located at the cathode surface, almost all its energy input is lost to the cathode due to ion impact, heat conduction and energy exchange. The cathode fall is given by:

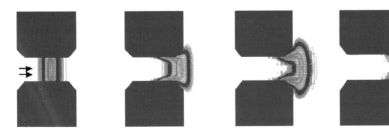

Fig. 1.15 Computed evolution of a glow plasma in a cross flow of 20 m/s in air at 1 bar and 300 K. Ignition system: 0.275 mJ in the breakdown phase with a subsequent glow current decreasing linearly within 1 ms from 100 mA to zero [7]

Table 1.2 Erosion rates of brass electrodes by Breakdown, Arc and Glow discharges measured in air at 300 K and at 1 bar pressure [6]

Device	Suppl. Energy E_{el}/mJ	Spark Gap l/mm	Erosion: 10^6 sparks mass/g	Erosion: 10^6 sparks mass/g mJ
TCI: Glow, 3.4 ms, 500 V	33	0.5	$0.89\ 10^{-4}$	$2.7\ 10^{-6}$
	37	1.0	$1.29\ 10^{-4}$	$3.5\ 10^{-6}$
	41	2.0	$1.43\ 10^{-4}$	$3.5\ 10^{-6}$
CDI: weak Arc, 100 μs	1.6	0.5	$0.25\ 10^{-4}$	$15.6\ 10^{-6}$
	1.76	1.0	$0.32\ 10^{-4}$	$18.2\ 10^{-6}$
	1.17	2.0	$0.29\ 10^{-4}$	$24.8\ 10^{-6}$
CDI: Arc, 100 μs, 200 V	278	0.5	$467\ 10^{-4}$	$168\ 10^{-6}$
	225	1.0	$442\ 10^{-4}$	$196\ 10^{-6}$
	236	2.0	$483\ 10^{-4}$	$204\ 10^{-6}$
Breakdown, 5 ns, 20 kV	40	0.5	$6.4\ 10^{-4}$	$16\ 10^{-6}$
	40	1.0	$4.8\ 10^{-4}$	$12\ 10^{-6}$
	40	4.0	$1.3\ 10^{-4}$	$3.3\ 10^{-6}$

$$U_c = 3\, U_i \ln(1 + 1/\Gamma)$$

which amounts to as much as $U_c = 400V$ for typical values of the ionization voltage of gases ($U_i = 14.534V$ for N_2) and the cathode efficiency $\Gamma \approx 10^{-4}$. The cathode fall depends therefore sensibly on the nature of the gaseous environment and even more so on surface layers on the cathode surface (Γ *of oxides, soot, hydrocarbons, etc. instead of the pure cathode material!*). Erosion is due to sputtering and erosion rates are low (see Table 1.2).

Because a glow discharge may last for up to several milliseconds, sensible energy fractions from the gas can be conducted into the cathode especially under quiescent conditions. Hence the glow discharge has the worst energy efficiency of all discharge modes. But the long discharge duration provides also benefits since the glow plasma may be transported away from the electrodes by cross-flows forming a hairpin shaped low loss high surface structure (see Figs. 1.15 and 1.21). If the voltage along the hairpin becomes larger than the restriking voltage directly across the electrodes − 2–3 kV *due to the already existing ionization at the electrodes* - a new short cut channel is formed and the process is repeated as long as sufficient electric energy is available. This combination of cross-flow and glow discharge can be utilized very favorably to improve the ignition probability since it generates effective gap spacings being up to 20-fold higher than the actual geometrical spacing.

1.2.2.5 Energy Transfer Efficiencies

Due to its specific plasma properties each discharge mode can transmit only fractions of the supplied electrical energy into the gas in the spark gap volume ($\approx 5mm^3$ *in a standard plug*). Since its duration is extremely short the breakdown is not dependent on the gas velocity and the quiescent data in Fig. 1.16 apply in all cases.

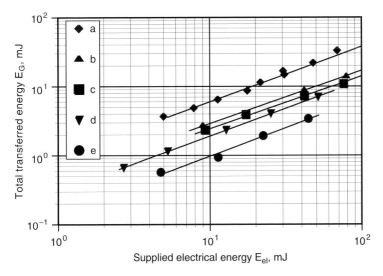

Fig. 1.16 Total energy transferred by the three discharge modes under quiescent conditions. For discharge intervals $< 10\mu s$ no influences by discharge duration or current level are observed. **a:** *Breakdown, gap 2* mm; **b:** *Arc, cathode diameter 0.2* mm; **c:** *Arc, c.d. 3* mm, **d:** *Glow, c.d. 0.2* mm; **e:** *Glow, c.d. 3* mm [11]

However, as discussed above the data for arc and glow show a strong effect on heat losses to the electrodes under quiescent conditions. Since arc and glow discharges are carried away by the gas flow in the gap region, the contact time with electrodes is reduced and hence the associated heat losses. This transport of the spark plasma by the flow improves significantly the ratio of column voltage to the sum of cathode and anode falls. Thus the transferable energy fraction to the gas increases markedly, albeit in a less concentrated form: the longer the discharge channel the smaller the effective plasma radius. This may be partially compensated by a better transport efficiency of heat and dissociated species from the axis of the spark to its surface.

In Fig. 1.17 the energy transfer data under cross flow conditions are presented approaching maximum values of 50% and 30% at 10–15 m/s for the arc and the glow, respectively. It has to be kept in mind, however, that the total transmitted energy is by no means a direct measure for the energy being finally available at the plasma surface. This is determined additionally by the temperature profile and the geometry of the plasma providing different ignition radii and reactive surfaces (see Section 1.2.3).

1.2.3 Flame Kernel Formation

Spectroscopically, chemical reactions can be observed already a few ns after spark onset i.e. already during the always preceding breakdown phase. However, since the plasma temperatures are still far too high to permit the existence of stable

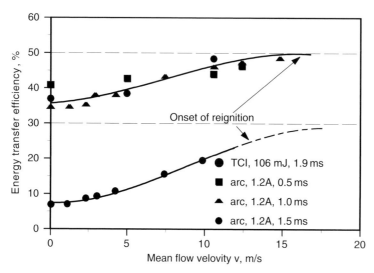

Fig. 1.17 Energy transfer efficiencies for arc and glow discharges under cross flow conditions in air at 300 K, 1 bar and a gap of 1 mm

combustion products - the plasma is dominated by dissociated molecules and excited or ionized atoms - these reactions occur at the plasma surface. The relatively small chemical energy of the fuel molecules inside the plasma kernel - compared to plasma energies per particle - is liberated instantaneously by the discharge and simply adds to its total energy.

Somewhere at the surface, however, there will always exist situations which provide most favorable conditions for initiating and supporting self-sustained flame propagation: sufficiently high temperatures, steep temperature gradients providing mass and heat transport and influxes of reactive species and radicals (N, O, H, C, etc.) which may carry many times their thermal energy in form of internal energy (at 8,000 K *a heavy plasma particle carries 0.7* eV *of thermal energy versus 5 eV of dissociation energy!*). Due to the high nitrogen content in combustible mixtures with air, early reactions will be dominated by the very energetic N radical forming species such as NO, NH and CN. Being unstable at lower temperatures these enter the conventional reaction scheme of hydrocarbons later and transfer their energy to other species as soon as the spark channel has cooled down to normal flame temperatures.

Hence reaction kinetics and combustion are started right at the very onset of the spark. The question is not whether the reactions are initiated by a spark but will the started reactions become self-sustained when the energetic support by the spark plasma ceases! This transition occurs at some $10 \mu s$ after spark onset when species such as OH, CH, C_2, CO etc. appear and indicate combustion processes as in stationary flames. As shown in Fig. 1.18 the plasma expansion velocity has dropped at $\approx 20 \mu s$ to values lower than the normal flame speed and the existence of self-sustained flames can be identified by the continued growth of the - initially:

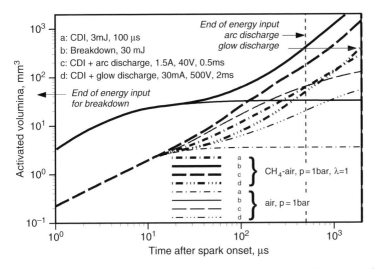

Fig. 1.18 Temporal evolution of plasma and flame kernels for the discharge modes Breakdown, Arc and Glow in a stoichiometric CH4-air mixture initially at 300 K, 1 bar and a 1-mm gap. The weak CDI is used to start the arc and glow discharges

spark kernel - now: flame kernel. The slopes in Fig. 1.18 indicate the same flame speeds for $t >> 100\mu s$ irrespective of the initiating discharge mode. Differences exist, however, in the size of the reacted volumes: initially larger spark kernels produce also larger flame kernels and higher amounts of burnt mass at any given time.

To measure flame kernel evolution consistently with adequate temporal and spatial resolution, simultaneous high speed Schlieren filming along two orthogonal axes has been proven to be the method of choice. The subsequent contributions are taken from the extensive and consistent work of Herweg [12, 13] covering both experimental and modeling aspects. A specially designed pancake type side chamber with a strong swirl was used to provide arbitrary values for mean flow U and turbulence levels u' simply by selecting appropriate spark plug locations. Propane-air mixtures were chosen to avoid additional effects as incomplete evaporation of fuel droplets or incomplete mixing. For further details see e.g. [13]. The electrical characteristics of the commercial TCI used in Herweg's experiments are given in Fig. 1.19. The linear display clearly shows re-striking due to flow effects: multiple jumps in the voltage trace. These jumps are caused by the lower restricting spark voltage of a new spark shortcutting the preceding hairpin structure. The voltage rise associated with the stretching of the spark channel due to a crossflow can conveniently be used to measure the momentary flow velocity at the spark plug location: the rate of increase in the spark voltage is directly proportional to the flow velocity [14].

Figure 1.20 provides a close-up view on flame kernel evolution in an operating engine. We notice a smooth spark dominated expansion up to to $200\mu s$ when the mean flow begins to push the spark kernel towards the ground electrode. Flame

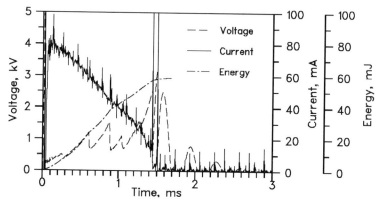

Fig. 1.19 Electrical parameters of the TCI at 1,000 rpm as used in the propane-air operated S.I. engine being referred to in the subsequent figures. Mean cross flow velocity in the spark gap: 12.5 m/s [13]

wrinkling starts around 330 μs and by 1 ms the discharge has been moved out of the gap. A more detailed insight into the important interactions between a spark plasma and a cross-flow is presented in Fig. 1.21 showing the wide range of transport and flame wrinkling effects by cross-flows.

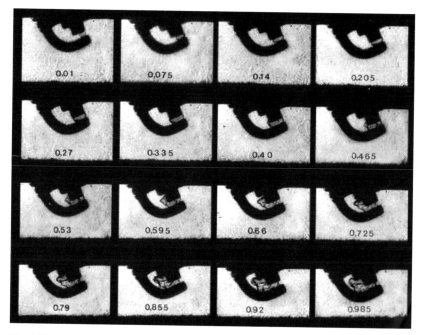

Fig. 1.20 High speed Schlieren film of ignition in a propane-air operated S.I. engine, $\phi = 0.77$, TCI, ignition timing 30° bTDC, pancake chamber, 1,500 rpm. Numbers indicate time in ms after spark onset. 1 ms corresponds to 9° crank angle [13]

Fig. 1.21 High speed
Schlieren film of ignition in a
Propane-air operated S.I.
engine under different flow
and turbulence conditions,
$\phi = 1$, TCI, 0.5 ms after
spark onset, pancake
chamber. The numbers
indicate the mean flow
velocity at the spark gap. The
corresponding turbulence
intensities are: 1,56; 0.78 and
0.52 m/s, respectively

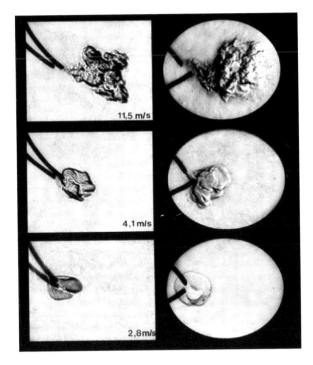

Figure 1.22 exemplifies that flame kernels in leaner mixtures exhibit a higher
degree of wrinkledness than in stoichiometric ones since the burn rates are lower
than the wrinkling rates due to turbulence. Therefore wrinkles are formed faster
than they can burn out. One notices also that the oldest part of the spark kernel *(head
of the cornucopia-like hairpin structure)* is reacting fastest because it contains the
accumulated energy of the breakdown, arc and early glow discharge phases. The
"legs" of the hairpin are generated by the late glow discharge alone at low energy
and power levels producing low temperatures and levels of dissociation *(i.e. less
reactive conditions)*.

Nevertheless ignition - especially of lean mixtures - benefits from the continued
late energy supply by a TCI whereas the short CDI spark is unable to create a self-
sustained flame kernel at high flow velocities (see Fig. 1.23).

1.2.4 Flame Kernel Modeling

As shown in the preceding sections, modeling of flame kernel formation requires
a simultaneous treatment of detailed spark physics, chemical kinetics and fluid
dynamic interactions. Since detailed modeling of the spark physics is still under
development, most published models are based on chemical kinetics - often using
only a one step reaction - and the first law of thermodynamics. Fluid dynamic
aspects are included more and more (e.g. [16, 17]) although still in an inadequate

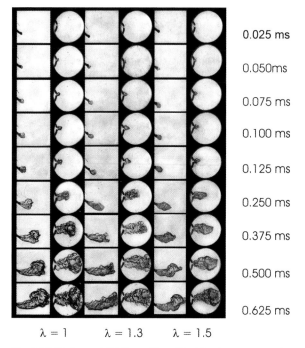

0.025 ms

0.050ms

0.075 ms

0.100 ms

0.125 ms

0.250 ms

0.375 ms

0.500 ms

0.625 ms

$\lambda = 1$ $\lambda = 1.3$ $\lambda = 1.5$

Fig. 1.22 High speed Schlieren film of ignition in a propane-air operated S.I. engine, $\lambda = 1/\phi$, TCI, ignition location near the cylinder walls, ignition timing $10°$ bTDC, pancake chamber, 1,000 rpm, part load. Time is in ms after spark onset. 1 ms corresponds to $6°$ crank angle

way. Turbulence *(i.e. u'-effects)* are sometimes cared for but rarely is the convection of the kernel by cross-flows accounted for nor the straining of the developing flame front. Also the importance of length and time scale effects is often neglected although it is obvious that the wrinkling of a small spark kernel can be accomplished only by scales being smaller than the kernel radius and - equally important - in time scales commensurate to flame kernel speeds. More and more progress is made, however.

Here we follow the approach of Herweg [15] which includes all important effects into a consistent model for flame kernel formation. The paper also gives an extensive literature review and many useful examples of possible treatments of the plasma physics and the fluid dynamics. The model is based on the thermodynamic system shown schematically in Fig. 1.24. For simplicity reasons a 1D formulation is used which, however, may easily be extended into a 3D version. Here only the general outline will be presented.

Fixing the coordinate system to the center of gravity of the flame kernel, the enthalpy change of the system is given by the change in internal energy U and the volume work of expansion:

$$\frac{dH_K}{dt} = \frac{d}{dt}(m_K h_K) = m_K \frac{dh_K}{dt} + h_K \frac{dm_K}{dt} = \frac{dU}{dt} + p \frac{dV_K}{dt} + V_K \frac{dp}{dt} \quad (1.1)$$

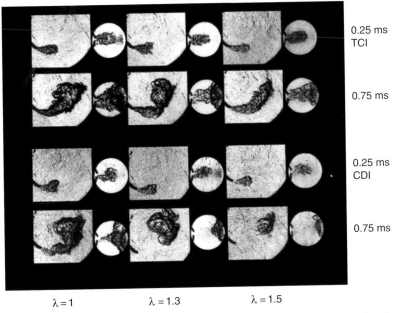

0.25 ms
TCI

0.75 ms

0.25 ms
CDI

0.75 ms

$\lambda = 1$ $\lambda = 1.3$ $\lambda = 1.5$

Fig. 1.23 High speed Schlieren film of ignition in a propane-air operated S.I. engine showing the effect of a TCI (60 mJ, 1.5 ms) and CDI (6 mJ, 0.1 ms) ignition system, $\lambda = 1/\phi$, ignition location near the cylinder walls, ignition timing 10° bTDC, pan cake chamber, 1,250 rpm, part load. Time is in ms after spark onset. 1 ms corresponds to 7.5° crank angle

The change in internal energy is given by:

$$\frac{dU}{dt} = \frac{dE_Z}{dt} + \frac{dQ_{Ch}}{dt} - \frac{dQ_{WV}}{dt} - p\,\frac{dV}{dt}, \qquad (1.2)$$

with E_{pl} the effective energy transferred into the plasma, Q_{ch} the heat of reaction and Q_{wv} the sum of all heat losses to electrodes, walls, etc. Since the flame speed develops on the moving plasma surface of the expanding plasma kernel the effective mass burn rate will be:

$$\frac{dm_K}{dt} = \frac{d(\rho_K V_K)}{dt} = \rho_K \frac{dV_K}{dt} + V_K \frac{d\rho_K}{dt} = \rho_u\, A_K\,(S_t + S_{Plasma}), \qquad (1.3)$$

and the effective expansion rate of the flame kernel:

$$\frac{dr_K}{dt} = \frac{1}{A_K}\frac{dV_K}{dt} = \frac{\rho_u}{\rho_K}\,(S_t + S_{Plasma}) + \frac{V_K}{A_K}\left[\frac{1}{T_K}\frac{dT_K}{dt} - \frac{1}{p}\frac{dp}{dt}\right] \qquad (1.4)$$

Rearranging and reordering results in 3 differential equations to be solved simultaneously:

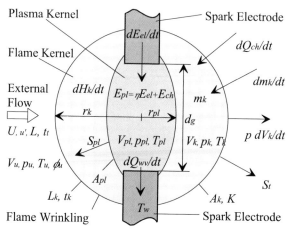

Fig. 1.24 Schematic of the thermodynamic system used for modeling ignition by a TCI under engine conditions. Unburned mixture properties: V_u, p_u, T_u, ρ_u; flow properties: U, u'; integral length and integral time scale: L, t_t; integral flame kernel length and time scale: L_k, t_k; flame kernel properties: V_k, p_k, T_k, m_k; Plasma properties: V_{pl}, p_{pl}, T_{pl}; Energy terms: E_{el}, E_{pl}, E_{ch}; energy/mass fluxes: dH_k/dt, dE_{el}/dt, dQ_{ch}/dt, dQ_{wv}/dt, pdV_k/dt, dm_k/dt; spark gap, flame kernel surface, strain rate, plasma speed, turbulent burn rate and wall temperature: d_g, A_k, K, S_{pl}, S_t, T_w

$$\frac{dh_K}{dt} = \frac{[h_b - h_K]}{\rho_K V_K} \rho_u A_K \, (S_t + S_{Plasma})$$
$$+ \frac{1}{\rho_K V_K} \left[\frac{dE_Z}{dt} - \frac{dQ_{WV}}{dt} \right] + \frac{1}{\rho_K} \frac{dp}{dt} \tag{1.5}$$

$$\frac{dV_K}{dt} = \frac{\rho_u}{\rho_K} A_K \, (S_t + S_{Plasma}) - V_K \frac{1}{\rho_K} \frac{d\rho_K}{dt}$$
$$= \frac{\rho_u}{\rho_K} A_K \, (S_t + S_{Plasma}) + V_K \left[\frac{1}{T_K} \frac{dT_K}{dt} - \frac{1}{p} \frac{dp}{dt} \right] \tag{1.6}$$

$$\frac{dr_K}{dt} = \frac{1}{A_K} \frac{dV_K}{dt} = \frac{\rho_u}{\rho_K} (S_t + S_{Plasma}) + \frac{V_K}{A_K} \left[\frac{1}{T_K} \frac{dT_K}{dt} - \frac{1}{p} \frac{dp}{dt} \right] \tag{1.7}$$

To provide the required data for the energy terms and the propagation speeds the following approaches were chosen. The time dependent increase in plasma energy is the sum of all energy inputs by breakdown, arc and glow phases as well as the chemical energy of the mixture mass contained inside the plasma kernel:

$$E_{Pl} = E_B + \eta_A \, E_A + \eta_G \, E_G + E_{ch}$$

The coefficients η_a, η_G account for the low energy transfer efficiency of the arc and glow and - for the sake of simplicity - also for the fraction of the total energy actually available in the plasma surface sheath. Since the TCI is predominantly a glow discharge its time dependent Gaussean temperature profile is calculated by the heat conduction equation in cylindrical form with the rate of plasma energy supply as an input:

$$\frac{\partial T_{Pl}}{\partial t} = a \, \Delta T + \frac{p}{\rho \, c_p} = a \, \Delta T + \frac{P}{\rho \, c_p \, V_{Pl}}, \tag{1.8}$$

$$\frac{\partial T_{Pl}}{\partial t} = \bar{a} \left(\frac{\partial^2 T_{Pl}}{\partial r^2} + \frac{n}{r} \frac{\partial T_{Pl}}{\partial r} \right) + \frac{\eta_{B.G} \, U(t) \, I(t)}{\bar{\rho} \, \bar{c}_p \, \pi \, r_E^2 \, d_{el}}, \tag{1.9}$$

The plasma speed S_{pl} is derived by differentiating the resulting $r(t)$ at the point of inflexion. The heat losses to the electrodes - later also the additional losses to the combustion chamber walls - are calculated from:

$$Q_{WV} = h \, F_{Wall} \, [T_K - T_{Wall}], \tag{1.10}$$

Following Bray [18] the strained turbulent burning velocity is:

$$S_t = I_0 \, S_L + 2 \, [D_T \, W]^{1/2}, \tag{1.11}$$

The strain I_0 is calculated for a Lewis number of 1 (typical for practical flames) based on the principles given by Law [19] to:

$$I_0 = 1 - \left(\frac{\delta_l}{15 \, L} \right)^{1/2} \left(\frac{u'}{S_l} \right)^{3/2} - 2 \, \frac{\delta_l}{S_l} \frac{1}{r_K} \frac{dr_K}{dt} \tag{1.12}$$

with the laminar burning velocity data for propane taken from Gülder [20] with the coefficients $S_{l0} = 48$ cms; $\alpha = 1.77$; $\beta = -0.2$; $f \approx 1$:

$$S_l = S_{l0} \left[\frac{T_u}{T_0} \right]^{\alpha} \left[\frac{p}{p_0} \right]^{\beta} (1 - f F), \tag{1.13}$$

The reaction rate W is calculated using a modified BML model [18] to account for moderate $(1 < u'/S_l < 10)$ and low $(0 < u'/S_l < 1)$ turbulence conditions prevailing under engine conditions since the standard formulations $(u'/S_l \longrightarrow \infty)$ for u'/S_l proved to be inappropriate. The reduced length and time scales at the kernel surface are computed from turbulent diffusion predicting as reasonable exponential correlation coefficients:

$$L_K = L \left(1 - exp\left(-\frac{r_K}{L}\right)\right) \tag{1.14}$$

$$t_K = t_t \left[1 - exp\left(-\frac{t}{t_t}\right)\right] = t_t \left[1 - exp\left(-\frac{\left(\bar{U}^2 + u'^2\right)^{1/2} + S_l}{L} t\right)\right] \tag{1.15}$$

Pulling the expressions together one obtains for the strained, size dependent turbulent flame speed of a developing flame kernel:
(with flame holder effect at the spark electrodes)

$$\frac{S_t}{S_l} = I_0 + I_0^{1/2} \left(\frac{\left[\bar{U}^2 + u'^2\right]^{1/2}}{\left[\bar{U}^2 + u'^2\right]^{1/2} + S_l}\right)^{1/2}$$

$$\left(1 - exp\left(-\frac{r}{L}\right)\right)^{1/2} \left[1 - exp\left(-\frac{t}{t_t}\right)\right]^{1/2} \left(\frac{u'}{S_l}\right)^{5/6} \tag{1.16}$$

(without flame holder effect at the spark electrodes)

$$\frac{S_t}{S_l} = I_0 + I_0^{1/2} \left(\frac{u'}{u' + S_l}\right)^{1/2} \left(1 - exp\left(-\frac{r}{L}\right)\right)^{1/2}$$

$$\left[1 - exp\left(-\frac{t}{t_t}\right)\right]^{1/2} \left(\frac{u'}{S_l}\right)^{5/6}. \tag{1.17}$$

where the terms from left to right account for strain, effective turbulence, integral kernel length scale, integral kernel time scale and for fully developed flame propagation.

Modeling results are presented in Figs. 1.25–1.32 and compared to experimental results. For all operating conditions: low and high loads, low and high engine speeds (and flow fields), long and short duration ignition systems (TCI, CDI), low and high ignition energies and stoichiometric or very lean mixtures an excellent agreement between model prediction and engine results has been found. This indicates that the presented flame kernel model captures virtually all relevant conditions occurring in S.I. engines. From spark onset to about $200\,\mu s$ the flame speed drops from very high values (plasma supported reactions) to a minimum. This minimum is the result of the opposing effects of the decaying support by the spark plasma and of the increasing influences of strain, flow / turbulence and laminar burning velocity. The influences of the flow field increase with time due to the flame speed dependence on the effective integral length and time scales controlling size and life time of the flame kernel. The data corroborate also the importance of including U to account for the hairpin structure of glow discharges in flows. For radii < 1 mm the flame

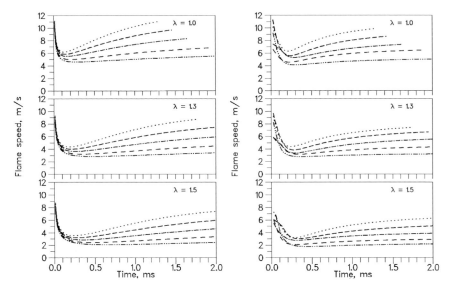

Fig. 1.25 **Central** ignition. Comparison of modeled (*left*) and measured (*right*) flame speeds **versus time** of **TCI** initiated flame kernels in a propane-air operated S.I. engine showing the effect of engine speed and equivalence ratio ($\lambda = 1/\phi$). High swirl pan cake chamber, ignition timing: $10°$ bTDC, engine speeds (top to bottom): 1,250, 1,000, 750, 500, 300 rpm

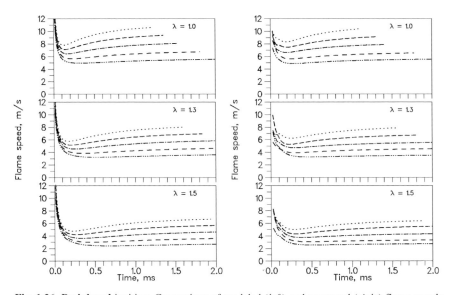

Fig. 1.26 **Peripheral** ignition. Comparison of modeled (*left*) and measured (*right*) flame speeds **versus time** of **TCI** initiated flame kernels in a propane-air operated S.I. engine showing the effect of engine speed and equivalence ratio ($\lambda = 1/\phi$). High swirl pan cake chamber, ignition timing: $10°$ bTDC, engine speeds (top to bottom): 1,250, 1,000, 750, 500, 300 rpm

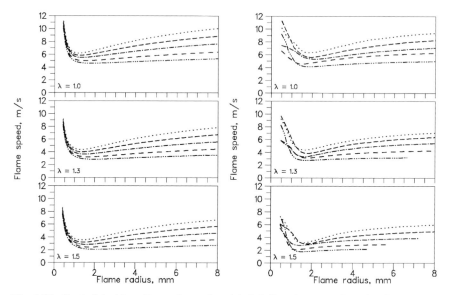

Fig. 1.27 Central ignition. Comparison of modeled (*left*) and measured (*right*) flame speeds **versus kernel radius** of **TCI** initiated flame kernels in a propane-air operated S.I. engine showing the effect of engine speed and equivalence ratio ($\lambda = 1/\phi$). High swirl pan cake chamber, ignition timing: 10° bTDC, engine speeds (top to bottom): 1,250, 1,000, 750, 500, 300 rpm

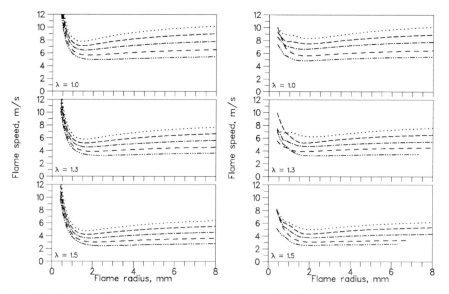

Fig. 1.28 Peripheral ignition. Comparison of modeled (*left*) and measured (*right*) flame speeds **versus kernel radius** of **TCI** initiated flame kernels in a propane-air operated S.I. engine showing the effect of engine speed and equivalence ratio ($\lambda = 1/\phi$). High swirl pan cake chamber, ignition timing: 10° bTDC, engine speeds (top to bottom): 1,250, 1,000, 750, 500, 300 rpm

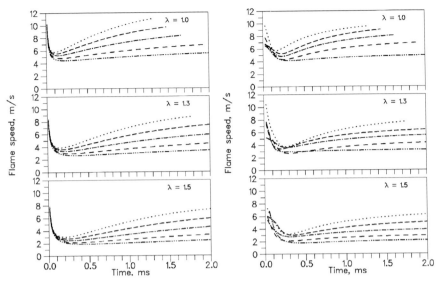

Fig. 1.29 Central ignition. Comparison of modeled (*left*) and measured (*right*) flame speeds **versus time** of **CDI** initiated flame kernels in a propane-air operated S.I. engine showing the effect of engine speed and equivalence ratio ($\lambda = 1/\phi$). High swirl pan cake chamber, ignition timing: $10°$ bTDC, engine speeds (top to bottom): 1,250, 1,000, 750, 500, 300 rpm

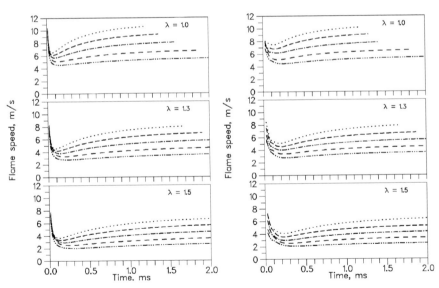

Fig. 1.30 Peripheral ignition. Comparison of modeled (*left*) and measured (*right*) flame speeds **versus time** of **CDI** initiated flame kernels in a propane-air operated S.I. engine showing the effect of engine speed and equivalence ratio ($\lambda = 1/\phi$). High swirl pan cake chamber, ignition timing: $10°$ bTDC, engine speeds (top to bottom): 1,250, 1,000, 750, 500, 300 rpm

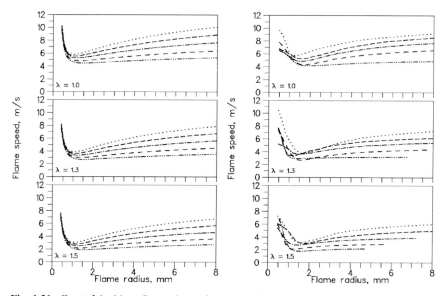

Fig. 1.31 Central ignition. Comparison of modeled (*left*) and measured (*right*) flame speeds **versus kernel radius** of **CDI** initiated flame kernels in a propane-air operated S.I. engine showing the effect of engine speed and equivalence ratio ($\lambda = 1/\phi$). High swirl pan cake chamber, ignition timing: 10° bTDC, engine speeds (top to bottom): 1,250, 1,000, 750, 500, 300 rpm

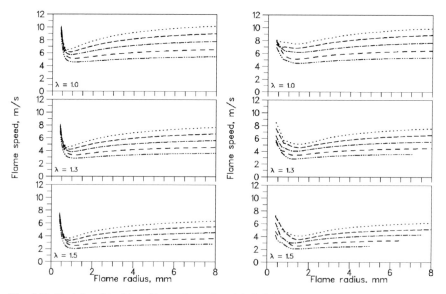

Fig. 1.32 Peripheral ignition. Comparison of modeled (*left*) and measured (*right*) flame speeds **versus kernel radius** of **CDI** initiated flame kernels in a propane-air operated S.I. engine showing the effect of engine speed and equivalence ratio ($\lambda = 1/\phi$). High swirl pan cake chamber, ignition timing: 10° bTDC, engine speeds (top to bottom): 1,250, 1,000, 750, 500, 300 rpm

kernel growth is dominated by the properties of the ignition system and its power input. The flame speed minimum occurs at $r \approx 2$ mm and is determined by a balance between S_l and the strain rate since turbulence effects are still small. For 2 mm < r < 10 mm the flame kernel acquires the properties of a fully developed flame. It is important to note that the local conditions at and around the spark gap determine flame kernel growth not the global ones!

1.2.5 Outlook for Spark Ignition

Increasingly more stringent exhaust emission regulations, the efficient implementation of which must be guaranteed by on-board-diagnostics and the need to improve the fuel efficiency still further impose more severe requirements on ignition systems for conventional well premixed S.I. engines than in the past. There will be a general demand for a higher performance at lower cost, less weight, less space requirements and preferably additional functions. Energy consumption should be minimal at 100% reliability and ignition probability without a need for maintenance or plug replacement. The outlook to meet this requirements is good as pointed out in a recent review by Maly [21]. Ignition systems for GDI engines will have to cope with incomplete mixing, presence of fuel droplets in the spark gap region and plug fouling due to soot formation. There is also a demand for higher ignition energies to meet lean mixture conditions better.

These contradicting demands may be met with new smart ignition systems exploiting still unused potentials in spark physics, combining at least: 1. specially designed long life spark plugs, 2. adaptive multiple sparking and 3. ion current sensing via the spark plug. Tendencies for such developments can be observed already in the open literature.

The life time of a spark plug may be increased by a: elimination of any arc phases, b: reduction of the peak currents and the discharge duration, c: diverting the discharge current to parts of the electrodes where erosion is permissible. The insert in Fig. 1.33 shows the principle of a long life plug. Small anchor spots (1) made of materials with high work function and low evaporation rates (e.g. Pt) sit on electrodes made of materials with inverse properties (e.g. CrNi) and determine ignition voltage and ignition location.

This combination of materials forces the initial breakdown to occur at the anchor spots and the subsequent discharge modes to move immediately over to the base electrode designed to have enough material to sustain continued erosion over the desired life span of the plug. Conductive deposits on the insulator may be cleaned by occasional surface sparks over the auxiliary gap (4). The benefits of such a design are clearly demonstrated in Fig. 1.33.

As shown schematically in Fig. 1.34 a conventional TCI always delivers the maximum ignition energy although the engine most often does need much less. If the total energy is split into a number of energy parcels, each delivering the

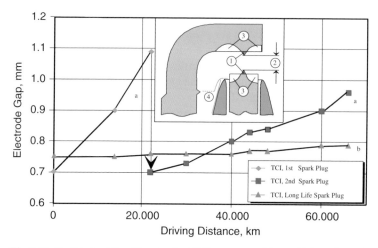

Fig. 1.33 Useful plug life with a standard TCI as measured in a vehicle under real road conditions. **a:** standard spark plug, **b:** specially designed long life spark plug. Insert shows schematically a long life plug design. 1: anchor spots, 2: spark gap, 3: planned erosion areas, 4: auxiliary surface spark gap for insulator cleaning [21]

minimum energy required for ignition (e.g. 5 mJ) an ignition concept with minimum weight and volume may be conceived combining minimum erosion with 100% ignition probability. After delivering the first energy parcel a suitable combustion sensor performs an inflammation check. If inflammation was unsuccessful another parcel will be delivered and so on until successful ignition and self-sustained flame

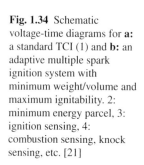

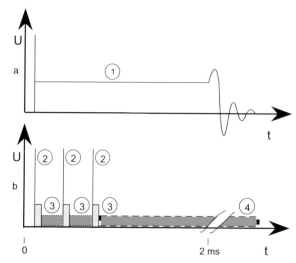

Fig. 1.34 Schematic voltage-time diagrams for **a:** a standard TCI (1) and **b:** an adaptive multiple spark ignition system with minimum weight/volume and maximum ignitability. 2: minimum energy parcel, 3: ignition sensing, 4: combustion sensing, knock sensing, etc. [21]

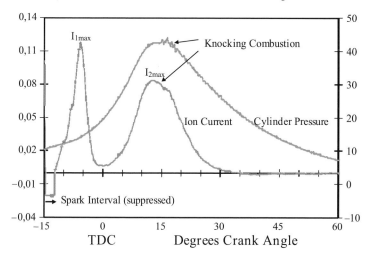

Fig. 1.35 Cylinder pressure and ion current signal as measured via the spark plug in a knocking production type S.I. engine. Ignition timing 15° bTDC. Ion current: around I_{1max} indicating flame properties, around I_{2max} indicating heat release properties [21]

propagation are confirmed. Thus it is guaranteed that each cylinder is individually ignited according to its specific demands under all operating conditions.

The spark plug itself may be used as a combustion sensor if ion current sensing is applied. Ion sensing can easily be integrated into a suitable ignition system providing detailed information on flame propagation and the thermodynamics of combustion in each cylinder [21–23]. As shown in Fig. 1.35 for a knocking engine, there is a first ion current peak characterizing ignition, inflammation and flame propagation, and a second peak containing details of thermodynamics and combustion including knock. The ion signals may be easily and consistently evaluated by modern electronic control units. Thus future ignition systems may provide multi-sensing capabilities aside from markedly improved ignition performance at no extra costs. In the contrary, sophisticated engine control becomes feasible requiring no extra sensors apart from an appropriate ignition system. It is believed therefore that the ignition requirements of future S.I. engines regarding reliability, performance and servicing can be met by new systems exploiting still unused potentials in spark physics. This will also lead to lower production and operational costs simultaneously with improvements in product quality and customer satisfaction.

1.3 Combustion in S.I. Engines

The homogeneous charge gasoline engine presents - at least in principle - the fewest challenges to understand its combustion process. It has been extensively studied and much progress has been made in recent years in its theoretical analysis, diagnostics and its modeling. Therefore, a wealth of information may be found in the literature. Here we will follow with preference the review of Maly [24] - providing also a detailed compilation of relevant literature - and the papers of Weller et al. [25] and Heel et al. [26] which cover the field consistently. Since flow and mixing of fuel and air are treated in Chapter 3 of this book, Section 1.3 will address the current status in S.I. combustion research as a continuation of the flame kernel phase being treated in Section 1.2.3. Although the charge is not truly perfectly mixed, it is safe to treat combustion in current S.I. engines as a homogeneous, premixed turbulent flame. Its properties are different, however, from burner flames due to the high pressure conditions in an engine and the instationarity of the whole combustion process.

1.3.1 Diagnostics

Since turbulent engine flames propagate under high pressure (O: 1–5 MPa) special diagnostic tools have been developed to access key features for the analysis of their properties. The results are used to optimize combustion schemes in transparent engines as well as to derive and validate better models. Aside from conventional visualization by still photos or high resolution movies using schlieren- or shadow-graphy [27] and standard test bench tools, dedicated non-intrusive laser-based imaging techniques are now available for measuring flame structure, temperatures, concentrations, radicals, flow fields and turbulence.

Cycle resolved flame *structures* are most easily obtained by Mie scattering from seeding particles being added to the intake air and a laser sheet technique with a copper vapor laser [28, 104]. Single images from individual cycles may be recorded by LIF of fluorescent HC molecules which happen to be present in the fuel [29], or more reliably, by LIF of oxygen-insensitive dopants e.g. ketones, added to the fuel [30–32]. Rayleigh measurements are also possible, but require, difficult to satisfy dust- and particle-free conditions [33]. Flame thickness and integral scales of wrinkling are easily retrieved by image processing techniques [34, 35] (see Fig. 1.36). Fractal analysis has been tried but produce excessive scatter in the data, especially in single shot applications, and no physical insight. Even in ensemble averaged measurements neither the postulated inner and outer cutoff scales nor the slope can be extracted unambiguously.

Direct *temperature* imaging in engines is still not possible. A 2-line OH-thermometry was proposed [36] but since this method is linked to the presence of OH it also will not be generally applicable. Indirect means as Rayleigh scattering can be useful under special conditions [33]. It is the only technique available, however, to measure - indirectly via density changes - 2D temperature fields in engines [37–39]. Single shot CARS is applicable for temperature and concentration measurements outside flame fronts but the lacking resolution below 1,000 K and the fact that it is a point measurement limits its use for engine applications severely.

Concentration of gases and fuel are easily imaged now by LIF techniques. If the laser is tuned to specific wavelengths, molecules as O_2 or *radicals* as OH [33], [44, 40], NO [37, 39], CH or intermediates as HCOH [41–43] can be readily measured qualitatively, in some cases also quantitatively. At high pressures as in engines ($p > 10$ bar) the NO cannot be excited through the flame front with wave-

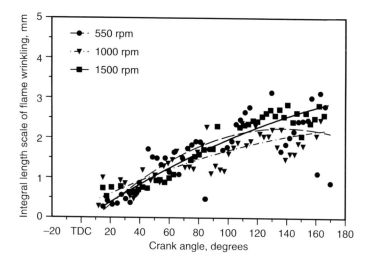

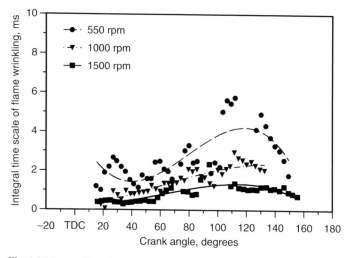

Fig. 1.36 Integral length and time scales of flame wrinkling measured in a fired optical S.I. engine. CR 10:1; propane-air; $\phi = 0.66$; quarter load [34, 35]

lengths below 210 nm because of strong absorption by reaction intermediates. Using a wavelength of 248 nm provides, however, good results under well mixed S.I. engine conditions. [38, 49].

Due to radiation quenching, absolute LIF data are difficult to obtain, but since LIF measurements are often used for relative model verification or for the analysis of relative spatial distributions, this is no general drawback. Used in combination with additional diagnostics or with models, LIF constitutes a very powerful tool for practical and theoretical work. In Figs. 1.37 and 1.38 examples are shown of

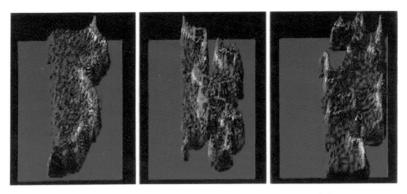

Fig. 1.37 LIF images of the OH distribution in a laser sheet through - left to right - a propagating ($\phi = 1$), a lean ($\phi = 0.77$) and a partially quenched flame ($\phi = 0.67$). 500 rpm, propane-air, quarter load. Quenching occurs last inside inlets [40, 44]

measured OH as well as NO and simultaneous temperature distributions in engine combustion.

Flow fields in fired engines are readily measured now by 2D PIV techniques either single shot [45, 46], cycle resolved using copper vapor lasers by a movie PIV [47] or even 3D [48]. The cross correlation technique with image shifting is

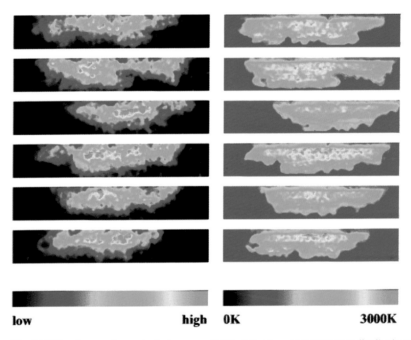

low   high   0K   3000K

Fig. 1.38 Simultaneously single shot measured NO - *left side* - and temperature distributions - *right side* - in an optical S.I. engine operated on propane-air mixtures. Quarter load, 1,000 rpm, $\phi = 1$, ignition timing 20° bTDC, measuring timing 8° bTDC [38]

to be preferred for its significantly higher accuracy and resolution. Thus detailed information on flow structure, time dependence and scales can be obtained under realistic conditions even in single shots. These 2D techniques favorably support the well established point measuring techniques LDV and PDA.

Although some diagnostic tools are still qualitative in nature all necessary instrumentation is now available for detailed combustion imaging and analysis especially if combined with suitable modeling efforts.

1.3.2 Reaction Kinetics

Knowledge in this field and computing power have progressed to a point where even large detailed kinetics may be used for engine combustion [50–52] so that reduced mechanisms with questionable accuracy can be avoided. However, although computing times are still unacceptably long for practical applications, fundamental studies may be carried out providing a deeper insight into pollutant formation processes. Detailed reaction mechanisms may be automatically generated for large molecules so that realistic fuels are becoming tractable [53]. There are many mechanisms available in the literature being updated regularly. Good results for NO and soot calculations have been obtained with the recent mechanisms of Bowman et al. [54] and Mauss [55].

To shorten computing times the detailed chemistry can be stored in libraries or tailored to demand by automatic reduction [53]. A promising approach appears to be the method of Intrinsically Low-Dimensional Manifolds [56, 57], however, little information is available on how these approaches perform in practical cases. Available evidence indicates a very limited applicability to engine conditions, difficulties in treating wide ϕ-ranges and very high computing costs. Since detailed chemistry is predominantly needed for computing formation and depletion of pollutants, simple or reduced chemistry may readily be used for heat release calculations. An overview on reduced mechanisms for hydrocarbons can be found in the book of Peters et al. [58]. An assessment of the applicability of the current reaction kinetics mechanisms to practical cases is a follows [59]:

- **fuels:** very well: paraffins, straight, branched and NOx chemistry, good: olefins, alcohols, fair: oxigenated hydrocarbons, ethers, epoxides, bad: aromatics,
- **processes:** well theoretically: laminar flames, ignition in shock tubes, flow reactors, static reactors, well practically: detonation, still difficult: pulsed combustion, turbulent flames, furnaces, burners, combustion in engines.

Aside from the need for fast but accurate kinetic schemes there are more problems to be solved for gasoline fuel kinetics: extension of the kinetics beyond C8 molecules, inclusion of olefins and aromatics and methods to handle multicomponent mixtures - containing at least some typical components. On the application side, time efficient and accurate models are needed for treating NOx-formation and -reburn, post flame and exhaust pipe UHC oxidation as well as kinetics of additives. Currently available mechanisms ending with Heptane/Octane can be used with acceptable results, however, until better mechanisms become available.

1.3.3 Turbulent Flame Propagation

In Fig. 1.20 it had been shown that a flame kernel develops a wrinkled surface structure which becomes the more wrinkled the larger the burned region is growing. The continuation of this process is presented in Fig. 1.39 showing now a fully developed propagating engine flame.

In Fig. 1.39 a special technique has been used to force a 2D representation of flame propagation: the clearance height of the combustion chamber was reduced to below 1 mm. The results obtained thereby support the commonly accepted picture of an engine flame as being a very thin discontinuity separating the burned from the unburned region. The width of this transition zone, i.e. the flame thickness δ_l of the flame front is thinner that the smallest turbulence eddies $\delta_l < l_K$ where $l_K =$ Kolmogorov scale, so it will not be influenced by turbulence directly. Therefore, the rate of converting unburned mass into burned mass can be described by the laminar burning velocity S_l which represents the combined effects of all laminar reaction processes occurring in the flame front (e.g. chemical reactions, transport processes, etc.). Since normally the characteristic time scale of laminar burning $t_l = l/S_l \approx 40\,\mu s$ is much smaller than the characteristic turbulence time $t_t = L/u' \approx 1\text{ms}$, with t_t rotation time for an eddy of the size of the integral length scale L, chemistry and turbulence effects may be decoupled. Chemistry is treated as being so fast that propagation of the turbulent flame front is governed by the properties of the flow field. Due to the temperature increase in the burned region the observable laminar flame speed is:

Fig. 1.39 Structure of a turbulent flame front measured in a pancake type S.I. engine operated on 80 RON gasoline. CR = 10:1, 1,000 rpm, full load, $\phi = 1$, ignition timing 14° bTDC. *Left frame*: 4.8° aTDC, *right frame*: 5.9° aTDC [42]

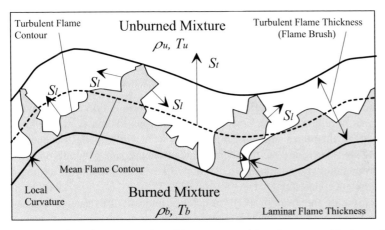

Fig. 1.40 Schematic representation of the structure and the relevant quantities in a turbulent engine flame

$$v_l = \frac{T_b}{T_u} = \frac{\rho_u}{\rho_b} S_l$$

Simple and accepted correlation functions are available for the *laminar burning velocity S_l* [see e.g. 80] which work well and are extensively used in simplified engine codes. A compilation of unstrained S_l data of many important hydrocarbons has been provided by Law [60]. In Fig. 1.40 a schematic structure of the turbulent engine flame front is presented. Relevant engine data have been compiled into Table 1.3.

The interactions of the turbulent flow field with this laminar flame front are:

- **wrinkling** (folding by the full or reduced range of eddy sizes),
- **straining** (lateral stretching of the flame front),
- **effective eddy sizes** (turbulent scales smaller than the dimension of the flame - e.g. for flame kernels and for flames approaching a wall - will wrinkle the front,

Table 1.3 Typical values (Orders of Magnitude) of quantities characterizing engine flames [62]

Property		Dimension
Turbulence Intensity	u'	2 m/s
Turbulent Reynolds Number	Re_l	300
Damköhler Number	Da	20
Karlovitz Stretch Factor	K	0.2
Integral Length Scale	L	2 mm
Taylor Micro Length Scale	l_T	0.7 mm
Kolmogorov Scale	l_K	0.03 mm
Laminar Flame Thickness	δ_l	0.02 mm
Laminar Flame Speed	S_l	0.5 m/s
Ratio u'/S_l		4
Ratio S_t/S_l		4
Mean Flame Radius of Curvature		2 mm
Length Scale of Flame Wrinkling		2 mm

whereas in the inverse case the larger scales will only convect the front without modifying its structure),
- **curvature** (increased / decreased losses at convex / concave sections),
- **rate of pressure change** dp/dt and
- **rate of wrinkling** (response time in instationary turbulence) [61].

The incorporation of all these effects into a full 3D formulation of the Navier Stokes equations in addition to the conservation equations for mass, momentum, energy and chemical species and its direct solution is still outside the current computer capacities. Therefore the turbulent flame propagation is generally viewed as a process propagating proportionally to the flame speed with simplifying assumptions about turbulence interactions.

In the simplest case the mean property turbulence intensity u' is used to characterize the turbulence effects on flame propagation which is quite convenient since it can be accessed by available measuring techniques so that correlation functions may be established readily. For thermodynamic purposes and heat release calculations, correlations of the type:

$$\frac{S_t}{S_l} = 1 + \left(\frac{u'}{S_l}\right)^n \quad n \approx 5/6...1 \text{ for unstrained cases and}$$

$$\frac{S_t}{S_l} = I_0 \left(1 + A\frac{u'}{S_l}\right)^n \quad I_0 = \text{strainrate}, \ A \approx 1...2.5 \text{ with straining.}$$

are widely used and handle practical combustion problems quite satisfactorily. This is especially true if parameter fitting and/or measured in-cylinder pressure traces are used in addition. Because of its simplicity it has been used in modeling flame kernel formation (Section 1.2.4), incorporating the first three effects listed above which are most important. More complex approaches will be treated in Section 1.3.4. In Fig. 1.41 an overview of published turbulent burning velocities is presented showing a very wide spread which is attributed to experimental problems. The flame kernel model predictions from Section 1.2.4 are included. Since these data have been extensively validated they are considered to be quite reliable. More recent results have been published by Bradley et al. [63].

1.3.4 Combustion Modeling

The main effort over the past decades was devoted to the development of multidimensional predictive modeling of flow, heat transfer and combustion in reciprocating engines. Basically, codes were implemented solving the conservation equations governing the physical processes with suitable numerical procedures to provide acceptable accuracy under engine conditions. On the numerical side good progress has been made so that more recently the main focus has shifted towards improving the modeling of the physics and chemistry. Although the turbulence modeling is still not satisfactory and it's improvement is a major topic in its own right, the

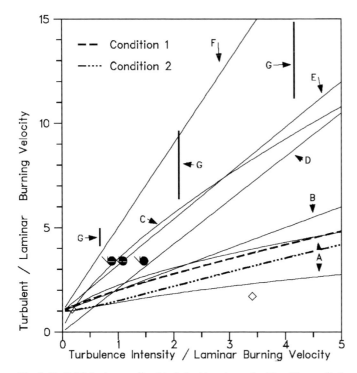

Fig. 1.41 Published normalized turbulent burning velocities. The predictions of the flame kernel model are included as broken curves: **Condition 1:** $U >> S_l$ and $I_0 = 0$; **Condition 2:** $U = S - l$ and $I_0 = 0$. **A:** Clavin and Williams [64], **B:** Tabaczynski et al. [65], **C:** Klimov [66], **D:** Pope and Annand [67], **E:** Witze at al. [68], **F:** Mattavi et al. [69], **G:** zur Loye and Bracco [30], **H:** Witze and Mendes-Lopez [70], **I:** Ho and Santavicca [71]

limiting factors in available codes are the combustion models. This applies both to the conventional ensemble averaged approaches as well as to the emerging LES and DNS activities. Numerous models are still being proposed awaiting validation under realistic engine conditions. Therefore, no extensive discussion of all the different models is in place but rather a more general look into the main issues.

It can be safely said that engine flows can be simulated now quite reliably in 3D ensemble averaged formulations using commercial well validated codes (e.g. KIVA, STAR-CD, FIRE, etc.). Realistic complex geometries with moving pistons and valves may be used. The number of grid points required for an acceptable accuracy is in the range of < 1 million which still can be handled although computing times are measured in days rather than hours. It is expected, however, that the rapid increases in computer performance will reduce these still unattractively huge running time expenses in the near future.

Mixing in S.I. engines can be handled quite satisfactorily for gaseous components so that modeling of premixed combustion is in a good shape as is shown in Figs. 1.42 and 1.43. Spray injection and spray combustion for the emerging GDIs, however, are still in a developmental stage with urgent need for better models for spray formation,

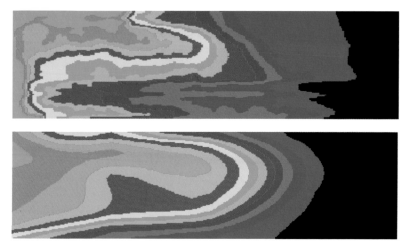

Fig. 1.42 Measured (top, LIF, acetone tracer in propane/air) and calculated (STAR-CD) fuel distribution in the square piston engine at 45° aTDC in the intake stroke [72, 118]

mixing of sprays and spray combustion. The main issues in modeling S.I. engine combustion are: combustion models, NO formation / depletion, formation / depletion of unburned hydrocarbons (UHCs) and knocking combustion. Topics which have seen major changes recently will be addressed below in more detail.

1.3.4.1 Premixed Combustion Models

The existing models fall in three categories: Eddy Break Up (EBU) [74], Thin Wrinkled Flame (TWF) and Flame Area Evolution (FAE) [118]. In the EBU class the local burning rate is assumed to be controlled by small scale turbulent

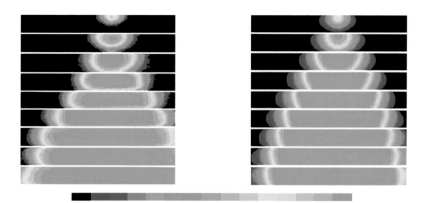

Fig. 1.43 Measured (left, LIF, 50 cycles averaged) and calculated (STAR-CD) flame propagation in the square piston engine seen through the cylinder wall. Ignition timing: 14° bTDC, Frames from the top: 11, 8, 5, 2° bTDC, 1,4,7,10,13° aTDC. black: unburned, red: burned [73, 118]

mixing (mixed is burned) with a characteristic time scale related to the turbulence dissipation scale. This assumption is incorrect and leads to concave flame shapes in engines, as shown in Fig. 1.44, in contrast to the correct convex flame front presented in Fig. 1.43.

For correct near wall behavior the characteristic burn time should not be equated to the turbulence $t_t = k/\epsilon$ time but derived from spectral descriptions of turbulence [75, 76] which account for the change in the spectrum of scales during flame kernel formation and on approaching a wall.

TWF models [77–79] account properly for the normally thin (of the order of the Kolmogorov scale) and wrinkled flame structure (see Fig. 1.45), the wrinkling being responsible for the increased burning rate over a smooth flame burning at laminar burning velocity. This laminar flamelet assumption has the advantage of allowing detailed chemistry and flame straining effects to be included economically. Unfortunately these models assume that the characteristic scale of wrinkling is proportional to characteristic turbulence scales leading to the same problems as in EBU modeling.

The more recently developed FAE models [80, 81] share some elements of the TWF models as the thin flame assumption and the laminar flamelet burning. The local wrinkled flame area is determined, however, by an evolution equation allowing incorporation of turbulence effects as well as effects generated by the flame itself and non-local effects. This type of modeling provides very reasonable results in simulating combustion in S.I. engines as shown in a recent paper [26] where combustion modeling was based on the Weller flame wrinkling model [25, 76]. It

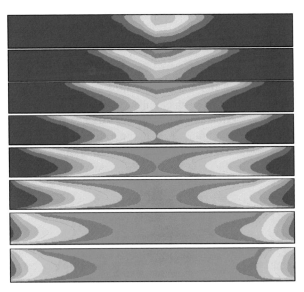

Fig. 1.44 Flame evolution in the same S.I. engine and under the same conditions as in Fig. 1.43 as predicted by the standard EBU model, using $t_c = t_t$ (characteristic combustion time (turbulence time) [25]

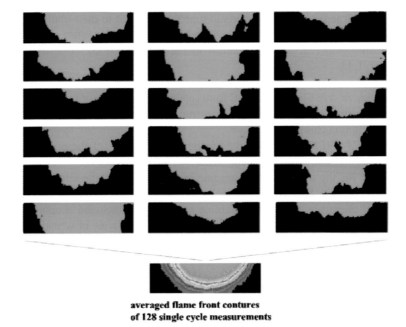

**averaged flame front contures
of 128 single cycle measurements**

Fig. 1.45 Single cycle flame contours in a S.I. engine viewed through the cylinder head. *Red*:
burned, *Black*: unburned. *Lower Frame*: Measured ensemble average over 50 cycle [38]

provides a transport equation for the spatial and temporal evolution of the wrinkling
factor Ξ. The wrinkling factor is defined by:

$$\Xi = \frac{A_t}{A} = \frac{S_t}{S_t^0}$$

where A_t and A are the true wrinkled and the smooth projected flame areas,
respectively. This ratio is set equal to the ratio of turbulent to laminar unstrained
burning velocity. In the most general 2-equation Weller model a transport equation
is provided for the spatial and temporal evolution of Ξ. In the results shown subsequently, a simplified 1-equation version was used assuming equilibrium between
the production and destruction of Ξ. In this case the local fuel consumption rate is
related to the wrinkling factor by:

$$\dot{\Omega} = \rho_u S_l \Xi \mid \nabla \tilde{b} \mid$$

where $\tilde{b}$ is the fuel regress variable (varying from 0 in the unburned to ρ_u in the
burned gases) and ρ_u is the unburned gas density. The equilibrium Ξ distribution is
obtained from the flame speed relation (4b) in the flame kernel model of Herweg in
Section 1.2.4. and the laminar flame speed correlation of Gülder [20].

More recent data on S_l can be found in a paper by Bradley et al. [63]. I_0 accounts
for the strain and the exponential terms for the turbulence length and time scale

Fig. 1.46 Cylinder pressure for base case and various equivalence ratios. *Symbols*: measurements, *Lines*: simulation. Lambda = $1/\phi$. M: measured, C: calculated [26, 118]

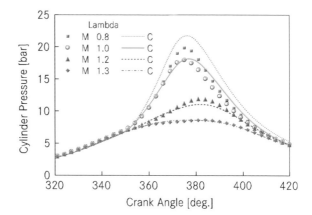

effects when the flame kernel (radius r_K) and life time are small relative to the integral turbulent scales L and t_t. For fully developed flames the exponential terms tend to 1 and the flame speed ratio becomes:

$$S_t/S_l = I_0[1 + A[u'/S_l)^n]\qquad\qquad A \approx 1...2.5, n = 5/6...1$$

Sample calculation results from Heel et al. [26] are presented in Figs. 1.46 – 1.48. showing generally a good accuracy when compared to detailed experimental results from an optically accessible S.I. engine. Mixing and flame propagation data for the same engine have been shown already in Figs. 1.42 and 1.43, respectively. In Fig. 1.49 a vertical view into the engine is shown to complement the data given in Fig. 1.43.

It should be noted that there are alternative wholly theoretically based ways of modeling these phenomena such as the spectral approach [81] which have also been successfully used with the 1-equation Weller model [25].

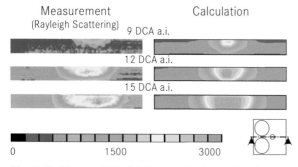

Fig. 1.47 Measured (Rayleigh scattering images) and computed density fields in the vertical bisector plane for base conditions [26, 118]. Insert: plane of interest between the two valves and through the spark plug seen through the cylinderhead of the square piston engine

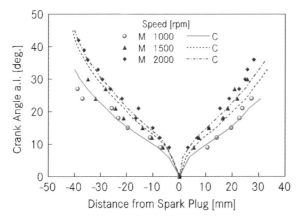

Fig. 1.48 Measured (symbols) and computed flame front evolution along the vertical bisector plane at various engine speeds. M: measured, C: calculated [26]

1.3.5 Pollutants

Apart from the CO_2 emission, the most critical pollutants in gasoline engines are NOx and Unburned Hydrocarbons (UHCs).

Although the engine-out emissions can be drastically reduced by the standard Three Way Catalyst (TWC) in-cylinder measures to reduce the fuel consumption usually cause higher engine-out NOx levels which must be reduced by a suitable control of the combustion process. The UHCs need a separate treatment since they originate predominantly in the cold start / warm-up phase [82, 83] where engine and catalyst are still cold, mixing is inefficient and aftertreatment is still inactive as shown in Fig. 1.50.

1.3.5.1 Nitric Oxides (NOx)

In NO modeling for engines it is now widely accepted that both thermal (extended Zeldovich) and prompt NO must be accounted for, although pure thermal modeling does provide reasonable results as shown in Fig. 1.51.

The shape of the NO curve is typical for engines operated at low speeds and at $\phi = 1$. The NO mass fraction starts to rise around TDC and reaches a maximum at the cylinder pressure maximum which corresponds to the highest temperatures in the engine. Subsequently the NO decreases with decreasing temperature until the temperature has fallen below $\approx 1,800$ K where the NO reduction freezes. The frozen level of NO $\approx 2,500$ ppm corresponds reasonably well with the exhaust data of $\approx 3,000$ ppm.

To include pollutant contributions from the flame front and from the burned gases it is attractive to include combined flamelet and PDF sub-models into multi-dimensional CFD codes; however, no practical results have been published so far. Known results indicate:

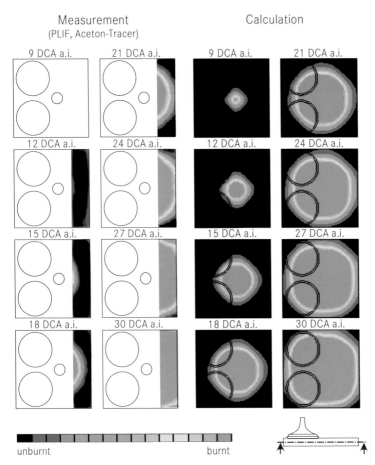

Fig. 1.49 Comparison of ensemble-averaged progress variable distribution in the horizontal bisector plane (seen through the cylinder head of the square piston engine) [27]. *Left*: LIF measurements, *right*: simulation [26]

1. prompt mechanisms play a significant role at the lower temperature end, i.e. below ≈1,800 K,
2. fuel chemistry affects NO production, albeit by unknown processes,
3. deposits may lead to higher NO levels due to higher charge temperatures,
4. highly turbulence-enhanced combustion produces comparable NO levels as conventional turbulent combustion.

To compute prompt NO requires very detailed kinetic schemes. Good results have been obtained when using the kinetics of Bowman et al. [54] even for computations combining thermal and prompt NO formation. In Fig. 1.52 an example is shown of formation and reduction of NO in the wake of a burning heptane droplet under simulated engine conditions. Since S.I. engines are not fully premixed (see

Fig. 1.50 UHC emissions
during the FTP 75 test.
Origin and levels of
maximum allowable
emissions for TLEV, LEV
and ULEV standards [82]

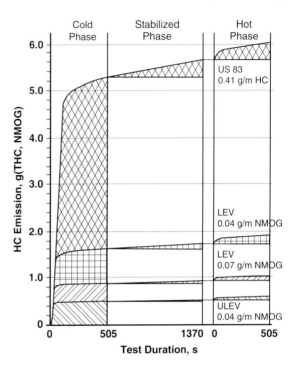

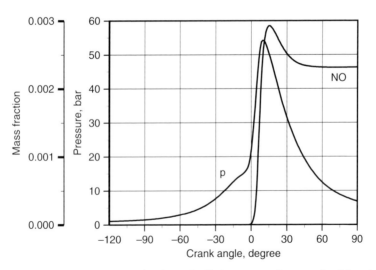

Fig. 1.51 Computed NO mass fraction and cylinder pressure of an operating S.I. engine. 2,000 rpm,
full load, $\phi = 1$, ignition timing: $13°$ bTDC [84, 85]

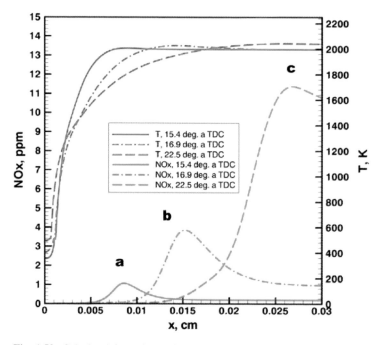

Fig. 1.52 Calculated formation and reburning of NO in the wake of a burning 25μm heptane droplet. Cross flow from left to right, droplet at 0:0, simulated engine conditions at the end of injection, 1% O_2. NO is formed in the downstream periphery whereas the NO formed earlier in a cycle (**a, b**) is reburned in the rich region near the droplet (**b, c**)

Chapter 1) and may have rich pockets, the incorporation of the prompt mechanisms appears to be mandatory for predictive NO modeling.

1.3.5.2 Unburned Hydrocarbons (UHC)

Direct numerical simulation [86, 87] has opened up new insights into bulk and wall quenching of engine flames. The results show that heat losses, curvature, viscous dissipation and transient dynamics have significant effects especially for small scale turbulence. Simplified models have been derived for wall quenching [88] facilitating modeling of UHC from boundary layers in engine codes. Oxidation of UHCs early in the power stroke can be handled by conventional kinetics, whereas late oxidation (crevice out-flows, quench layers, burn-out in the exhaust pipe) occurs at such low temperatures and in a non-premixed oxygen deficient environment where kinetics are still unreliable or not available yet at all.

Results from engine tests [90–95] indicate for:

- *engine-out emissions:*

  1. During warm-up over the first 30 s about 75% of the emitted UHCs is unreacted fuel dropping to a steady state value of 53% after $\approx$60 s.

2. the main components during warm-up are unburned fuel species and reaction products consisting of C2-C4 olefins, methane and acetylene,
3. the specific reactivity right after cold start is due to unburned fuel,
4. the percentage of C2-C4 olefins with high reactivity, and the specific reactivity increase as warm-up proceeds although the total UHC decrease drastically.

- *catalyst-out emissions:*

  1. before light-off the C2-C4 olefins, after light-off the unburned fuel species dominate the specific reactivity,
  2. hydrocarbons with boiling points lower than the momentary catalyst temperature are hardly present, hydrocarbons with higher boiling points are almost totally absorbed during catalyst warm-up,
  3. after catalyst light-off, hydrocarbons with a higher reactivity are oxidized more readily. The ranking of the conversion efficiency is: C2-C4 olefins > alkylbenzenes >> paraffins and benzenes,
  4. the equivalence ratio has a significant influence on specific reactivity and mass of toxic air pollutants.

The current understanding of the mechanisms leading to UHC emissions from SI engines has been compiled into a complete flow chart presented in Fig. 1.53 clarifying the origins and subsequent processes being responsible for UHC's from a warmed-up SI engine. Although most of the mechanisms are not rigidly known yet, the chart is an extremely useful guide for the needs both in practical and research work. It is noteworthy that about 9% of the fuel escapes the normal combustion process and causes a loss of 6% in IMEP! Although these figures may vary somewhat in different engines and at other operating conditions they are generally supported by other engine work such as, for example:

- in-cylinder measurements of crevice flow and wall layers which are corroborated by LIF images from an optical engine (see Figs. 1.54 and 1.55),
- fuel absorption-desorption in oil films (see Figs. 1.56 and 1.57),
- effects by piston temperature [94] and exhaust valve leakage [95].

There are also encouraging attempts to model *UHC oxidation* [97] and effects of *engine design parameters* on UHC emission [98]. If calibrated for a single operating condition, reasonably good results are obtained for other speeds, loads, air-fuel ratios, EGR rates and spark timings. The model predicts that:

- bore to stroke ratio has a small effect,
- an decrease in displacement per cylinder increases UHC,
- increase in compression ratio or in crevice volume per displacement increases UHC. These results agree well with the proposed mechanisms in Fig. 1.50.

The post-oxidation of UHCs is a very complex process being not well understood. There is a real need for more theoretical and experimental work in these areas

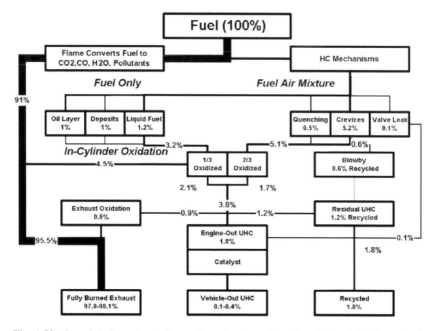

Fig. 1.53 Complete flow chart of normal combustion of gasoline fuel and UHC mechanisms. Numbers indicate the UHC emission index in % fuel entering each engine cylinder, redrawn after [90]

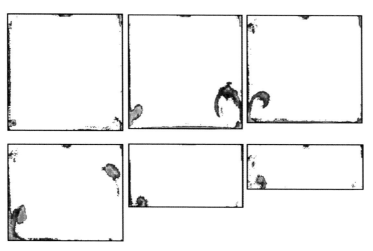

Fig. 1.54 LIF images of unburned hydrocarbons. Top (topland crevice outflow): 0, 20, 40° aBDC, bottom (scraping-off of wall quench layer): 60, 100, 120° aBDC [72]

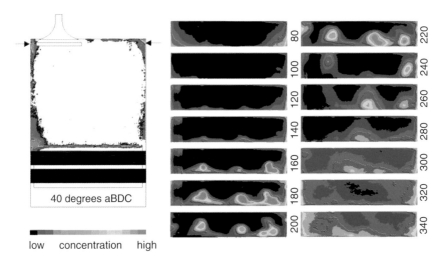

Fig. 1.55 LIF images of unburned hydrocarbons from head gasket and piston rings in the square piston engine. **Left**: vertical cross section, outflow from the head gasket and the piston ring section at 40° aBDC, **Center** and **Right**, top to bottom: vertical cross section, outflow from the piston ring section. Frames are taken at indicated degrees bBDC in the expansion storke [96]

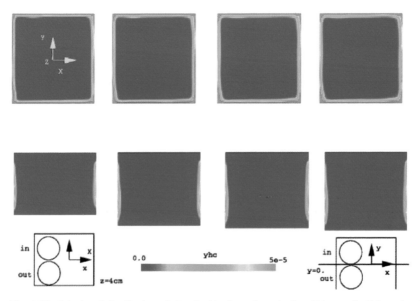

Fig. 1.56 Calculated distribution of absorbed hydrocarbons in the oil layer of a S.I. engine. *Top row*: horizontal (4 cm below the cylinder head), *bottom row*: vertical cross section through the center of the square piston engine. 1,000 rpm, 2.5 BMEP, $\phi = 1$, spark timing 20° bTDC. Left to right: 100, 110, 120 and 130° aTDC [99]

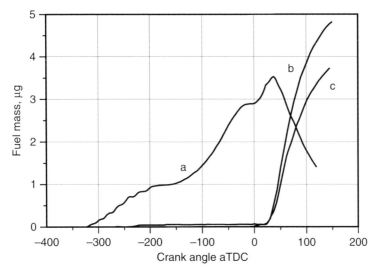

Fig. 1.57 Calculated hydrocarbon quantities obtained by integrating over the entire combustion chamber of the engine. **a:** evolution of the total mass absorbed in the oil layer, **b:** total desorbed mass from the oil layer, **c:** oxidized fraction of the desorbed mass [99]

to provide reliable, physically based sub-models. The predicted desorbed fraction oxidized in Fig. 1.57 is too high ($\approx 75\%$) when compared to the 30% being estimated on the basis of measured engine data. Nevertheless the crude model used for simulating the contribution from oil layers [99] provides at least some insight into the spatial distribution and a rough estimate on the temporal evolution of UHC sources.

1.3.5.3 Fuel Effects

Generally there is little effect of the fuel type on heat release in engines if differences in heat of evaporation, heat of combustion, ignition delay and burning velocities are balanced out by operating conditions of the engine. However, there is a direct effect on pollutant formation and emissions: The fuel properties (structural and compositional characteristics) are still not integrated into the optimization process of engine performance. Ideally, for optimum performance, engine and fuel must be co-optimized simultaneously as a single system. The pressure on reducing pollutants has stimulated practical work in this area prompting also joint approaches in the US and European Auto-Oil programs.

Results from this work indicate that the major components in engine-out emissions are linearly related to the concentrations in the fuel, only the minor fractions arise from partially combusted fuel components. There is an effect of fuel structure on combustion chemistry but only a minor effect arising from the engine design. Although results are not always consistent, in summary, the fuel effects on vehicle emissions are:

- UHC is reduced by: an addition of MTBE, a reduction of the aromatic content or a reduction in T90. The UHC increase if the olefin fraction is increased,
- CO is reduced by: an addition of MTBE or a reduction in the aromatics content,
- NO is reduced by: an addition of MTBE or a decrease in the olefin or aromatic content. It increases if T90 is lowered or toluene is added. Deposits increase NO because they cause higher charge temperatures.

The ambient temperature during the test also affects the results. The percentage increase between 25°C and −7°C for the whole FTP cycle and for the cold transient portion alone are [100]:

	HC/%	CO/%	NOx/%	CO_2/%
FTP Cycle	286	396	16	10
Cold Transient	411	590	34	14

The relative contribution of the first cold transient phase to the whole FTP cycle is even more dominant at lower temperatures:

	HC/%	CO/%	NOx/%	CO_2/%
-7°	87	88	48	22
25°	68	65	44	21

The specific ozone reactivity (SOR), a major concern in atmospheric chemistry, is influenced strongly by the composition of the fuel since aromatics and olefins constitute about 80% of the SOR with aromatics arising predominantly from unreacted fuel. The engine-out SOR is higher than for the unreacted fuel and may be reduced by lowering T90 and the amount of aromatics and olefins in the fuel.

No detailed knowledge is yet available about the specific processes causing these changes but they are obviously more affected by design and operational characteristics of the vehicle than by differences in fuel composition. Nevertheless fuel structure and composition affect the formation of pollutants to a significant degree so there is an urgent need to clarify the details.

1.3.6 Knocking Combustion

Knock, an irregular combustion feature, limits the maximum compression ratio and hence the efficiency of the engine. It is promoted by end-gas temperature, fuel structures with long chains and residuals carried over from previous cycles. Knock is a complex interaction of auto-igniting exothermic centers and their mutual fluid dynamic responses in the inhomogeneous end-gas.

The auto-ignition part is extensively treated and modeled in the literature, and is well understood [53, 101–103]. The low temperature chemistry (< 900 K) is started by the RO2 isomeration processes producing OH radicals by QOOH decomposition consuming some fuel and liberating some heat at low rates. Thus slowly a radical pool is built up leading eventually to hot ignition as soon as temperatures above 900 K are reached. This process may exhibit a distinct two-stage ignition behavior and a negative temperature coefficient (NTC) region in the overall reaction. The effect of additives is by way of influencing the OH radical pool: additives increasing the OH production accelerate hot ignition, while additives which remove OH inhibit it. Thus Anti-Knocks (i.e. Anti-Auto-Ignitions) can act either by retarding low temperature oxidation and chain branching or by inhibiting the high temperature HO_2-dominated hot ignition. MTBE and ETBE act in both regimes. In binary mixtures the olefins may act in the low temperature region as radical scavengers and retard the activities of paraffin's, whereas the paraffin may act as a radical scavenger in the NTC region and retard the activity of olefins. Experimental high pressure data on mixtures are becoming available more and more so it can be expected that realistic fuels may soon be modeled.

This chemically based model does not explain, however, origin, strength or variability of the pressure waves in knocking combustion nor the reasons for material damage associated with it. The pressure signals, commonly used as indicators for knock intensity, correlate only very poorly with end-gas temperature - although this should be the prime controlling parameter - and even less so with the mass of unburned fuel at onset of auto-ignition [104]. The reasons have been clarified in recent studies by the fluid dynamic part following auto-ignition [42, 105, 106]: contrary to general assumptions, auto-ignition occurs not uniformly throughout the end-gas region but in localized spots embedded in the inhomogeneous end-gas - in *exothermic centers* - with specific induction (time to ignition) and excitation times (time of heat release) [107–109]. These centers arise from incomplete mixing and constitute temperature and/or compositional heterogeneities. Transition from auto-ignition into knock is controlled by the properties of these exothermic centers: size, gradients and spatial distribution in the end-gas [42].

In Figs. 1.58 and 1.59 the transition of low to high temperature chemistry in the end-gas is shown by the formation and disappearance of formaldehyde. The reactions start irregularly in the inhomogeneous end-gas but approach an apparent "homogeneous" concentration (not temperature!) distribution in the negative temperature coefficient (NTC) region since the production rate of centers with lower initial temperature may exceed those being already in the NTC region.

Depending on local conditions, some exothermic centers may transit into the hot ignition regime as visualized by the burn-out of the formaldehyde. In Fig. 1.60 data from simultaneous LIF and ultra-high speed schlieren reveal the spatial and temporal evolution of exothermic centers into a spectrum of slow and fast burning centers each showing an individual behavior. In summary there are three different modes of knocking combustion [42, 106]:

- *deflagration,* the most common mode being characterized by small exothermic centers, steep gradients and very weak pressure oscillations. After hot ignition flame propagation is similar to normal flame propagation without causing any knock damage,

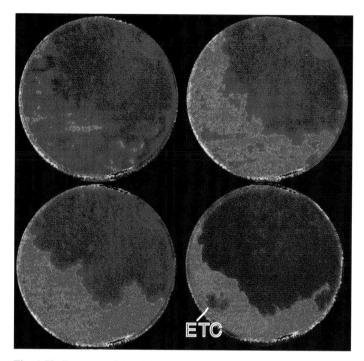

Fig. 1.58 Sequence of LIF images of formaldehyde (*green color*) formed in the end-gas of a S.I. engine during transition from normal (*blue color*) to knocking combustion showing end-gas inhomogeneity and formation of exothermic centers. Sequence (left to right, top to bottom, timing in degrees aTDC): 2°: random beginning of cool flame reactions producing HCOH, 4°: spot-wise formation of HCOH (= endgas inhomogeneity), 7°: apparently "homogeneous" HCOH-distribution in the NTC regime, and 9°: formation of hot flames (burning up HCOH) at exothermic center ETC [72]

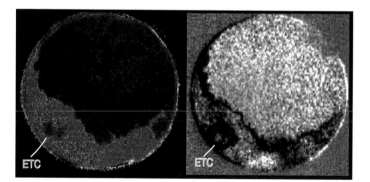

Fig. 1.59 Simultaneously recorded LIF (*left*) and Schlieren (*right*) graphs of formaldehyde. The LIF image shows the formaldehyde concentration formed in the cool flame regime (*green*) and its burnout in exothermic centers. The schlieren image shows the associated density and/or temperature changes [72]

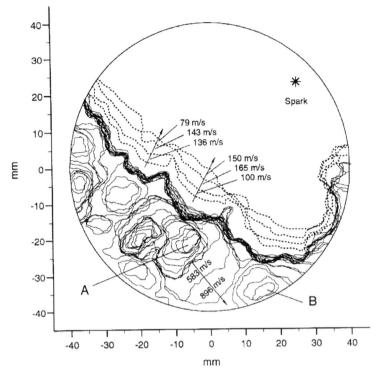

Fig. 1.60 Evaluated ultra-high speed schlieren movie (750.000 fps) showing spot-wise formation of exothermic centers (ETCs) in the endgas. Most centers (e.g. B) develop into deflagrations whereas the lower right hand side of A transits into a developing detonation. Note the random distribution of ETCs and the pushing back of the regular flame front by the fast pressure rise in the endgas [73]

- *thermal explosion*, being characterized by large centers with flat gradients. Pressure oscillations are moderate and knock damage is very light if any at all.
- *developing detonation*, the most violent mode occuring rarely but causing intense pressure oscillations and rapid surface damage to materials. It is characterized by medium sized centers having critical intermediate gradients which support the gradual growth of strong pressure waves. This mode is detrimental to the engine.

In practice all modes are present simultaneously. During the short excitation time they emit pressure waves which heat surrounding exothermic centers and thus shorten their induction time drastically. In consequence initially mild knock modes may drive other centers into a developing detonation mode directly or by a sort of avalanche effect depending on the statistical temporal and spatial distribution of the exothermic centers in the end-gas. Thus auto-ignition is controlled by temperature, but engine knock and its intensity is controlled by the properties, distribution and fluid dynamic interactions of the exothermic centers in the end-gas. They in turn are consequences of preceding mixing processes between residual exhaust, fuel, air as well as cold and hot gases due to heat transferred from chamber walls and valves. The developing detonation mode of knock is associated with very high rates of heat transfer ($>$ 100 MW/m^2) to the walls causing thermo-shock ($\Delta T_{Surface} > 150$ K) which is the prime cause for knock damage. In crevices (e.g. topland region) pressure piling occurs which under special circumstances may lead to excessive mechanical wall loadings causing also knock damage in combination with high heat transfer rates.

This combined auto-ignition and fluid dynamic description of knock explains consistently all features known in knocking combustion. All relevant processes are understood and can be modeled satisfactorily. There is a need, however, for reducing the sub-models for detailed chemistry and pressure wave interactions so they can be incorporated time efficiently in 3D engine codes.

1.3.7 Outlook

The Gasoline engine of the future will most probably be somewhat different from the engine we know today. The pressing needs to reduce fuel consumption and emissions to ever lower values, are stimulating new, more sophisticated approaches. Since there are several options available it is not clear how these engines will look like. Emerging, very interesting approaches are: GDI and concepts using downsizing, homogeneous lean burn and/or variable valve timing. They have in common that they improve the part load fuel efficiency and need dedicated engine control systems. In combustion the crucial issues remain fuel injection and fuel-air mixing - a field with a serious lack of physical-chemical based models - a fact being well known from Diesel engines - and exhaust aftertreatment. Subsequently a brief outlook on these key issues will be given.

1.3.7.1 Spray Combustion

Whereas the numerical aspects of spray modeling are relatively well developed, serious deficiencies exist in modeling the atomization of fuels and the fluid dynamic interactions of sprays with the cylinder charge. The most widely accepted atomization mechanism assumes that liquid fuel is issued from the nozzle as a liquid jet on whose surface instability waves form being amplified and eventually broken up by aerodynamic forces caused by the high relative velocity between liquid and gas. Penetration lengths may be predicted reasonably well whereas droplet sizes are still in poor agreement. Recent studies show, however, that the fuel is already disintegrated when it leaves the nozzle hole. This is assumed to be due to cavitation inside the nozzle. Therefore there is a need for better spray models.

Nevertheless the current models may be used - with caution - to promote progress in spray combustion research. As an example in Fig. 1.61 a typical result of simulating Diesel combustion with a modified KIVA II code is shown. The qualitative

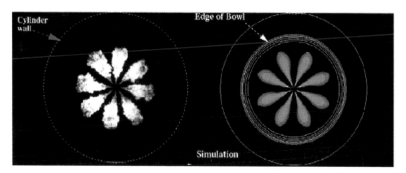

Fig. 1.61 Measured (*left*, 2D 2color method) and calculated (*right*, KIVA II) soot radiation in a modern Diesel engine [110]

agreement with experimental data is acceptably good and similar results should be expected for Gasoline injection.

However, a replacement of the current spray modeling by improved models is required before an acceptable performance can be expected. To assist in this effort new approaches are necessary to provide a better basis for model development. Since predictive spray models have to be rigidly based on the relevant physics and chemistry there is a still unresolved trade-off between the degree of rigidity needed and the computing resources required to provide it. Different attempts exist in the literature to reduce the problems in spray combustion to various degrees of simplification. A very promising approach is the group combustion concept being briefly presented here.

Group Combustion Model for Soot and NO Formation

The group combustion model [111] addresses the specific problems in reactive sprays - the interactions of droplets with each other and the gas which have a direct impact on self-ignition, partially premixed and diffusive combustion and, most important, on pollutant formation as NO_x and soot. It provides the necessary insight into the controlling processes needed for deriving improved modeling schemes for future engines. The group combustion concept is based on the following principles:

1. The space and time dependent characteristics of the real spray with its hundreds of millions of droplets is replaced by a small representative droplet group (e.g. 20 droplets). This group represents the droplet size distribution, the spatial distribution of the droplets and the interactions with the complex flow field at any location of interest in the spray plume.
2. The complex flow field around and inside a spray is replaced by a local single mean flow vector and an appropriate turbulence intensity which is feasible for a small group of e.g. 20 droplets having a characteristic group size of only 0.2–0.5 mm under engine conditios.

3. Heat transfer to and evaporation of each individual droplet are calculated by detailed physics. General transport equations are used both for the liquid and the gas phase which are solved simultaneously.
4. Reactions in the gas phase are treated in two ways: for overview runs a fast one step global chemistry is used for ignition and heat release. For detailed calculations the CHEMKIN [112] code and a modified Heptane kinetic code [55, 113], [114] are applied. The kinetic has been adapted to handle pressures of up to 100 bar [115] and the NO production / depletion by prompt and thermal processes [54].

In Fig. 1.62 a schematic representation of a group combustion model is shown. The interactions within the group are first computed in stage A using a 3D model with detailed droplet properties (variable sizes, locations, speeds and composition), detailed fluid dynamics, detailed droplet heating and evaporation using a one step global chemistry for ignition and combustion [113]. Combustion and pollutant formation are calculated then in stage B by a 1D single droplet model with detailed reaction kinetics to provide combustion and pollutant specifics with initial conditions provided by stage A. Details of the model may be found elsewhere [111], [116].

In Fig. 1.63 model predictions are presented showing that temperatures in a droplet group may vary over wide ranges depending both on time and location in the spray and hence evaporation and combustion are heavily depending on the local conditions. Ignition starts in the wake of the droplet cluster slightly on the lean side where evaporated fuel had the longest time to react. The different inertia of the droplets rapidly leads to a disintegration of the initial droplet structure and generates a complex distribution of the fuel vapor. Therefore widely varying individual conditions for pollutant formation exist at each location in the spray.

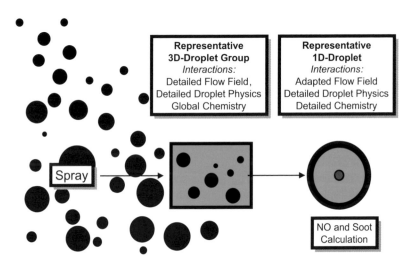

Fig. 1.62 Concept of the group combustion model [116]

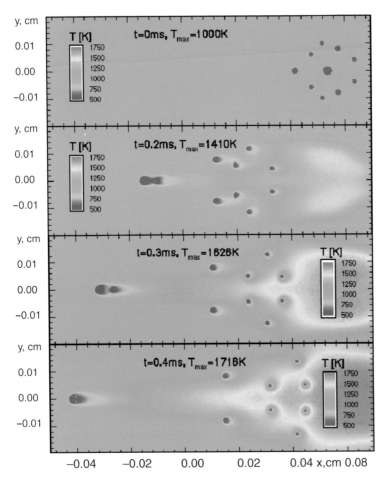

Fig. 1.63 Calculated 3D temperature distribution within a group of 10 different heptane droplets (6×15, 3×20 and $30\,\mu m$) at 0, 0.2, 0.3 and 0.4 ms after injection. Initial droplet velocity: 5 m/s right to left, incoming gas flow: 1.25 m/s left to right [111]

The widely different local combustion conditions can provide quite attractive features as shown in Fig. 1.64 where the rich mixture region behind a burning heptane droplet can provide reburning of the NO having been formed earlier in combustion. The figure shows that on approaching the droplet from the right hand side the assumed background level of 2,000 ppm NO is first reduced to NO_2 (near 0.01 cm), then new thermal NO is form around the peak temperature region and then all NO is fully reduced very close to the droplet surface and temperatures below about 1,800 K. The same kinetics has been applied successfully also to new denoxing schemes [116, 117]. Since, at present, insights into such details of combustion and pollutant formation can not be gained by other means making the group combustion highly attractive for fundamental studies and derivation of improved spray sub-models.

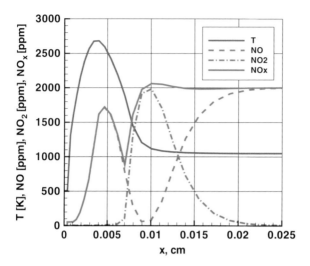

Fig. 1.64 Computed NO reburning in the wake of a 25 μm heptane droplet in a gas environment with 2000 ppm NO. Engine conditions with 1%O_2. Air flow: 5 m/s left to right. The droplet is at the origin

1.3.7.2 Exhaust Aftertreatment

The current most promising approach to reduce the engine-out NOx emissions of future lean burn concepts is the NOx storage catalyst [116, 117]. NOx is adsorbed during the normal lean operation of the engine which is subsequently reduced by a short rich phase. Fresh catalysts show highly attractive conversion efficiencies above 90%. However, its practical application requires a drastic reduction of the fuel sulfur at least below 10 ppm and an improvement of the thermal durability of the catalyst. It is believed that these problems will be overcome by joint efforts of car, catalyst and oil industry so very clean and fuel efficient engines can be realized in the future.

References

1. J.B. Heywood, "Internal Combustion Engine Fundamentals", McGraw-Hill, New York (1988).
2. S.S. Penner, B.P. Mullins, "Explosions, Detonations, Flamability and Ignition", Pergamon Press, London (1959).
3. B. Lewis, G. von Elbe, "Combustion, Flames and Explosions of Gases", 2nd edn., Academic Press, New York (1961).
4. H. Müller, S. Rhode, G. Klink, "Gemischbildung, Verbrennung und Abgas im Ottomotor, Fachbibliographie mit Referaten bis 1965", Universität Braunschweig, Braunschweigh (1972).
5. G. Konzelmann, "Über die Entflammung des Kraftstoff-Luftgemisches im Motor, Bosch Techn. Berichte 1, 6, 297–304 (1966).
6. R.R. Maly, "Spark Ignition, its Physics and Effect on the Internal Combustion Process", in Fuel Economy: Road Vehicles Powered by Spark Ignition Engines, ed. by J.C. Hilliard, G.S. Springer, Plenum Press, New York 91–148 (1984).

7. J. Köhler, W. Lawrence, M. Schäfer, R. Schmidt, W. Stolz, "Spark Plasma Modeling", Final Report "Engine and Fuel interactions in Real Engines", ed. by R. Maly, Daimler-Benz AG, CEC-Daimler-Benz Project, CEC Contract JOU 2-CT 92-0081, Brussels (1995).

8. M. Schäfer, "Der Zündfunke, ein Beitrag zur Modellierung der motorischen Verbrennung", PhD Thesis, University of Stuttgart (1997).

9. H. Albrecht, W.H. Bloss, W. Herden, R.R. Maly, B. Saggau, E. Wagner, "New Aspects of Spark Ignition", SAE Paper 770853 (1977).

10. R.R. Maly, M. Vogel, "Initiation and Propagation of Flame Fronts in Lean CH-Air Mixtures by the Three Modes of the Ignition Spark", 17th Symposium (Int) on Combustion, The Combustion Institute, Pittsburgh, 821–831 (1978).

11. G.F.W. Ziegler, "Entflammung magerer Methan/Luft-Gemische durch kurzzeitige Bogen- und Glimmentladungen", PhD Thesis, University of Stuttgart (1991).

12. R. Herweg, G.F.W. Ziegler, "Untersuchung der Flammenkernbildung im Ottomotor", Abschlußbericht Vorhaben 349 (AIF-Nr. 6359), FVV, Frankfurt (1988).

13. R. Herweg, "Die Entflammung brennbarer, turbulenter Gemische durch elektrische Zündanlagen - Bildung von Flammenkernen", PhD Thesis, University of Stuttgart (1992).

14. R.R. Maly, H. Meinel, "Determination of Flow Velocity, Turbulence Intensity and Length and Time Scales from Gas Discharge Parameters", 5th Int. Symposium on Plasma Chemistry, Edinburgh, 552–557, (1981).

15. R. Herweg, R.R. Maly, "A Fundamental Model for Flame Kernel Formation in S.I. Engines", SAE Paper 922243 (1992).

16. D. Bradley, F.K.K. Lung, "Spark Ignition and the Early Stages of Turbulent Flame Propagation", Combust. Flame, 69, 71–93 (1987).

17. Th. Mantel, "Three-Dimensional Numerical Simulations of Flame Kernel Formation Around a Spark Plug", SAE Paper 920587 (1992).

18. K.N.C. Bray, "Studies of the Turbulent Burning Velocity", Report CUED/A-Thermo/Tr.32, Cambridge University, Eng, Dept., England (1990)

19. C.K. Law, D.L. Zhu, G. Yu, "Propagation and Extinction of Stretched Premixed Flames", 21th Symposium (Int.) on Combustion, 1419–1426 (1986).

20. Ö.L. Gülder, "Correlations of Laminar Combustion for Alternative S.I. Engine Fuels", SAE-Paper 841000 (1984).

21. R.R. Maly, "Die Zukunft der Funkenzündung", MTZ, 59, Nr. 7–8, XXVIII–XXIII (1998), English version: "The Future of Spark Ignition", MTZ Worldwide 7–8/98, Supplement, 37–41 (1998).

22. P. Hohner, "Ein adaptives Zündsystem mit integrierter Motorsensorik", PhD Thesis, University of Stuttgart (1998).

23. Wilstermann, "Wechselspannungszündung mit integrierter Ionenstrommessung als Sensor für die Verbrebbungs- und Motorregelung", PhD Thesis, University of Karlsruhe (1999), Fortschritts-Berichte VDI, Reihe 12, Nr. 389.

24. R.R. Maly, "State of the Art and Future Needs in S.I. Engine Combustion", Invited Topical Review, 25th Symposium (Int) on Combustion, The Combustion Institute, Pittsburgh (1994).

25. H. Weller, S. Uslu, A.D. Gosman, R.R. Maly, R. Herweg, B. Heel, "Prediction of Combustion in Homogeneous-Charge Spark Ignition Engines", Proc. COMODIA'94, JSME, Tokyo, 163–169 (1994).

26. B. Heel, R.R. Maly, H.G. Weller, A.D. Gosman, "Validation of S.I. Combustion Model Over Range of Speed, Load, Equivalence Ratio and Spark Timing", Proc. COMODIA'98, JSME, Tokyo, 255–260 (1998).

27. R.R. Maly, G. Eberspach, W. Pfister, "Laser Diagnostics for Single Cycle Analysis of Crank Angle Resolved Length and Time Scales in Engine Combustion", Proc. COMODIA'90, Kyoto, Japan, 399–404 (1990).

28. G.F.W. Ziegler, R. Herweg, P. Meinhardt, R.R. Maly, "Cycle-Resolved Flame Structure Analysis of Turbulent Premixed Engine Flames", Proc. XXIII FISITA Congress, Paper 905001, Turin, Italy (1990).

29. E. Winklhofer, H. Phillip, G. Fraidl, H. Fuchs, "Fuel and Flame Imaging in SI Engines", SAE Paper 930871 (1993).
30. A.O. zur Loye, F.V. Bracco, "Two-Dimensional Visualization of Premixed-Charge Flame Structures in an I.C. Engine", SAE Paper 870454 (1987).
31. J. Mantzaras, F.G. Felton, F.V. Bracco, "3-D Visualization of Premixed Charge Engine Flames: Islands of Reactants and Products; Fractal Dimensions and Homogeneity", SAE Paper 881633 (1988).
32. F. Lawrenz, J. Köhler, F. Meier, W. Stolz, R. Wirth, W.H. Bloss, R.R. Maly, E. Wagner, M. Zahn, "Quantitative 2D LIF Measurements of Air/Fuel Ratios During the Intake Stroke in a Transparent S.I. Engine", SAE Paper 922320 (1992).
33. A. Orth, V. Sick, J. Wolfrum, R.R. Maly, M. Zahn, "Simultaneous 2D-Single Shot Imaging of OH Concentrations and Temperature Fields in a S.I. Engine Simulator", 25th Symposium (Int) on Combustion, The Combustion Institute, Pittsburgh (1994).
34. R.R. Maly, K.N.C. Bray, T.C. Chew, "An Integral Time Scale of Evolution for Non-Stationary Turbulent Premixed Flames", Combus. Sci. Technol., 66, 139–147 (1989).
35. K.N.C Bray, T.C. Chew, R.R. Maly, "Quantitative Evaluation of Length and Time Scales of Turbulent Engine Combustion from 2D Laser Sheets", Proc. 7th Symposium Turbine Shear Flow, Stanford University, (1989).
36. J. Wolfrum, Private communication (1994).
37. A. Bräumer, V. Sick, J. Wolfrum, V, Drewes, M. Zahn, R.R. Maly, "Quantitative Two-Dimensional Measurements of Nitric Oxide and Temperature Distributions in a Transparent Square Piston S.I. Engine", SAE Paper 952462 (1995).
38. V. Drewes, "Zweidimensionale Konzentrations- und Temperaturfeld-bestimmung in einem Forschungsmotor mit quadratischem Querschnitt", PhD Thesis, University of Heidelberg (1995).
39. C. Schulz, V. Sick, J. Wolfrum, V. Drewes, M. Zahn, R.R. Maly, "Quantitative 2D Single Shot Imaging of NO Concentrations and Temperatures in a Transparent SI Engine", 26th Symposium (Int.) on Combustion, The Combustion Institute, Pittsburgh, 2597–2604 (1996).
40. H. Becker, P.B. Monkhouse, J. Wolfrum, R.S. Cant, K.N.C. Bray, R.R. Maly, W. Pfister, G. Stahl, R. Warnatz, "Investigation of Extinction in Unsteady Flames in Turbulent Combustion by 2D-LIF of OH Radicals and Flamelet Analysis", 23rd Symposium (Int.) on Combustion, The Combustion Institute, Pittsburgh (1990).
41. G. Blessing, "Klopfuntersuchungen mit Hilfe der Formaldehyd-LIF", in Report "Gas/Surface Interactions and Damaging Mechanisms in Knocking Combustion", ed. by R. Maly, Daimler-Benz AG, CEC-Daimler-Benz Project, CEC Contract JOUE 0028-D-(MB), Brussels (1993).
42. G. König, "Auto-ignition and Knock Aerodynamics in Engine Combustion", PhD Thesis, Mech. Dept., Leeds University (1993).
43. G. König, R.R. Maly, S. Schüffel, G. Blessing, "Effect of Engine Conditions on Knock", in Final Report "Gas/Surface Interactions and Damaging Mechanisms in Knocking Combustion" , ed. by R. Maly, Daimler-Benz AG, CEC-Daimler-Benz Project, CEC Contract JOUE 0028-D-(MB), Brussels (1993).
44. H. Becker, A. Arnold, R. Suntz, P.B. Monkhouse, J. Wolfrum, R.R. Maly, W. Pfister, "Investigation of Flame Structure and Burning Behavior in an IC Engine by 2D-LIF of OH Radicals", Appl. Phys. B, 50, 473–478 (1990).
45. D.L. Reuss, "Two-Dimensional Particle Image Velocimetry with Electrooptical Image Shifting in an Internal Combustion Engine, Proc. SPIE, Vol. 2005, p. 413–424, Optical Diagnostics in Fluid and Thermal Flow, Soyoung S. Cha; James D. Trolinger; Eds., Bellingham WA 98227-0010 USA (1993).
46. E. Nino, B.F. Gajdeczko, P.G., Felton, "Two-Color Particle Image Velocimetry in an Engine with Combustion", SAE Paper 930872 (1993).

47. W. Stolz, J. Köhler, F. Lawrenz, F. Meier, W.H. Bloss, R.R. Maly, R. Herweg, M. Zahn, "Cycle Resolved Flow Field Measurements Using a PIV Movie Technique in a S.I. Engine", SAE Paper 922354 (1992).

48. D.H. Barnhart, R.J. Adrian, G.C. Papen, "Phase Conjugate Holographic System for High Resolution Particle Image Velocimetry", Appl. Optics 33, 30, 7159–7170 (1994).

49. V. Drewes, H. Häcker, B. Heel, R. Herweg, R.R. Maly, M. Zahn, "NO and UHC in S.I. Engines", in Final Report "Engine and Fuel interactions in Real Engines", ed. by R. Maly, Daimler-Benz AG, CEC-Daimler-Benz Project, CEC Contract JOU 2-CT 92-0081, Brussels (1995).

50. J.C. Keck, "Rate-Controlled Constrained-Equilibrium Theory of Chemical Reactions in Complex Systems", Prog. Energy Combust. Sci., 16, 125–154 (1990).

51. S. Hochgreb, L.F. Dryer, "A Comprehensive Study on CH_2O Oxidation Kinetics", Comb. Flame, 91, 257–284 (1992).

52. J. Warnatz, "Resolution of Gas Phase and Surface Combustion Chemistry into Elementary Reactions", 24th Symposium (Int.) on Combustion, The Combustion Institute, Pittsburgh, 553–579 (1992).

53. C. Chevalier, W.J. Pitz, J. Warnatz, C.K. Westbrook, H. Melenk, "Hydrocarbon Ignition: Automatic Generation of Reaction Mechanisms and Application to Modeling of Engine Knock", 24th Symposium (Int.) on Combustion, The Combustion Institute, Pittsburgh, 93–101 (1992).

54. C.T. Bowman, R.K. Hanson, D.F. Davidson, W.C. (Jr.) Gardiner, V. Lissianski, G.P. Smith, D.M. Golden, M. Frenklach, M. Goldenberg, "Gri-Mech 22.11", www.me.berkeley.edu/gri˙mech (1997).

55. F. Mauss, "Entwicklung eines kinetischen Modells der Rußbildung mit schneller Polimerization", PhD Thesis, RWTH Aachen (1997).

56. U. Maas, S.B. Pope, "Implementation of Simplified Chemical Kinetics Based on Intrinsic Low-Dimensional Manifolds", 24th Symposium (Int.) on Combustion, The Combustion Institute, Pittsburgh, 24–103 (1992).

57. U. Maas, S.B. Pope, "Laminar Flame Calculations Using Simplified Chemical Kinetics Based on Intrinsic Low-Dimensional Manifolds", 25th Symposium (Int.) on Combustion, The Combustion Institute, Pittsburgh (1994).

58. N. Peters, B. Rogg, "Reduced Kinetic Mechanisms for Applications in Combustion Systems", Springer Verlag, Berlin, Heidelberg (1993).

59. C.K. Westbrook, "Combustion Chemistry Modeling in Engines", Engineering Foundation Conference on Present and Future Engines for Automobiles, St. Barbara, August 25–30, 1991.

60. C.K. Law, "A Compilation of Experimental Data on Laminar Burning Velocities", in Reduced Kinetic Mechanisms for Applications in Combustion Systems, ed. by N. Peters, B. Rogg, Springer Verlag, Berlin, Heidelberg, 15–26 (1993).

61. P. Cambray, G. Joulin, "Length Scales of Wrinkling of Weakly-Forced, Unstable Premixed Flames", Combust. Sci. Tech., 97, 405–428 (1994).

62. J.B. Heywood, "Combustion and its Modelling in Spark Ignition Engines", Proc. COMODIA'94, JSME, Tokyo, 1–15 (1994).

63. D. Bradley, R.A. Hicks, M. Lawes, C.G.W. Sheppard, R. Wooley, "The Measurement of Laminar Burning Velocities and Markstein Numbers for Iso-octane-Air and Iso-octane-n-Heptane-Air Mixtures at Elevated Temperatures and Pressures in an Explosion Bomb", Comb. Flame, 115, 126–144 (1998).

64. P. Clavin, F.A. Williams, "Theory of Premixed Flame Propagation in Large Scale Turbulence", J. Fluid Mech., 90, 589 (1979).

65. R.J. Tabaczynski, C.R. Ferguson, K. Radhakrishnan, "A Turbulent Entrainment Model for Spark Ignition Engine Combustion", SAE Paper 770647 (1977).

66. A.M. Klimov, "Premixed Turbulent Flames - Interplay of Hydrodynamics and Chemical Phenomena", in Flames, Lasers and Reactive Systems, Prog. Astr. Aero. Vol. 88, Ed. American Institute of Aeronautics and Astronautics, Inc., New York, NY, USA (1983).

67. B. Pope, M.S. Anand, "Flamelet and Distributed Combustion in Premixed Flames", 20th Symposium (Int.) on Combustion, The Combustion Institute, Pittsburgh (1985).

68. P.O. Witze, J.K. Martin, C. Borgnakke, "Measurements and Predictions of the Pre-Combustion Fluid Motion and Combustion Rates in a Spark Ignition Engine", SAE Paper 831697 (1983).

69. J.N. Mattavi, E.G. Groff, F.V. Matekunas, "Turbulence, Flame Motion and Combustion Chamber Geometry - Their Interactions in a Lean-Combustion Engine", Proc. Conference on Fuel Economy and Emissions of Lean Burn Engines, IMechE, London, C100/79 (1979).

70. P.O. Witze, J.M.C. Mendes-Lopez, "Direct Measurements of the Turbulent Burning Velocity in a Homogeneous-Charge Engine", SAE Paper 861531 (1986).

71. C.M. Ho, D.A. Santavicca, "Turbulence Effects on Early Flame Kernel Growth", SAE Paper 872100 (1987).

72. R.R. Maly, "Applied Flow and Combustion Diagnostics for I.C. Engines", Invited Paper, Proc. Computational Fluid Dynamics Conference '94, Stuttgart, John Wiley & Sons (1994).

73. R.R. Maly, "State of the Art and Future Needs in S.I. Engine Combustion", Invited Topical Review, 25th Symposium (Int.) on Combustion, The Combustion Institute, Pittsburgh (1994).

74. T.W. Kuo, R.D. Reitz, "Computation of Premixed-Charge Combustion in Pancake and Pentroof Engines", SAE Paper 860670 (1986).

75. H. Weller, Uslu, A.D. Gosman, R.R. Maly, R. Herweg, and B. Heel, "Prediction of Combustion in Homogeneous-Charge Spark Ignition Engines", Proc. COMODIA'94, JSME, Tokyo, 163–169 (1994).

76. H.G. Weller, "The Development of a New Flame Area Combustion Model Using Conditional Averaging", Thermo-Fluids Section Report TF/9307, Dept. Mech. Eng., Imperial College London (1993).

77. K. Boulouchos, T. Steiner, P. Dimopoulos, "Investigation of Flame Speed Models for the Flame Growth Period during Premixed Engine Combustion", SAE Paper 940476 (1994).

78. R. Borghi, B. Argueyrolles, S. Gauffie, P. Souhaite, "Application of a Presumed pdf Model of Turbulent Combustion to Reciprocating Engines", 21th Symposium (Int.) on Combustion, The Combustion Institute, Pittsburgh 1591–1599 (1996).

79. K.N.C. Bray, "Studies of the Turbulent Burning Velocity", Proc. Royal Soc., London, A431 (1990).

80. P. Boudier, S. Henriot, T. Poinsot, T. Baritaud, "A Model for Turbulent Flame Ignition and Propagation in Spark Ignition Engines, 24th Symposium (Int.) on Combustion, The Combustion Institute, Pittsburgh (1992).

81. H.G. Weller, C.J. Marooney, A.D. Gosman, "A New Spectral Method for Calculation of Time-Varying Area of a Laminar Flame in Homogeneous Turbulence", 23rd Symposium (Int.) on Combustion, The Combustion Institute, Pittsburg (1990).

82. G.K. Fraidl, F. Quissek, E. Winklhofer, "Improvement of LEV/ULEV Potential of Fuel Efficient High Performance Engines", SAE Paper 920416 (1992).

83. G. Almkvist, S. Eriksson, "An Analysis of Air to Fuel Ratio Response in a Multi Point Fuel Injected Engine Under Transient Conditions", SAE Paper 932753 (1993).

84. J.M. Duclos, C. Griard, A. Torres, T. Baritaud, "Numerical Modeling of a Stratified Combustion Chamber", in Final Report "Gasoline Engine with Reduced Raw Emissions", ed. by R. Maly, Daimler-Benz AG, CEC-Daimler-Benz Project, CEC Contract TAUT-CT 92-0003, Brussels (1997).

85. J.M. Duclos, G. Bruneau, T. Baritaud, "3D Modeling of Combustion and Pollutants in a 4-valve S.I. Engine; Effect of Fuel and Residuals Distribution and Spark Location", SAE Paper 961959 (1996).

86. C. Meneveau, T. Poinsot, "Stretching and Quenching of Flamelets in Premixed Turbulent Combustion", Comb. Flame, 86, 311–332 (1991).
87. T. Poinsot, D. Veynante, S. Candel, "Quenching Process and Premixed Turbulent Combustion Diagrams", J. Fluid Mech., 228, 561–606 (1991).
88. T. Poinsot, D.C. Haworth, G. Bruneau, "Direct Simulation and Modeling of Flame-Wall Interaction for Premixed Turbulent Combustion", Comb. Flame, 95, 118–132 (1993).
89. W.K. Cheng, D. Hamrin, J.B. Heywood, S. Hochgreb, K. Min, M. Norris, "An Overview of Hydrocarbons Emissions Mechanisms in Spark-Ignition Engines", SAE Paper 932708 (1993).
90. P.G. Brown, W.A. Woods, "Measurements of Unburned Hydrocarbons in a Spark Ignition Combustion Engine during the Warm-Up Period", SAE Paper 922233 (1992).
91. S. Kubo, M. Yamamoto, Y. Kizaki, S. Yamazaki, T. Tanaka, K. Nakanishi, "Speciated Hydrocarbon Emissions of SI Engine During Cold Start and Warm-Up", SAE Paper 932706 (1993).
92. R.G. Nitschke, "Reactivity of SI Engine Exhaust under Steady-State and Simulated Cold-Start Operating Conditions", SAE Paper 932704 (1993).
93. R.M. Frank, J.B. Heywood, "The Effect of Piston Temperature on Hydrocarbon Emissions from a Spark-Ignited Direct-Injection Engine", SAE Paper 910558 (1991).
94. P.R. Meernik, A.C. Alkidas, "Impact of Exhaust Valve Leakage on Engine-Out Hydrocarbons", SAE Paper 932752 (1993).
95. V. Drewes, H. Häcker, B. Heel, R.R. Maly, M. Zahn, "NO and UHC in S.I. Engines", in Report 3/95 "Engine and Fuel Interactions in Real Engines", ed. by R. Maly, Daimler-Benz AG, CEC-Daimler-Benz Project, CEC Contract TAUT-CT 92-0003, Brussels (1995).
96. T. Tamura, S. Hochgreb, "Chemical Kinetic Modeling of the Oxidation of Unburned Hydrocarbons", SAE Paper 922235 (1992).
97. F.H. Trinker, J. Cheng, G.C. Davis, "A Feedgas HC Emission Model for SI Engines Including Partial Burn Effects", SAE Paper 932705 (1993).
98. C. Huynh, T, Baritaud, "Modeling Absorption / Desorption in Oil Films", in Final Report "Engine and Fuel Interactions in Real Engines", ed. by R. Maly, Daimler-Benz AG, CEC-Daimler-Benz Project, CEC Contract TAUT-CT 92-0003, Brussels (1995).
99. W.R. Leppard, J.D. Benson, J.C. Knepper, V.R. Burns, W.J. Koehl, R.A. Gorse, L.A. Rapp, A.M. Hochhauser, R.M. Reuter, "How Heavy Hydrocarbons in the Fuel Affect Exhaust Emissions: Correlation of Fuel, Engine-Out, and Tailpipe Speciation - The Auto/Oil Air Quality Improvement Research Program", SAE Paper 932724 (1993).
100. C. Chevalier, P. Louessard, U.C. Müller, J. Warnatz, "A Detailed Low-Temperature Reaction Mechanism of n-Heptane Auto-Ignition", COMODIA'94, JSME, Tokyo, 93–97 (1990).
101. W.J. Pitz, C.K. Westbrook, W.R. Leppard, "The Autoignition Chemistry of Paraffinic Fuels and Pro-Knock and Anti-Knock Additives: A Detailed Chemical Kinetic Study", SAE Paper 912314 (1991).
102. W.R. Leppard, "The Autoignition Chemistries of Primary Reference Fuels, Olefins/Paraffin Binary Mixtures, and Non-Linear Octane Blending", SAE Paper 922325 (1992).
103. G. König, C.G.W. Sheppard, "End Gas Autoignition and Knock in SI Engines", SAE Paper 902135 (1990).
104. R.R. Maly, R. Klein, N. Peters, G. König, "Theoretical and Experimental Investigation of Knock Induced Surface Destruction", SAE Paper 900025 (1990).
105. G. König, R.R. Maly, D. Bradley, A.K.C. Lau, C.G.W. Sheppard, "Role of Exothermic Centres on Knock Initiation and Knock Damage", SAE Paper 902136 (1990).
106. A.K. Oppenheim, "Dynamic Features of Combustion", Phil. Trans. R. Soc. London A, 315, 471–508 (1985).
107. A.E. Lutz, "Numerical Study of Thermal Ignition", Sandia Report, SAND 88-8228.UC4 (1988).

108. A.E. Lutz, R.J. Kee, J.A. Miller, H.A. Dwyer, A.K. Oppenheim, "Dynamic Effects of Auto-ignition Centers for Hydrogen and C-1,2-Hydrocarbon Fuels", 22nd Symposium (Int.) on Combustion, The Combustion Institute, Pittsburgh (1988).
109. P. Dittrich, F. Wirbeleit, J. Willand, K. Binder, "Multi-Dimensional Modeling of the Effect of Injection Systems on DI Diesel Engine Combustion and NO-Formation", SAE Paper 98FL-512 (1998).
110. P. Stapf, R.R. Maly, H.A. Dwyer, "A Group Combustion Model for Treating Reactive Sprays in I.C. Engines", 27th Symposium (Int.) on Combustion, The Combustion Institute, Pittsburgh (1998).
111. R.J. Kee, F.M. Rupley, J.A. Miller, "CHEMKIN II, A FORTRAN Chemical Kinetics Package for the Analysis of Gas-Phase Chemical Kinetics", Sandia Report, SAND89-8009 (1990).
112. C.K. Westbrook, F.L. Dryer, "Chemical Kinetic Modeling of Hydrocarbon Combustion", Prog. Energy Combust. Sci., 10, 1–57 (1984).
113. M. Nehse, J. Warnatz, C. Chevalier, "Kinetic Modeling of the Oxidation of Large Aliphatic Hydrocarbons", 26th Symposium (Int.) on Combustion, The Combustion Institute, Pittsburgh (1996).
114. R.R. Maly, P. Stapf, G. König, "Neue Ansätze zur Modellierung der Rußbildung", in Dieselmotorentechnik 98, ed. by U. Essers, 553, Technische Akademie Essingen (1998).
115. R. R. Maly, P. Stapf, G. König, "Progress in Soot Modeling for Engines", Key Note Paper, Proc. COMODIA'98, JSME, Tokyo, 25–34 (1998).
116. B. Krutzsch, G. Wenninger, M. Weibel, P. Stapf, A. Funk, D.E. Webster, E. Chaize, B. Kasemo, J. Martens, A. Kiennemann, "Reduction of NO˙x in Lean Exhaust by Selective NOx-Recirculation (SNR-Technique) Part I: System and Decomposition Process", SAE Paper 982592 (1998).
117. N. Fekete, R. Kemmler, D. Voigtländer, B. Krutzsch, E. Zimmer, G. Wenninger, W. Strehlau, J.A.A. van den Tillaart, J. Leyrer, E.S. Lox, W. Müller, "Evaluation of NOx Storage Catalysts for Lean Burn Gasoline Fueled Passenger Cars", SAE Paper 970746 (1997).
118. R.R. Maly, "Progress in Combustion Research", IMechE Prestige Lecture, IMechE, London (1998).

Chapter 2
Flow, Mixture Preparation and Combustion in Direct-Injection Two-Stroke Gasoline Engines

Todd D. Fansler and Michael C. Drake

2.1 Attractions and Challenges of Two-Stroke Engines

In the mid-1980's, two-stroke-cycle gasoline engines featuring direct injection (DI) of fuel into the combustion chamber began to receive attention worldwide as passenger-car powerplants [1, 2, 3, 4, 5, 6]. This intense interest came despite the fact that emissions regulations and customer-acceptance issues (e.g., poor fuel economy, smoky and malodorous exhaust, and the need to mix lubricating oil with the fuel) had driven carbureted two-stroke-powered Saab and Suzuki vehicles from the North American and Western European markets in the late 1960's. Automotive two-strokes also suffered from the unsavory reputation of the East German Trabant and Wartburg vehicles, whose engines survived largely unchanged for fifty years [7]. In the mid-1990's automotive two-stroke engine efforts waned in North America [8, 9, 10, 11], while development continued for a while in Europe [12, 13, 14, 15] and Australia [16]. However, DI two-stroke (DI2S) engines have been introduced commercially for recreational marine engines [17, 18, 19, 20], motorcycles [21], and personal watercraft [19, 22, 23], and they are under development for scooters [24] and snowmobiles [25]. A DI retrofit kit has also been developed to reduce excessive exhaust emissions and fuel consumption by carbureted 2S engines on light vehicles such as auto-rickshaws in Southeast Asia [26]. Research on DI2S engines continues, with emphasis on emissions mechanisms and control (e.g., [27, 28, 29]).

In this chapter, we first discuss the attractions of DI2S engines and then delineate the major development challenges. The bulk of the chapter describes the systematic application of optical and other diagnostic techniques to address these challenges and summarizes the resulting insights.

The primary attractions of two-stroke engines are their excellent specific power, comparatively simple design, and low cost. Compared to a four-stroke-cycle engine

Todd D. Fansler
General Motors Research & Development Center, 30500 Mound Road, Warren, MI 48090-9055, USA

Michael C. Drake
General Motors Research & Development Center, 30500 Mound Road, Warren, MI 48090-9055, USA

C. Arcoumanis, T. Kamimoto (eds.), *Flow and Combustion in Reciprocating Engines*,
DOI: 10.1007/978-3-540-68901-0_2, © Springer-Verlag Berlin Heidelberg 2009

of the same displacement and combustion stability, two-stroke-cycle engines can provide substantially more output power and smoother operation because they fire twice as often (i.e., once per crankshaft revolution).[1] Alternatively, for a given output, a two-stroke engine can be substantially smaller and lighter than its four-stroke counterpart (hence their popularity in motorcycles and snowmobiles as well as in power tools such as chain saws, weed cutters and snow throwers). Smaller, lighter power plants are also attractive for automobiles as fuel-economy and exhaust-emissions requirements become increasingly stringent and as aerodynamics and aesthetics make lower hoodlines desirable. Engines of the type discussed in this chapter – ported two-stroke engines with crankcase-compression scavenging – offer additional advantages of reduced light-load pumping losses, mechanical simplicity, and lower friction compared to conventional four-stroke spark-ignition (SI) engines. The last two advantages result from operation without the usual overhead-valve and lubrication hardware (a small amount of lubricating oil is atomized into the intake air, avoiding the need for a conventional oil pump and oil-control piston rings). The dry-sump lubrication also allows the engine to be mounted in any orientation.

To provide a rationale for the studies summarized in this chapter, we first describe key differences between direct-injection two-stroke engines and conventional premixed-charge two- and four-stroke SI engines in terms of flow, mixture preparation and combustion. For details on premixed-charge two-stroke engines, see general engine texts (e.g., [30, 31]), Blair's two-stroke treatises [32, 33] and Heywood and Sher's book [34]; for details on conventional premixed-charge four-stroke engines (see [30, 31] and Chap. 1 of this volume).

Scavenging is the term used (particularly in the context of two-stroke engines) for the gas-exchange process that removes combustion products from the cylinder and brings in fresh air (or air and fuel in premixed-charge engines). Scavenging is important in all internal-combustion (IC) engines, but it is especially critical in two-stroke engines of the type discussed here because they require this gas exchange to be accomplished without the aid of overhead valves and in roughly half the time available in a four-stroke engine operating at the same speed.[2]

Figure 2.1 shows a schematic of a two-stroke engine with *crankcase-compression* scavenging. As the piston rises toward its top-dead-center (TDC) position to compress the charge in the cylinder (Fig. 2.1a), the volume of the crankcase increases (each cylinder has a separate, sealed compartment). The reduced crankcase pressure allows a set of reed valves to open, and fresh air (or air and fuel) is drawn into the crankcase. As the piston passes through TDC during combustion (Fig. 2.1b), the reed valves close, and the descending piston compresses the air trapped in the crankcase. (Typical *crankcase* compression ratios are only about 1.3–1.5,

---

[1] The factor-of-two advantage that one might expect at first glance is not achieved in practice for several reasons, including the lower effective compression ratio at which two-stroke engines typically operate [32, 34].

[2] Intake and exhaust tuning effects are very important in two-stroke engines at high speeds and loads and in multicylinder configurations [32, 34], but will not be discussed here.

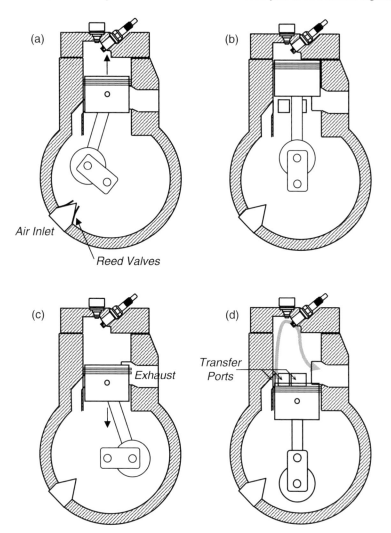

Fig. 2.1 Schematic diagram illustrating the scavenging process in a crankcase-compression two-stroke engine

which is why these engines have lower light-load pumping losses than four-stroke SI engines.) As the piston descends past the top of the exhaust port (Fig. 2.1c), typically around mid-stroke, the hot cylinder gases blow down into the exhaust port. Shortly thereafter, the compressed gas in the crankcase flows into the cylinder as the descending piston uncovers the *transfer ports* that are arrayed around the bottom of the cylinder. The transfer ports are usually designed to induce a looping flow, as sketched in Fig. 2.1d. As the piston ascends after bottom-dead-center (BDC), it covers first the transfer ports and finally the exhaust port, after which the charge in the cylinder undergoes compression, ignition, combustion and expansion.

Even with careful transfer-port design, the scavenging process is generally less effective in crankcase-compression two-stroke engines than in four-stroke engines. Four principal factors contribute: (1) Some of the incoming air (or air and fuel, if the engine is operated on a premixed charge) is *short-circuited* directly to the exhaust port. (2) The gases from the crankcase typically enter the cylinder as highly transient, near-sonic jets which can strongly mix the incoming and residual gases. (3) The incoming jet flows may interact unstably with each other and with the chamber surfaces. (4) At part load, too little air is admitted to the crankcase to remove all the cylinder contents, so that the engine necessarily operates with a high residual fraction (sometimes >50%) except at heavy load.[3] In turn, these factors lead to a number of problems. (1) If the engine operates with a premixed charge (e.g., a carbureted 2S engine), so that the air and fuel both enter the cylinder together from the crankcase, *substantial amounts of fuel will short-circuit directly to the exhaust*, increasing fuel consumption and producing high unburned-hydrocarbon (HC) emissions and smoke. As much as a third of the inducted fuel can escape to the exhaust in this way. (2) Mixing of incoming gases with the cylinder contents exacerbates the effects of short-circuiting. (3) Flow instabilities can be a source of cycle-to-cycle variation in the composition and spatial distribution of the cylinder contents and hence a source of cycle-to-cycle variation in combustion and emissions. (4) Dilution of the fuel-air mixture by the intrinsically high residual fraction can lead to very poor combustion stability at part load (including "four-stroking," in which firing and misfiring cycles alternate). Note also that the scavenging process in a crankcase-compression two-stroke engine does not scale even roughly with engine speed,[4] unlike the piston-driven gas-exchange process in four-stroke-cycle engines.

Short circuiting of fuel to the exhaust can be eliminated almost completely[5] by scavenging the engine with air alone and by *directly injecting* the fuel into the combustion chamber sufficiently late in the cycle so that no fuel can escape before the exhaust port is closed off by the piston [24, 32]. This leads to major reductions in fuel consumption, smoke, and engine-out HC emissions compared to premixed-charge two-stroke engines, and is the major reason for the revival of interest in DI2S engines. Furthermore, direct fuel injection permits *stratified-charge* operation at part load, in which the fuel is concentrated to form an ignitable cloud near the spark plug even when the quantity of fuel per cycle is so small that a homogeneous fuel-air-residual mixture would be beyond the dilute ignition limit. In the two-stroke context,

---

[3] Throttling a crankcase-compression two-stroke engine reduces the amount of air that is admitted to the crankcase and hence that is available to scavenge the cylinder, but throttling only slightly affects the density and hence the mass of the gas that is trapped in the cylinder when all the ports have been covered. Thus throttling primarily alters the residual fraction (and the amount of incoming fuel in a premixed-charge engine).

[4] This need not be an issue if the two-stroke engine is externally scavenged with a blower driven from the crankshaft. Blower-driven scavenging also permits use of a conventional one-piece crankshaft and wet-sump lubrication.

[5] Short circuiting can contribute substantially to DI2S engine-out HCs at the heaviest loads where (as discussed momentarily) injection must be advanced substantially [35].

stratified-charge operation is attractive for three principal reasons: (1) it permits stable operation across the speed-load range (including idle) with load controlled by the quantity of injected fuel rather than by throttling, (2) it permits operation under highly diluted conditions that reduce NOx formation by reducing peak temperatures [35], and (3) it lowers hydrocarbon emissions by reducing or eliminating the need to overfuel the engine for the first cycle (or first few cycles) of a cold start.

But stratified-charge combustion, whether in two-stroke or-four-stroke engines, brings its own problems. Over eighty years ago, Ricardo [35] remarked that:

> working with a stratified charge . . . is possible and the high efficiency theoretically obtainable from it can be approached. The worst feature about it is that, *if not just right, it may be very wrong;* a small change in form or dimension may upset the whole system. (italics added)

In Ricardo's day, engine developers were primarily concerned about part-load efficiency, combustion stability, and full-load performance. Today, stringent emissions regulations make stratified-charge operation even more challenging (see Chap. 3 and [37, 38]). Stratified-charge SI engines are generically susceptible to excessive HC emissions at light load (thought to be caused primarily by fuel-air mixture on the periphery of the injected fuel cloud that is too lean to burn[6]) and to high NOx emissions (because some fraction of the stratified charge burns at the stoichiometric air-fuel ratio that maximizes the local temperature and hence NOx production). In-cylinder control of HC and NOx formation is especially critical in stratified-charge engines because their lean, low-temperature exhaust makes catalytic after-treatment more difficult.

Furthermore, the wide speed-load range imposed by vehicle driving schedules poses problems of combustion control that are considerably more complex for stratified-charge SI engines in general, and for DI2S engines in particular, than they are for more conventional four-stroke premixed-charge engines. At light load, late injection is needed to concentrate the fuel in a burnable cloud around the ignition site, as already mentioned. At full load, where output power is limited by the engine's air throughput capacity, the engine needs to operate with a uniform stoichiometric or slightly rich mixture, which requires as much mixing time as possible between fuel injection and ignition. The basic strategy adopted in the experimental program described in this chapter was to vary the charge stratification continuously with load between these extremes. Recent direct-injection four-stroke DI4S engines take a different approach, operating in either highly stratified or homogeneous modes, with the latter including a region of homogeneous lean operation at intermediate loads [37, 39, 40, 41, 42].

In this chapter, we focus on part-load, stratified-charge DI2S operation. The principal challenges for flow, mixture preparation and combustion are then to:

---

[6] By contrast, in premixed-charge four-stroke engines (see chapter 1), the dominant HC emissions mechanism is the storage and release of fuel trapped in combustion-chamber crevices, in particular the piston top-ring-land crevice [30].

1. Scavenge the engine with maximum efficiency and minimum cyclic variation.
2. Control the fuel spray and its interaction with the piston and combustion-chamber surfaces and with the in-cylinder air motion to produce a readily ignitable fuel-air-residual mixture at the spark gap every cycle, while minimizing the wetting of combustion-chamber surfaces with fuel, which can produce particulate emissions [43].
3. Optimize combustion by burning the stratified mixture as completely as possible in order to minimize HC emissions and with optimal phasing relative to piston motion in order to maximize fuel economy.
4. Minimize NOx production primarily by reducing the flame temperature using the maximum charge dilution by residual gases that is consistent with stable combustion and acceptable HC emissions.[7]

The sensitivity of stratified-charge combustion to small perturbations (particularly in mixture composition at ignition, which will be illustrated in Sect. 2.5.8.1) imposes performance and emissions penalties if these conditions are not met for *every* engine cycle. This is considerably more difficult than optimizing the multi-cycle average operation. Furthermore, speed-load transients require carefully controlled transitions in charge stratification and residual-gas content. Except for the first item listed, these same challenges confront four-stroke DI stratified-charge engines, which have once again become the focus of intense development activity (see Chap. 3) and, unlike their prototype predecessors [36, 44, 45, 46, 47, 48], have gone into commercial production [39, 40, 41, 42, 49, 50, 51, 52].

In the remainder of this chapter, we describe a series of primarily laser-based measurements that address these issues and summarize the resulting insights. Specifically, we discuss the application of the following techniques:

1. Laser-Doppler velocimetry (LDV)
2. Exciplex laser-induced fluorescence (LIF) imaging of liquid and vapor fuel
3. High-speed planar Mie scattering imaging
4. High-speed spectrally resolved imaging of combustion luminosity
5. LIF of commercial gasoline and of isooctane doped with fluorescent markers
6. Heat-release analysis of cylinder-pressure data
7. Individual-cycle exhaust hydrocarbon (HC) emission measurements using a fast-response flame-ionization detector
8. LIF imaging of NO together with conventional engine-out NOx concentration measurements.

These experiments were performed in a series of optically accessible research engines that operated progressively more realistically and that also reflected progress

---

[7] In automotive applications, *in-cylinder* NO formation tends to be intrinsically lower in a two-stroke engine because, for a given *vehicle* power requirement, it operates at roughly half the *engine* load as its four-stroke counterpart, and NO formation is a strongly non-linear function of temperature and hence of engine load [35].

in the development of experimental three-cylinder and V6 DI2S engines intended for passenger-car application.

2.2 Optical Two-Stroke Research Engines

Over the course of several years, GM's two-stroke engine program was directed first towards development of a three-cylinder engine for compact passenger cars and later towards development of a V6 engine (Fig. 2.2) as an enhanced-performance option (220 brake horsepower at 5600 rpm) for a mid-size sports sedan. Combustion-chamber design, fuel-injection systems and porting all underwent substantial evolution during the course of the program.

As is often the case in development programs, the optical research engines did not always fully reflect the current state of the developmental hardware. This situation was due both to hardware availability and to the need to provide optical access to the cylinder, which requires additional design and fabrication effort and which always entails some compromise in terms of engine geometry and operation.

The first optical engine to be discussed here (Fig. 2.3) operated only under motored conditions with highly simplified cylinder heads mounted on a commercially available three-cylinder two-stroke crankcase and block whose "Schnürle" porting arrangement [32, 53] was similar to that used in the early stages of the three-cylinder engine development effort. For LDV measurements of scavenging flows (Sect. 2.3), whose primary emphasis was on the flows entering the cylinder and the formation of the characteristic scavenging loop pattern (sketched in Fig. 2.1), the combustion-chamber geometry was simplified to a right circular cylinder with a flat piston crown and a flat quartz window as the cylinder head (Fig. 2.3a). Backscatter LDV through this window permitted all three velocity components to be measured without modifying the ports to provide optical access (Sect. 2.3.2.1).

Fig. 2.2 GM experimental CDS2 V6 direct-injection two-stroke engine

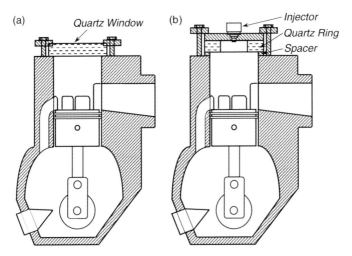

Fig. 2.3 Schematics of first two optical engines. (**a**) Flat-head version used for scavenging-flow measurements. (**b**) Version with central cylindrical bowl in cylinder head used for liquid/vapor fuel visualization

For spectrally discriminated visualization of liquid- and vapor-phase fuel using exciplex fluorescence (Sect. 2.4), the same engine was reconfigured with a centrally located bowl-in-head combustion chamber that was formed by clamping a quartz ring between the cylinder block and a metal plate in which the fuel injector was mounted (Fig. 2.3b). A metal spacer ring below the quartz ring was used to vary compression ratio. Extensive studies of fuel sprays from a variety of injectors were also carried out with simple off-axis combustion chambers using planar laser Mie scattering together with high-speed film and video imaging.

In order to make the most direct use of optical diagnostics in engine development – especially for such design-specific issues as cyclic variability and hydrocarbon emissions – the optical experiments should simulate "real" engine operation as closely as possible. Fuel sprays, mixture preparation, combustion, and emissions formation under realistic firing conditions were studied in a single-cylinder optical research engine (Fig. 2.4a) whose cylinder block, piston, rings, fuel-injection system and lubrication system were all taken from one of the experimental V6 engines. Note that this design follows the *spray-guided* DISI engine concept (see Chap. 3), in which charge stratification is achieved through close spacing of the fuel spray and ignition in both space and time.

To permit high-quality imaging, the rounded contours of the V6 engine's bowl-in-head combustion chamber were modified to a nearly rectangular shape (Fig. 2.4b) and fitted with flat ultraviolet-grade quartz windows that formed two orthogonal sides. The combustion-chamber volume, the chamber-entrance area, the TDC piston-to-head clearance, and the location of the fuel injector relative to the spark plug were the same in the optical and unmodified versions. As discussed in Sect. 2.5.3.2, the optical engine achieved realistic part-load operation as gauged by indicated work, combustion stability, and exhaust emissions.

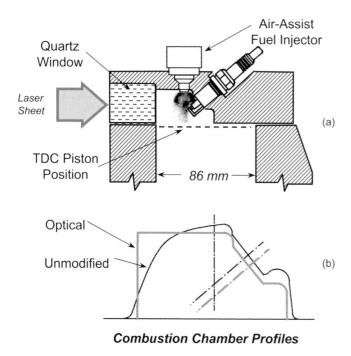

Combustion Chamber Profiles

Fig. 2.4 (**a**) Cross-sectional diagram of combustion chamber of realistic Mod-3 optical single-cylinder research engine used for mixture preparation, combustion and HC emissions measurements. (**b**) Combustion chamber profiles of unmodified and optical engines on enlarged scale

2.3 Scavenging Flow Measurements Using Laser Velocimetry

Prior to this work, comparatively little *quantitative* experimental data were available on flow fields in crankcase-compression two-stroke engines. Systematic laser-Doppler-velocimetry (LDV) measurements (velocity–crank-angle histories of three vector components at nearly 200 spatial locations) were therefore undertaken in a representative engine to characterize in detail both the flow entering the cylinder through the transfer ports and the resulting in-cylinder flow field. Full details can be found in [54, 55]. Like all of the diagnostic work discussed in this chapter, the principal goals of the LDV program were to obtain physical insight and engineering guidance, as well as to obtain an extensive data set under well characterized conditions for assessing computational-fluid-dynamics (CFD) calculations [56, 57]. Early CFD work also used the port-efflux LDV data to establish boundary conditions for calculations of the in-cylinder flow field and combustion without modeling the flow through the transfer passages that connect the crankcase to the cylinder [58].

2.3.1 Engine Configuration and Operating Conditions

As mentioned in Sect. 2.2, the LDV measurements were performed in one cylinder of a commercially available two-stroke engine. To simulate moderate-load, moderate

Table 2.1 Specifications of two-stroke optical research engine used for scavenging-flow measurements

Number of cylinders	1
Displacement	0.5 liter
Bore	86 mm
Stroke	86 mm
Connecting-rod length	191.5 mm
Wrist-pin offset	1.0 mm
Compression ratio (trapped)	6.5

speed passenger-car operation, the engine was motored at a speed of 1600 rpm and a delivery ratio[8] of 0.5. The nominally mirror-symmetric arrangement of the exhaust port and the six transfer ports is illustrated schematically in several of the flow-field representations in Sect. 2.3.3. For engine specifications, see Table 2.1.

2.3.2 Photon-Correlation Laser-Doppler-Velocimetry

Both the LDV system and the data-analysis procedures have been described previously in detail [54, 55, 59]. The LDV apparatus (Fig. 2.5) uses conventional

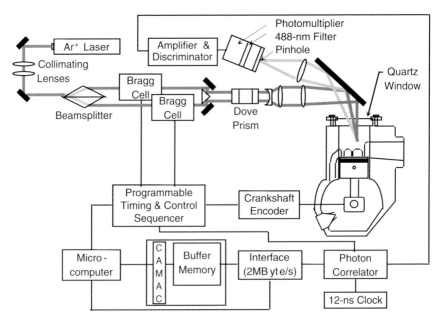

Fig. 2.5 Block diagram of photon-correlation LDV system

---

[8] *Delivery ratio,* a measure of overall engine air flow through a two-stroke engine, is defined as the fraction of a displacement volume of fresh air at ambient conditions that is delivered to the cylinder each cycle.

discrete optical components together with photon-correlation signal processing, which permits the desired flow-velocity information to be extracted from very weak (single-photon) signals despite significant stray-light noise (Sect. 2.3.2.2).

2.3.2.1 LDV Hardware

As in all dual-beam LDV systems, two coherent laser beams (here from an argon-ion laser at 488 nm) are focused to intersect and form a small measurement volume in a flowing fluid. Minute tracer particles ($\sim$0.5-μm silicone oil droplets) scatter light as they cross the interference fringes formed in the measurement volume. The scattered light intensity is therefore modulated at a frequency (the difference between the Doppler-shift frequencies associated with each laser beam) that is proportional to the velocity of the particles normal to the bisector of the intersecting beams.

The LDV apparatus (Fig. 2.5) is a single-velocity-component system that uses a backscatter configuration, dual-Bragg-cell frequency shifting to resolve the velocity-direction ambiguity, rapid data transfer to a large buffer memory, and automated, off-line fast-Fourier-transform (FFT) data reduction of the scattered-light intensity autocorrelation functions (correlograms) formed by a real-time digital correlator. The transmitting optics also include collimating lenses to place the laser beam waists precisely at the beam intersection, a Dove prism to rotate the plane of the intersecting beams and thereby change the direction of the velocity measurement, and a beam expander that reduces the final focal diameter in order to increase the signal intensity and improve spatial resolution. To minimize stray-light acceptance and to reduce the effective length of the measurement volume, scattered light is collected about 15° off axis and imaged onto a pinhole spatial filter. A photo-multiplier tube and signal-conditioning electronics produce single-photon detection pulses of uniform amplitude and duration (10 ns). The entire optical system was mounted on a milling-machine base.

With the beam arrangement shown in Fig. 2.5, measuring the x and y velocity components normal to the cylinder axis was straightforward within the interior of the cylinder. To obtain the mean axial velocity $\langle U_z \rangle$ without optical access through the cylinder wall, LDV data were also collected with the incident laser beams rotated 30° away from the cylinder axis. The mean velocity component measured with this beam orientation was a linear combination of $\langle U_z \rangle$ and $\langle U_x \rangle$ or $\langle U_y \rangle$, from which $\langle U_z \rangle$ was extracted. Velocity measurements at the port-cylinder interface required that the beams used to measure the x and y components be tilted 5° in order to avoid obstruction by the cylinder wall. The beam plane was tilted an additional 30° to measure $\langle U_z \rangle$. With careful alignment, useful port-efflux velocity data were obtained within $\sim$1–2 mm of the piston surface.

2.3.2.2 Photon-Correlation Signal Processing

The attraction of photon-correlation LDV is that (at the cost of non-trivial elec-tronics and software effort) it achieves (1) maximal sensitivity by processing the scattered-light signals at the single-photon level to form their intensity autocorrelation function (ACF) in real time and (2) maximal efficiency in extracting the desired velocity information by (off-line) Fourier transforming the ACF to obtain the power spectrum of the scattered light. This permits velocity measurements under adverse signal-to-noise conditions where conventional LDV frequency trackers and counters fail. For instantaneous velocity measurements from single particle transits (*single-burst* operation), performance comparable to photon-correlation processing has recently become available in the latest generation of commercial Fourier-transform and autocorrelation-based real-time processors [60, 61]. However, the photon ACF can also be accumulated over many successive scattering-particle transits, permitting an ensemble-averaged velocity probability density function (PDF) to be recovered from the power spectrum even when the scattered light is too weak for single-burst operation [59, 62].[9] This is the data-acquisition mode used here, where high flow velocities (>200 m/s) and unfavorable light-collection efficiency often combined to produce detected signal levels of less than one photon per interference-fringe crossing [55, 59].

2.3.2.3 Data Acquisition and Analysis

All the velocity data were acquired as ensemble averages over many engine cycles at specified engine crank angles. Experiment control and synchronization to the engine cycle is achieved by a custom-built programmable sequencer together with a 0.5°-resolution crankshaft encoder and a microcomputer, which also performs off-line data reduction.

The photon correlator (Malvern K7026) uses a shift-register delay-line architecture to form a real-time digital approximation to the autocorrelation function of the scattered light intensity $I(t)$:

$$G(\tau) = \int I(t)I(t+\tau)dt \cong G(k\Delta t) = \sum_{i=0}^{N_W} n_i n_{i+k},$$

where the delay time $\tau = k\Delta t; k = 1, 2, 3, \ldots, 64; n_i$ is the number (typically <1 on average) of photons detected in a brief sampling interval Δt (=12 ns here); and N_W is the number of samples in each data-acquisition window.

The raw data were taken as a series of short-duration "correlograms" by gating the correlator on at each desired measurement crank angle for a brief time (1° or 2°), latching the result into the correlator's output buffer, and transferring the data into buffer memory in parallel with accumulation of the next correlogram. When all the results from the stipulated number of engine cycles (>500 here) had been stored, ensemble-averaged correlograms were constructed by summing all the short-duration correlograms associated with each specified crank angle window.

---

[9] With this mode of data acquisition, photon-correlation LDV is also intrinsically corrected to first order for velocity-sampling bias (more fast particles cross the measurement volume per unit time than slow ones) [62].

When the correlogram is accumulated at fixed crank angle over the transits of many scattering particles during many engine cycles, the coherent Doppler signal emerges clearly from the statistical noise. Each correlogram (e.g., Fig. 2.6a,b) exhibits damped oscillations at the mean Doppler-difference frequency superimposed on a Gaussian pedestal that is due to the laser-beam intensity profile and on a flat (uncorrelated) stray-light background. The damping of the oscillations reflects the distribution of velocities due to turbulence and any other source of flow variation during the accumulation of the correlogram. Fourier transformation of the correlogram yields the power spectrum of the scattered light, in which the velocity PDF appears as a peak (Fig. 2.6c,d). The Gaussian pedestal contributes a broad, near-dc peak to the spectrum which here obscures velocities below −250 m/s, as indicated by the shaded bands.

For any velocity component, the ensemble-mean velocity $\langle U \rangle$ and the ensemble-rms velocity fluctuation u' were evaluated as the first and second moments of the PDF $P(U)$, with the limits of integration taken as the velocities on either side of the Doppler peak at which $P(U)$ had fallen to the estimated noise floor of the spectrum. PDF broadening by the Gaussian beam profile was corrected [55, 59] to eliminate overestimation of u'.

Figure 2.6 demonstrates that this *multi-burst* photon-correlation scheme works well with mean signal levels of substantially less than one photon per Doppler

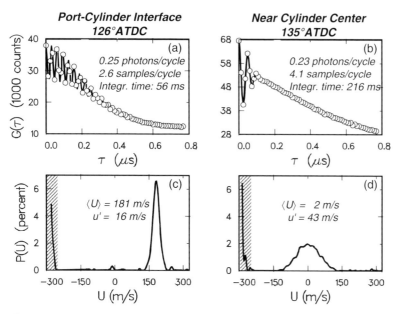

Fig. 2.6 Correlograms (**a,b**) and resulting velocity PDFs (scattered-light power spectra) (**c,d**) for measurement locations at the port-cylinder interface and near the cylinder center. The cross-hatched regions in **c** and **d** indicate the near-DC contribution of the Gaussian beam profile to the Doppler spectra

cycle[10] (fringe crossing) and at frequencies approaching the Nyquist limit of two samples per Doppler cycle, even in the presence of significant stray-light noise (note the uncorrelated background level in the correlograms; in Fig. 2.6a, the LDV measurement volume is 1.3 mm above the piston surface). With an interference-fringe spacing of 15 μm, an applied frequency difference of 20 MHz between the laser beams, and the 12-ns correlator sample duration (Nyquist maximum Doppler frequency of 41.67 MHz), a usable velocity range of −250 to +325 m/s was obtained. This was sufficient to cope with the high flow velocities, strong flow reversals, and high turbulence intensities encountered in the engine.

2.3.3 Scavenging Flow-Field Survey

Figure 2.7 shows mean-velocity vector maps of the flow entering the cylinder through the transfer ports at crank angle 135°ATDC, roughly 15° after they have been uncovered by the piston. The large exhaust port and the six smaller transfer ports are nominally mirror symmetric about an axis (defined here as the x axis) that is offset 60° from the crankshaft axis. The port-efflux velocity profiles are fairly uniform circumferentially and approximately follow the port layout's nominal symmetry with respect to the x axis, although there are some deviations (e.g., flow attachment) at the edges of some of the ports. The angles by which the port-efflux velocities are tilted up out of the x–y plane (not shown) display a similar degree of circumferential uniformity along each port and are mirror symmetric to within a few degrees [54]. As illustrated by the velocity–crank-angle plots in Fig. 2.8, the port-efflux velocity magnitude falls rapidly (an order of magnitude in 2–3 ms) as the crankcase discharges into the cylinder. This behavior parallels the rapid decline in the crankcase-to-cylinder pressure difference (Fig. 2.8a) that drives the scavenging

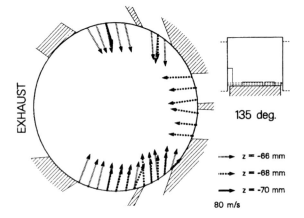

Fig. 2.7 Vector plot showing port efflux velocities at crank angle 135° (overlay of velocity vectors from three planes at distances of 66, 68 and 70 mm below the cylinder head)

135 deg.

EXHAUST

z = -66 mm
z = -68 mm
z = -70 mm
80 m/s

---

[10] Even with photon correlation, usable *single-burst* LDV measurements require at least 1 detected photon per Doppler cycle.

Fig. 2.8 Variation with crank angle of driving pressure difference between crankcase and cylinder (*upper*) together with resulting mean (*middle*) and rms (*lower*) port-efflux (radial) velocities for two points on opposite sides of the nominal port symmetry axis. Solid and dashed arrows indicate measurement locations and corresponding velocity curves

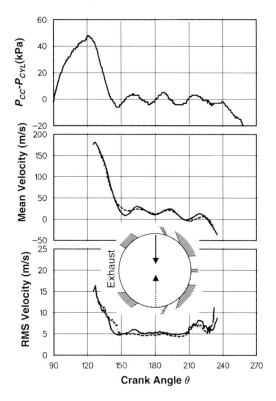

process. Little time is needed to discharge the crankcase because of the comparatively low delivery ratio at this simulated part-load test condition.[11] During this comparatively brief interval (roughly between $120°$–$140°$ ATDC), the port-efflux flows are, in essence, highly transient but well behaved turbulent slot jets, with rms fluctuation intensities $u' \lesssim 10\%$ of the local ensemble-mean velocity, as typified by the radial-inflow PDF of Fig. 2.6c.

Despite the near-symmetry and modest rms velocities of the port-efflux jets, the interaction of these jets with each other and with the chamber surfaces produces an in-cylinder flow field characterized by strong asymmetry and very high rms velocities ($u'/\langle U \rangle \sim 30$–$40\%$, as illustrated by Fig. 2.6d), as well as locally high mean velocities (Fig. 2.9). As the strong inflow diminishes and the crankcase and cylinder pressures equilibrate, the scavenging-loop vortex becomes recognizable in the axial (symmetry) plane (Fig. 2.10), but the flow at the port-cylinder interface becomes complex (note, the spatially non-uniform backflow into two of the transfer ports). By BDC (Fig. 2.11), the overall velocity level has fallen considerably, and the scavenging-loop vortex has become very well defined. A residual effect of the asymmetry in the in-cylinder flow is the development of unintentional swirl, as seen most

---

[11] This also implies a high residual fraction at part load, as mentioned earlier.

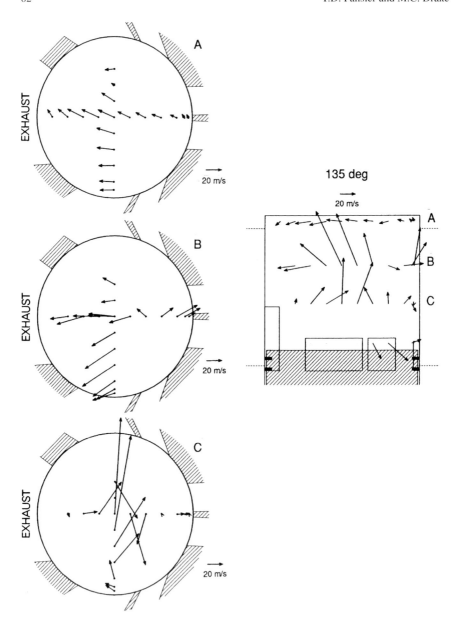

Fig. 2.9 In-cylinder velocity field at CA = 135°

clearly in planes B and C. As the piston compresses the scavenging-loop vortex, causing it first to spin up and then (in this pancake-chamber geometry) to break down into turbulence shortly before TDC, the swirl remains as the only organized motion near TDC [54].

Fig. 2.10 In-cylinder velocity field at 150° ATDC showing both axial plane of nominal symmetry (*upper*) and a diametral cross section just below the tops of the transfer ports (*lower*).

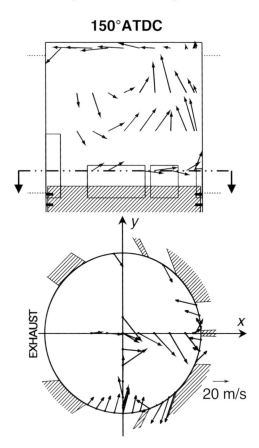

2.3.4 Discussion

2.3.4.1 Comparison to Four-Stroke Engines

To a good approximation, the mean and rms velocities in four-stroke spark-ignition and Diesel engines scale linearly with engine speed because the intake and compression processes are driven by the rate of change of the cylinder volume, which is proportional to engine speed [63]. The mean piston speed $\overline{V}_P = 2NS$ (where N is the engine speed in revolutions per second and S is the engine stroke) therefore provides a useful normalizing parameter that allows meaningful comparison of the valve-efflux and in-cylinder velocities between different engines irrespective of engine speed or volumetric efficiency. In two-stroke engines of the type studied here, however, the gas-exchange process is driven primarily by the discharge of compressed air from the crankcase into the cylinder when the intake ports are uncovered by the piston, and hence, as mentioned in Sect. 2.1, the gas velocities do *not*

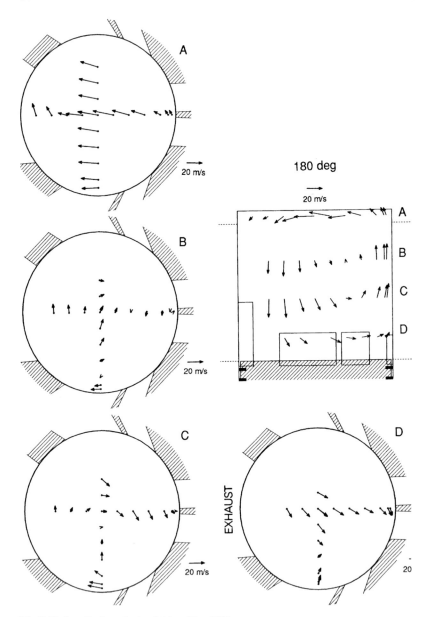

Fig. 2.11 In-cylinder velocity field at CA = 180°

scale in any simple way with engine speed.[12] This is probably why velocities in two-stroke engines have generally not been expressed in terms of mean piston speed. Two alternative reference velocities for two-stroke engines are discussed briefly in footnotes below.[13,14]

Normalization by the mean piston speed is nevertheless an instructive way to place the range of velocities observed in this study in the context of four-stroke SI engines [63], bearing in mind that the normalized velocities will vary with engine speed and delivery ratio. For the present engine, $\overline{V}_P = 4.59$ m/s at the operating condition of $N = 1600$ rpm and delivery ratio $\eta_{del} = 0.5$. The largest mean velocities, which occur at the port exits during scavenging, are in the range of about 40–$50\overline{V}_P$ (Fig. 2.6b), while the largest rms velocities are about 8–$10\overline{V}_P$ and are found near the cylinder center (Fig. 2.6d). These values are large on the scale of four-stroke SI engines, where the maximum mean and rms velocities during intake are typically in the ranges of 3–$10\overline{V}_P$ and 1–$2\overline{V}_P$, respectively. In four-stroke engines with simple open chambers and in the absence of intake-generated tumble and strong squish flows, the rms velocities decay after intake to values of about 0.3–$0.5\overline{V}_P$ at typical ignition times shortly before compression TDC. Intake-generated tumble, which occurs naturally in four-valve pent-roof engines, leads to essentially the same vortex spin-up/breakdown process during compression as that mentioned in Sect. 2.3.3 for the two-stroke scavenging-loop flow. This process enhances turbulence, leading to rms velocities just before TDC of about 0.8–$1\overline{V}_P$ in four-stroke engines, values which are only slightly less than the rms velocities of about 1–$1.2\overline{V}_P$ measured just before TDC in the two-stroke engine studied here at the stated operating condition [54].

2.3.4.2 Implications for DI2S Engine Design and Operation

The impulsive character of the crankcase discharge into the cylinder and the high initial port-efflux velocities (~ 200 m/s) are intrinsic to crankcase-compression

---

[12] In-cylinder velocities in two-stroke engines with external scavenging driven by a blower connected to the crankshaft do scale with engine speed.

[13] A reference velocity based on, say, delivered air mass per cycle, ambient air density and total port area, i.e., $\overline{V}_{ref} = \dot{M}_{del}/\rho_{amb}A_{port}$, might appear to be preferable to $\overline{V}_P$ for two-stroke engines. Note, however, that *any* reference velocity that involves air mass per engine cycle implicitly involves engine speed and is hence proportional to mean piston speed. For the example here, the mass of air delivered per cycle $\dot{M}_{del} = \rho_{amb}\eta_{del}V_D N$, where η_{del} is the delivery ratio, and V_D is the displacement volume. But $\overline{V}_D = A_{bore}S$, where A_{bore} is the bore area, and hence $\overline{V}_{ref} = \eta_{del}(A_{bore}/A_{port})\overline{V}_P/2$. For the present work, $A_{port} = 19.9$ cm^2, so $\overline{V}_{ref} = 3.35$ m/s.

[14] The ideal (discharge coefficient = 1) port-efflux velocity when the transfer ports are first uncovered could be evaluated from the orifice-flow equations and used as a reference velocity that is independent of engine speed. This approach requires conditions in the crankcase at port opening to be assumed, measured or calculated. For this study, the crankcase pressure data yield $\overline{V}_{port,ideal} \approx 250$ m/s. Such a reference velocity might be useful for comparing flows between two-stroke engines, but it seems unlikely to be helpful for comparing in-cylinder velocities between two- and four-stroke engines.

scavenging, although they could be eliminated by external (blower) scavenging. The observed flow asymmetry, including the appearance of appreciable swirl, diminishes scavenging effectiveness by increasing short circuiting and fresh-residual gas mixing [32, 53]. CFD calculations [56, 57] were able to reproduce many of the major features of the LDV data, but predicted much less asymmetry, even though the calculations included a digitized representation of the transfer passages and ports. The CFD results are very sensitive, however, to small (sub-mm) geometric changes such as a difference in the heights of the transfer ports on opposite sides of the symmetry plane.[15] The sensitivity to small perturbations, the very high ensemble-rms velocity fluctuations measured within the cylinder (which are consistently underpredicted by CFD), and the bimodal velocity PDFs observed at some locations ([54]; not shown here) all suggest that reliance on ostensibly symmetric colliding-jet flows to establish the desired scavenging-loop flow structure may lead to large-scale instabilities from one cycle to another. Finally, the observed flow asymmetry, the sensitivity to small geometric variations (such as might easily occur in manufacturing), and the possibility of large-scale cyclic flow-field variation are all undesirable in emissions-regulated applications and require an improved port design that is less susceptible to these problems.

2.4 Liquid/Vapor Fuel Visualization Using Exciplex Fluorescence

For an initial study of air-assist fuel sprays and charge stratification, the exciplex laser-induced fluorescence (LIF) technique [64, 65, 66, 67] was applied to simulated (non-firing) near-idle conditions in the same engine as used for the LDV study, but with a centrally located bowl-in-head combustion chamber (Fig. 2.3b). The major attraction of the exciplex technique is that it enables spectrally separated imaging of liquid and vapor phases of the fuel. The technique has become well known and will only be discussed briefly, together with a set of representative engine results. Additional details, results of exciplex imaging in an atmospheric-pressure test rig, and CFD modeling of the sprays can be found in [68]. Differences between the exciplex technique, other approaches to LIF imaging of fuel, and Mie-scattering spray imaging are discussed later in Sect. 2.5.2.2.

2.4.1 Exciplex Liquid/Vapor Visualization

With the exciplex technique, small quantities of organic dopants are added to the fuel. When excited by ultraviolet (UV) light, the dopants produce different-colored fluorescence from the liquid and vapor phases. Following [66, 67], we use

---

[15] The LDV data suggest that jets from opposing ports arrive at the symmetry plane at slightly different times, which could be due to small differences in port heights. A slightly tilted piston would have much the same effect [54].

decane fuel doped with naphthalene and TMPD (n,n,n',n'-tetramethyl-p-phenylene diamine) in the proportion 89:10:1 by weight.

Exciplex-visualization photophysics is discussed in detail in [64, 65, 66, 67]. In brief, a TMPD molecule can be driven from its ground state to an electronically excited state (denoted TMPD*) by absorption of an ultraviolet photon. In the gas phase, the TMPD* molecule can return to the ground state either by emission of a longer-wavelength fluorescence photon or by collisional quenching (i.e., collisional de-excitation without emission of light, which is especially severe when oxygen is present). In the liquid mixture, however, TMPD* can bind with naphthalene to form an *exci*ted-state com*plex*, or *exciplex*, a molecule which is bound only in an excited state and which has no stable ground state. Relative to fluorescence from isolated TMPD*, the fluorescence from the TMPD*-naphthalene exciplex is red-shifted by the binding energy of the exciplex. The concentrations of TMPD and naphthalene in the two phases can be arranged to ensure that essentially all the TMPD* molecules bind to naphthalene in the liquid and essentially none bind in the gas phase. The net result is spectrally separated fluorescence from the two phases: purple emission from the vapor and green emission from the liquid.

2.4.2 Planar Exciplex Imaging System

Figure 2.12 is a schematic diagram of the experiment. The output beam of a pulsed, frequency-tripled Nd:YAG laser (355 nm wavelength, 35–40 mJ/pulse energy) was formed into a thin vertical sheet ($\sim$0.5 $\times$ 40 mm) and directed through the transparent quartz head of the engine. Fluorescence excited in a planar slice along the axis of the fuel spray was collected at 90° to the incident light sheet. Note that the light sheet was parallel to the crankshaft axis and was therefore at an angle of 60° to the nominal symmetry plane of the ports.

Simultaneous single-laser-shot liquid- and vapor-phase images were obtained by forming two images of the combustion chamber with a commercial glass biprism [66], isolating the liquid (green, 500-nm) and vapor (purple, 400-nm) fluorescence with narrow-band filters, and recording both filtered images with a cooled, image-intensified vidicon camera system.[16] Neutral density filters were sometimes needed to attenuate the liquid fluorescence. The laser pulse duration (10 ns) and the fluorescence lifetime ($<$100 ns) were so brief that all motion was frozen. Because the fluorescence is strongly quenched by collision of the excited molecules with oxygen, the cylinder of interest was motored on nitrogen, while the other two cylinders were motored on air.

Ideally (i.e., with an optically thin spray, a spatially uniform laser light sheet, and perfect co-evaporation of the fuel and dopants), the fluorescence intensity incident

---

[16] This vidicon system was vastly inferior to the slow-scan, intensified CCD camera used later for the fuel LIF imaging described in §4, and we will therefore not try the reader's patience with its idiosynchrasies. For details, consult [68].

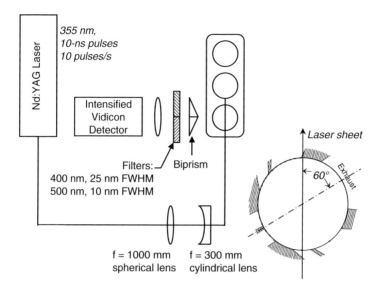

Fig. 2.12 Block diagram of exciplex LIF experiment to visualize liquid and vapor fuel distributions in one cylinder of a two-stroke engine. The port diagram at the lower right shows the 60° offset between the port layout's nominal symmetry plane and the laser sheet, which propagates parallel to the crankshaft

on each detector pixel is proportional to the mass of liquid or vapor fuel present in the corresponding physical volume element of the test section. However, several limitations complicate the exciplex image interpretation [69, 70]:

1. The TMPD-naphthalene exciplex system matches the vaporization characteristics of hydrocarbon fuels considerably less volatile than most gasoline constituents. Decane was therefore used as the base fuel in these experiments, and the intake nitrogen was heated to raise the crankcase temperature to 40°C. Exciplex dopants which better match the vaporization of gasoline have been developed more recently [71, 72, 73].

2. The short-wavelength tail of the liquid's fluorescence spectrum overlaps the pass band of the vapor-fluorescence filter [66, 67]. This crosstalk allows liquid to masquerade as vapor at the level of some 10% of the liquid fluorescence intensity, as estimated by imaging a large drop of liquid through the vapor-fluorescence filter. We ignore this effect here.

3. Even under ideal circumstances, the images here provide *relative* rather than absolute mass distributions. (Substantial effort has been devoted to calibration techniques for absolute liquid and vapor concentration measurements [74, 75, 75], which have proven difficult to perform reliably in engines, due in large part to the temperature dependence of the absorption and emission spectra of the dopants [69, 70, 74, 77]). Furthermore, we have not attempted to "flat-field" correct the images for either the spatially non-uniform sensitivity of the detection system or for shot-to-shot variations in the overall laser intensity or its spatial

distribution. Multiple-scattering effects, detector saturation and image blooming can also compromise the interpretation of the images.

4. The laser pulse repetition rate (10 Hz) and vidicon camera readout time were too slow to permit more than one image to be recorded during a single injection event. All the images were recorded in different engine cycles.

2.4.3 Engine Configuration and Operating Conditions

The cylinder-head geometry was shown earlier in Fig. 2.3b. A cylindrical bowl-in-head combustion chamber was formed by clamping a quartz ring with a 48-mm diameter by 20-mm high opening on top of a 6.5-mm-high aluminum spacer ring to one cylinder of the engine block. The effective compression ratio was 4.0.

The measurements were carried out at 600 rpm engine speed and a delivery ratio of 0.5. To promote vaporization of the decane fuel and TMPD marker, the nitrogen was heated to produce a gas temperature of 40°C in the crankcase, producing an estimated temperature in the cylinder at the start of injection of 100°C.

An experimental electronically controlled air-assist fuel injection system was used. The injector had a solenoid-driven outwardly-opening poppet with an included seat angle of 80°, and the injector and cylinder axes coincided. The injection pressure was 540 kPa, and the injection duration was 4 ms.

2.4.4 Exciplex Liquid/Vapor Imaging Results

Figure 2.13 contains examples from a series of exciplex-LIF liquid/vapor image pairs recorded in the motored engine from the start of injection through typical light-load ignition times. The outlines of the cylindrical combustion bowl and the injector tip are indicated. The liquid and vapor fluorescence intensities are encoded on linear 7-level gray scales from low (light gray) to high (dark gray). The boundary between white and the lightest gray is a threshold chosen to suppress vidicon detector noise. The spatial resolution of the images is approximately 1 mm/pixel.

The liquid spray initially emerges from the injector with an included angle greater than the poppet seat angle of 80°. Asymmetry is observed immediately, with liquid and vapor penetrating farther on the right side (crank angles 61° and 59° BTDC). Because the laser sheet travels from left to right in the images, this asymmetry is not an artifact due to attenuation of the incident light by the spray. As injection proceeds, the liquid spray and the vapor distribution both collapse and fill in along the axis. As has been known for many years [78], hollow-cone sprays collapse because entrainment of air into the spray produces a low-pressure region downstream of the injector. Towards the end of injection (the poppet is expected to close at 47° BTDC), the vapor distribution becomes distorted toward the right (exhaust-port) side of the chamber. This distortion is consistent with vapor transport by the remnant of the scavenging-loop flow.

Fig. 2.13 Series of
simultaneous images of
instantaneous liquid and
vapor fuel distributions in
motored DI2S engine
obtained by exciplex LIF. The
outlines indicate the
boundaries of the cylindrical
bowl and the injector tip.
Note that the bottom two
rows of images have been
rescaled

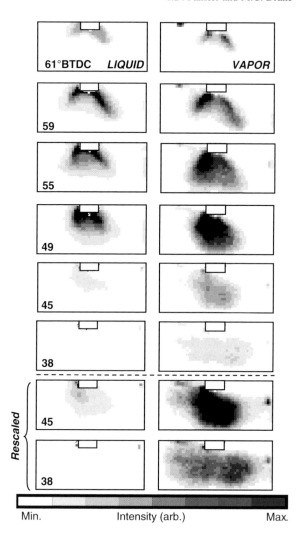

The liquid fluorescence intensity declines rapidly after the end of injection. The last two pairs of images in Fig. 2.13 are rethresholded and rescaled versions of the images at 45° and 38° BTDC. The liquid fluorescence signals are at most a few percent of the maximum signal level in the earlier images. These results suggest that *nearly all the liquid in the combustion chamber has evaporated by typical light-load ignition times* (30°–40° BTDC). Given the low compression ratio here (4.0) and the comparatively high boiling points of the dopants (TMPD: 265°C; naphthalene: 218°C) and of the decane base fuel (174°C), vaporization would be expected to be even more rapid for gasoline injection at full compression ratio (typically 6–7 for a two-stroke SI engine, as in our other experiments).

Large-scale transport and small-scale mixing cause the vapor distribution to broaden in extent and diminish in fluorescence intensity after the end of injection (considerable volume is available below the field of view). An important observation is that *steep concentration gradients* (twofold changes in fluorescence intensity over distances of 2–4 mm) *exist at the edges of the fuel-vapor cloud at typical light-load ignition times*. This suggests that cycle-to-cycle variation in the fuel-vapor concentration (especially in the local concentration at the spark gap) could be quite important to combustion stability at light load, where (as here) injection is delayed to concentrate the fuel around the spark gap and ignition is typically timed near the end of injection.

2.5 Mixture Preparation and Combustion in a Realistic Optical Two-Stroke Engine

In the remainder of this chapter, we discuss systematic measurements of stratified-charge mixture preparation and combustion in the optical research engine of Fig. 2.4, which accurately reproduced part-load operation of an experimental V6 passenger-car engine [79, 80, 81, 82]. Given the complex phenomena and coupled processes involved in mixture preparation and combustion, our approach has emphasized the systematic use of simultaneous, complementary diagnostics that provide both quantitative and qualitative information. Specifically, conventional cylinder-pressure measurements and analysis (which yield indicated work, mass-burning rate, mass-burned fraction, etc., for each engine cycle) are coupled with continuous, high-speed imaging (planar Mie scattering from liquid fuel; combustion luminosity) or with planar LIF imaging of fuel (liquid and/or vapor) or nitric oxide. The use of conventional and fast exhaust-hydrocarbon (HC) sampling in conjunction with high-speed imaging is also described.

2.5.1 Optical Engine Hardware

2.5.1.1 Engine Configuration

Figure 2.14 is a slightly simplified cross-sectional schematic that supplements Fig. 2.4 with additional details of the engine and its instrumentation. Table 2.2 summarizes engine specifications.

2.5.1.2 Fuel Injection

Although several different types of fuel injectors have been examined in this engine, the remainder of this chapter presents results obtained with two air-assist fuel injectors. Both were outwardly opening poppet injectors; their only significant difference was that their seat geometries were designed to create wider or narrower sprays, as illustrated by the planar Mie-scattering images in Fig. 2.15. These images are

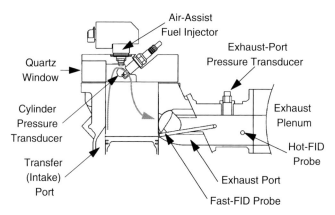

Fig. 2.14 Mod-3 optical two-stroke schematic, including fast-FID probe, hot-FID probe and pressure transducers

Table 2.2 Specifications of realistic two-stroke optical two-stroke research engine used for mixture-preparation and combustion measurements

Number of cylinders	1
Displacement	0.5 liter
Bore	86 mm
Stroke	86 mm
Connecting-rod length	191.5 mm
Wrist-pin offset	1.0 mm
Compression ratio (trapped)	6.5
Port timings (nominal) referred to piston top	
Exhaust port opening (EPO)	90° ATDC
Exhaust port closing (EPC)	270° ATDC
Intake[28] port opening (IPO)	120° ATDC
Intake port closing (IPC)	240° ATDC
Fuel	Amoco 91 RON

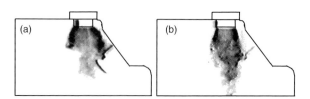

Fig. 2.15 Planar Mie-scattering images of fuel sprays from the (**a**) wide-spray and (**b**) narrow-spray injectors at approximately 40° BTDC. The images are digitized from a high-speed movie taken with the optical engine operated at Test Point 1 (see Table 2.3)

digitized from high-speed movies that were taken with the optical engine operating under the conditions of Test Point 1, as defined below. The different spray widths and penetrations led to different combustion stability and average HC emissions at light loads, as shown in Sect. 2.5.3.3.

With both injectors used here, compressed air was continuously supplied to the injector body at a pressure of 540 kPa. The fuel to be injected was pressurized to 620 kPa and was admitted into the injector body by a conventional port-fuel injector. This fuel-metering event ended 3 ms before the start of fuel injection into the cylinder.

2.5.1.3 Probe Diagnostics for Cylinder Pressure and Exhaust Emissions

Piezoelectric pressure transducers mounted in the rear wall of the combustion chamber, in the crankcase and in the exhaust port were used to record pressure-time data (0.5° crank angle intervals; 14-bit resolution). The cylinder-pressure data were analyzed with a two-stroke formulation of the single-zone analysis approach of Gatowski et al. [83].

Time-resolved exhaust-HC concentrations were measured with a Cambustion HR200 fast-response flame-ionization detector (FID), whose probe was close-mounted to minimize mixing and sampling response time ($\sim$2 ms here). *Time-averaged* HC concentrations were measured with a conventional heated-line FID located further downstream in the exhaust passage. Full details of the exhaust-sampling techniques, including the use of cylinder- and exhaust-pressure data to calculate the individual-cycle exhaust mass flow rate and thereby obtain cycle-resolved HC mass data, can be found in [82].

In conjunction with LIF imaging of in-cylinder NO (Sections 2.5.2.3 and 2.5.7.2), time-averaged exhaust NOx concentrations were measured in a few experiments using a Thermo-Electron chemiluminescent analyzer.

2.5.2 Optical Diagnostics

Figure 2.16 is a schematic diagram of the laser-based imaging diagnostics. Essentially the same layout was used for high-speed film/video imaging and for fuel LIF and NO LIF imaging, but with different lasers and cameras.

2.5.2.1 High-Speed Film and Video Imaging

The light source for high-speed imaging was a copper vapor laser (CVL) operating at wavelengths of 511 and 578 nm simultaneously and producing pulses of 2–3-mJ energy and $\sim$30-ns duration at rates up to 20000 pulses/s. Cylindrical lenses transformed the round laser beam into a thin ($\sim$0.5 mm thick $\times \sim$40 mm high) laser sheet, and mirrors directed it through one of the combustion chamber's quartz windows along the injector axis.

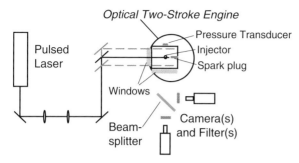

Fig. 2.16 Schematic diagram of laser-based imaging apparatus. Translating the final beam-steering mirror moves the laser sheet through the combustion chamber

Laser light that was Mie-scattered by the liquid fuel, together with flame luminosity, was collected at right angles through the second quartz window and imaged onto either a high-speed 16-mm film camera (Redlake HyCam) or, using a beam splitter, onto a pair of image-intensified high-speed video cameras (Kodak EM1012 with UVX imager). Both camera systems used quartz 105-mm focal length, f/4.5 Nikkor ultraviolet lenses. The cameras were typically operated at 4000 frames/s, with the CVL synchronized to fire once per film or video frame.

Spectrally resolved combustion imaging was sometimes performed by placing interference filters (10-nm bandpass) in front of each video camera to isolate combustion luminosity from selected regions of the UV and visible spectrum. Typical filter wavelengths were 310 nm (which passed ultraviolet chemiluminescence from the OH radical at 306 nm), 430 nm (CH-radical chemiluminescence and CO-O recombination continuum emission), and 520 nm (C_2-radical emission at 516 nm and CO-O and soot continuum emission). Images of the fuel spray were recorded by slightly tilting the 520-nm filter, thereby permitting an adjustable fraction of the 511-nm laser light that was Mie-scattered from the liquid droplets in the spray to reach the camera.

These techniques provided a continuous record over many consecutive engine cycles of the liquid fuel spray (Mie scattering), of ignition and combustion (flame emission, either broadband or spectrally resolved), and of condensed-phase constituents of crevice flows (Mie scattering). Note that since Mie scattering occurred only in the laser sheet, it provided a two-dimensional "slice" image through the liquid fuel distribution. The combustion luminosity, however, was integrated from the front to the rear of the combustion chamber due to the depth of field of the camera lens.

High-speed film imaging provides excellent spatial resolution and wide dynamic range with a nonlinear intensity response function that mimics the response of the human eye [84], but color film does not respond well to the UV (OH) and deep-blue (CH) emissions of lean and stoichiometric premixed combustion. High-speed video has the advantage of essentially instant turnaround since there is no need to develop film. The UV sensitivity and image intensification of the high-speed video cameras are also necessary for spectrally resolved imaging. With the equipment used here, these capabilities came at the price of limited dynamic range ($\sim$6 bits) and spatial resolution (240 × 192 pixels full-frame at 1000 frames/s and reduced stripwise to achieve higher framing rates). Newer high-speed CMOS digital imaging systems provide much higher frame rates ($\sim$100000 frames/s), spatial resolution, and dynamic range (10–12-bits). Compatible image intensifiers are available, although they always degrade spatial resolution and dynamic range somewhat.

Two variations on the theme of high-speed planar Mie imaging that we have applied but will not illustrate in any detail here deserve a brief mention. First, performed in an "injector du jour" mode of operation, planar Mie scattering has proven to be a rapid and effective way to obtain practical information under realistic in-cylinder conditions (flow field and gas density) on spray structure and penetration, cycle-to-cycle and injector-to-injector repeatability, secondary injections, dribble, etc. The results observed in motored and fired engines have sometimes differed significantly from those obtained in atmospheric-pressure and pressurized-vessel tests. Second, to a lesser extent, we have used planar Mie scattering with high-speed film and video to visualize early flame-kernel growth by recording light scattered from minute silicone-oil droplets seeded into the intake air so that the scattered light marks the *un*burned gas [85, 86, 87].

2.5.2.2 Fuel LIF Imaging

Liquid- and vapor-phase fuel have been imaged using a frequency-quadrupled Nd:YAG laser (266-nm wavelength, 10-ns pulse duration, $\sim$10–40-mJ pulse energy, $\sim$10-Hz pulse repetition rate) to excite fluorescence either from undoped commercial gasoline or, in a few auxiliary experiments (Sect. 2.5.5.3), from high-purity isooctane doped with a ketone tracer (acetone, 3-pentanone or 5-methyl-2-hexanone) [79, 80, 81]. The gasoline fluorescence emission spectrum spans the range 300–400 nm, with a peak at about 340 nm, whereas the ketone fluorescence extends to longer wavelengths (about 300–550 nm, peaking around 420 nm). In contrast to exciplex LIF (Sect. 2.4), this approach does not distinguish spectrally between the liquid and vapor phases, but the measurements can be performed more readily under normal firing conditions.

The optics for LIF imaging are very similar to those for high-speed imaging, except that quartz lenses are used to form a $\sim$0.5 × 50-mm laser sheet. The central 25-mm portion of the sheet height, over which the intensity profile is most uniform, is selected by cropping with an iris before directing it into the engine through one of the UV-grade quartz windows. Light emitted or scattered perpendicular to the laser beam exits the second UV quartz window and is imaged by a quartz camera lens (Nikkor UV lens, 105-mm, f/4.5) onto a digital computer-controlled intensified camera [Princeton Instruments (PI) ICCD-576]. Light at the laser wavelength (from specular or diffuse reflections off solid surfaces, Mie scattering and Rayleigh scattering) was strongly attenuated by a long-pass glass filter (WG-305 or WG-345).

A colored glass bandpass filter (BG-25 or 7-51) was sometimes also used to reject combustion luminosity at wavelengths longer than the fuel-fluorescence spectrum.

The PI camera has high dynamic range (10–12 bits), low noise (cooled detector to minimize dark noise and slow readout to minimize readout noise), wide spectral sensitivity (180–800 nm), high quantum efficiency ($\geq$12% at 300 nm), adjustable gain (from $\leq$1 to 80 counts/photoelectron), approximately linear intensity response at a given gain setting, and adjustable electronic gating (3.5 ns–80 ms). One disadvantage of the PI camera system is its relatively modest spatial resolution. The detector array has 576 $\times$ 384 pixels, but the fiber-optic-coupled intensifier degrades the resolution by about a factor of two. To reduce the data-transfer and -storage demands, we therefore bin the pixels 2 $\times$ 2 and acquire partial-frame images (230 $\times$ 135 binned pixels).

The engine crank angle, laser pulse timing and camera intensifier gate duration (0.2 μs) are all electronically synchronized, and the camera gain was adjusted for each engine condition. Typically either three 16-laser-shot average images or up to 100 single-shot images were taken. Data acquisition was usually completed within 40 s, but one experiment spanning a six-minute period of continuous firing was conducted (Sect. 2.5.8.2).

Fuel LIF is a very useful complement to laser Mie scattering and combustion-luminosity imaging. LIF can reveal the vapor fuel distribution, which Mie scattering and flame luminosity cannot. Furthermore, under certain conditions (to be discussed in Sect. 2.5.5.3), the LIF intensity incident on each detector pixel is proportional to the mass of fuel (whether liquid or vapor) and the intensity of the laser light sheet within the physical volume element corresponding to that pixel. In contrast, the Mie-scattering intensity is a complicated function of drop size, shape, and number density, which are not usually known with certainty in in-cylinder experiments.

In contrast to exciplex LIF, the fluorescence from gasoline (or from tracers such as ketones that can be added to non-fluorescing single-component fuels such as isooctane; see Sect. 2.5.5.3) does not distinguish spectrally between the liquid and vapor phases. Under the conditions of the experiments described here, however, vaporization is very rapid. Indeed, as illustrated in Fig. 2.17, comparison of

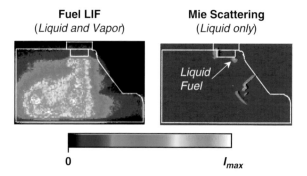

Fig. 2.17 Comparison of planar LIF (*left*) and Mie-scattering (*right*) images of fuel at 1° before ignition at TP3

Mie-scattering images (which are sensitive indicators of the presence of liquid) with fuel-LIF images shows that essentially all of the LIF at $1°$ before ignition is typically due to vapor-phase fuel, provided ignition does not occur during injection.

2.5.2.3 Nitric Oxide LIF Imaging

NO LIF imaging in a firing engine is considerably more difficult than fuel LIF imaging. Reasons include the need for tunable UV laser light, the effects of pressure and temperature on the NO absorption and emission spectra, and interfering absorption and fluorescence from other species.

Diatomic molecules absorb and emit light only at specific narrowly defined wavelengths, in contrast to the complex organic species used for fuel LIF imaging, which have broad, continuous absorption and emission spectra. The tunable UV light needed to excite NO molecules from the lowest vibrational level of the ground electronic state [written as $X^2\prod(v''=0)$ in spectroscopic notation] to the first electronic excited state [$A^2\sum(v'=0)$] was generated by using the second harmonic of the Nd:YAG laser to pump a tunable dye laser that operated on a mixture of Rhodamine 610 and 590 dyes. The dye-laser output at 574 nm was then frequency doubled to 287 nm and mixed with the residual 1064-nm Nd:YAG laser fundamental to obtain ~1-mJ pulses of tunable UV light near 226 nm. The laser beam was formed into a $\sim0.5 \times 12$ mm laser sheet and directed along the mid-height of the combustion chamber.

Fluorescence from the NO $A^2\sum(v'=0)$ excited state was isolated from the laser light and combustion luminosity with filters (a UG5 colored-glass filter and a 10-nm-bandpass interference filter centered at 249 nm) and recorded by the same image-intensified CCD camera that was used for the fuel LIF imaging. A preliminary experiment in which the LIF spectrum from the firing engine was recorded with a diode-array spectrometer showed only emission from the NO $A^2\sum(v'=0)$ level to the $X^2\prod(v''=1-4)$ vibrational levels, thus ruling out significant interference by fluorescence from polycyclic hydrocarbons or from vibrationally excited (i.e., hot) O_2, which has many absorption lines near 226 nm [88].

At room temperature and atmospheric pressure, the NO absorption lines are narrow [comparable to the laser line width (0.4–0.6 cm^{-1})], but at high pressures they shift in wavelength and broaden substantially [88]. The shifts are large enough that a laser tuned exactly to the center of an isolated absorption line at atmospheric pressure might scarcely excite NO molecules at ~20 atm. To minimize this problem, the laser was tuned to a region [specifically, the low-frequency (i.e., long-wavelength) side of the Q_1 bandhead at 226.0 nm] where several closely spaced rotational lines overlap even at low pressure.

NO fluorescence quenching cross sections vary strongly with the species of the collision partner and with temperature [89, 90], but calculations indicate that, fortuitously, *total* (all species) quenching cross sections do not vary greatly for premixed flames and engine combustion [91, 92].

At the low laser-sheet intensity used here, the NO LIF intensity should be proportional to the local NO concentration and the local laser intensity. To obtain more accurate images of relative NO concentration, a correction for the non-uniform laser-sheet intensity was performed by dividing the LIF images by a multi-shot-average "flat-field" reference image that was acquired under static conditions with the combustion chamber filled with a uniform mixture of 240 ppm NO in N_2 at atmospheric pressure and room temperature. Because of the shorter wavelengths involved, laser and fluorescence absorption by deposits on the combustion-chamber windows was significantly stronger for NO LIF than for fuel LIF, requiring the windows to be cleaned more frequently and flat-field reference images to be recorded for essentially every test.

2.5.3 Optical Engine Performance

2.5.3.1 Engine Test Conditions

The optical and other diagnostic measurements to be described subsequently were performed at the first and third of a set of six steady-state speed-load conditions that our GM colleagues devised to simulate key portions of the US FTP driving cycle for a 3.0-L V6 DI2S engine in a mid-size sports sedan. Table 2.3 lists the baseline operating conditions of the optical engine with the wide-spray injector at the two selected test points in detail; the values listed for the engine airflow and the fuel-injection and ignition timings yielded the minimum coefficient of variation (COV) of indicated mean effective pressure (IMEP) during steady, warmed-up operation.

Test Point 1 (TP1: 900 rpm, 100 kPa IMEP), the lowest speed-load condition of the six-point FTP simulation, is of particular interest because HC emissions from stratified-charge engines tend to be dominated by light-load operation. Specifically, TP1 accounted for 46% of the total engine-out HC emissions on the six-point FTP simulation but only 4% of engine-out NOx. *Test Point 3* (TP3: 1500 rpm, 180 kPa IMEP), representing a moderate cruise condition, was in the transition regime between light loads (characterized by late injection and highly stratified

Table 2.3 Engine operating conditions for mixture-preparation and combustion measurements

Test point	1	3
Engine speed (rpm)	900	1500
Coolant temperature (°C)	90	90
Intake air flow (kg/hr)	7.6	12.5
Spark timing (°BTDC)	30	39
Nominal injection advance (°BTDC)	59	95
Injection duration (ms)	4.0	5.0
Inj. fully opened (°BTDC)	52.0	83.4
Inj. fully closed (°BTDC)	29.4	36.7
Avg. fuel injected (mg/cycle)	4.5	6.4
Avg. air injected (mg/cycle)	3.7	6.2
Delivery ratio	0.23	0.25
Equivalence ratio (trapped)	0.44	0.69
Residual fraction	0.67	0.65
IMEP (kPa)	100	180
COV(IMEP), typical	7%	2.5%
Combustion efficiency, typical	92%	98%

combustion) and heavier loads (characterized by earlier injection and more premixed combustion). TP3 generated approximately 10% of the total engine-out HCs and 18% of engine-out NOx emissions on the simulated FTP cycle. TP1 and TP3 each accounted for about 20% of the fuel consumed in the FTP simulation.

For all tests, the engine was continuously fired (usually on commercial gasoline), and the engine block and head were typically heated to 90°C before beginning an experiment. To reduce window fouling, the engine was typically fired only 30–60 s before beginning optical measurements and therefore approached but did not achieve fully warmed-up operation. Continuous firing was required by the very high residual fractions (approximately 2/3 at our test conditions!) that are characteristic of two-stroke engines at light load. Skip firing, which is often used in optical-engine experiments to reduce window fouling and thermal/mechanical stresses, would have led to completely unrealistic temperatures and concentrations of fuel, air and residuals in the cylinder.

2.5.3.2 Comparison of Optical and Unmodified Engine Performance

To assess how well the optical experiments could approximate "real" engine operation, we tested our single-cylinder engine under steady-state, fully warmed-up conditions at TP1 and TP3 both with the optical-access cylinder head and with an unmodified, all-metal cylinder-head sectioned from a V6 two-stroke engine. Fig. 2.18 compares the results for engine-out time-averaged HC emissions and

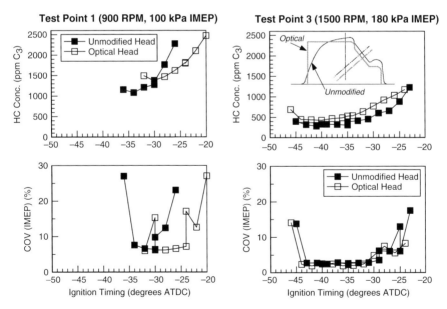

Fig. 2.18 Comparisons of engine performance with unmodified and optical cylinder heads at TP1 (*left*) and TP3 (*right*) using the wide-spray injector. Unburned HC concentration measured by conventional heated FID (*upper*) and COV(IMEP) (*lower*) are plotted against ignition timing at fixed injection advance

combustion stability [COV(IMEP)] as a function of ignition timing for fixed injection timing. The results indicate that the modifications to obtain optical access did not seriously alter engine operation. The single-cylinder results also agreed with data from V6 engines and other non-optical single-cylinder engines under the same test conditions.

2.5.3.3 Effects of Fuel Sprays on Combustion and HC Emissions

As mentioned in Sect. 2.5.1.2, there were notable differences in engine performance with the wide- and narrow-spray injectors, as illustrated by the engine-out HC concentration data as a function of spark timing in Fig. 2.19 for Test Point 1.

Three points are of particular interest here. First, the minimum engine-out HC concentration with the narrow-spray injector is lower by a factor of two at this test condition. Second, for fixed injection timing, optimum performance [minimum HC and COV(IMEP)] with the narrow-spray injector requires substantially earlier ignition (during mid-injection rather than at the end of injection). Third, the results for both injectors lie along essentially a single curve of HC concentration vs. spark advance, with the minimum delay from injection to ignition (consistent with stable combustion) leading to the minimum HCs. This is classic evidence that a major HC emission mechanism in stratified-charge SI engines is over-mixing of the injected fuel to form regions too lean to burn [44, 93, 94], a mechanism that is directly supported by LIF and combustion-luminosity imaging (Sect. 2.5.6).

Unfortunately, developing combustion systems for engines is rarely as simple as a glance at Fig. 2.19 might suggest. In fact, despite leading to lower HC emissions at the lightest loads, the narrow-spray injector was inferior to the wide-spray injector in terms of light-load combustion stability (specifically in the incidence of sporadic misfires) and full-load power. Understanding these differences is a principal motivation for employing the diagnostic techniques discussed here.

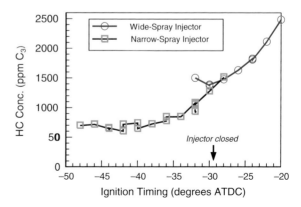

Fig. 2.19 Spark-timing sweep (for fixed injection timing) comparing exhaust HC concentrations with wide- and narrow-spray injectors in the optical cylinder head at TP1

2.5.4 Overview of Mixture Preparation and Combustion

Figures 2.20 and 2.21 provide an overall picture of the preparation and combustion of a stratified mixture in this engine by bringing cylinder-pressure, heat-release and time-resolved exhaust-HC data (Fig. 2.20) together with optical imaging results (Fig. 2.21). These are all typical results at Test Point 1, although not all were

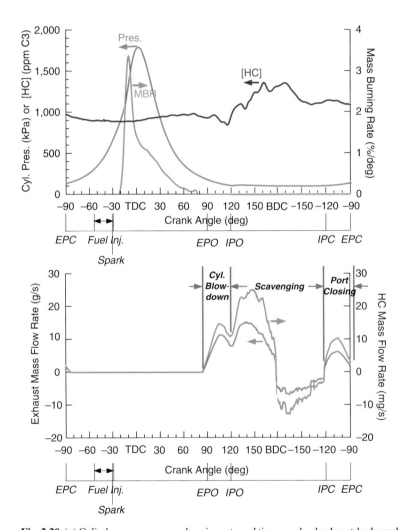

Fig. 2.20 (**a**) Cylinder pressure, mass-burning rate and time-resolved exhaust-hydrocarbon concentration vs. crank angle during a typical engine cycle at Test Point 1 (wide-spray injector). Port openings and closings, fuel-injection, and ignition timings are indicated along the top of the graph. (**b**) Exhaust mass flow rate and HC mass flow rate evaluated from cylinder and exhaust-port pressure data. The three main phases of the scavenging process are labeled

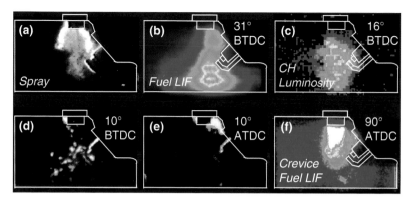

Fig. 2.21 High-speed color film, LIF, and spectrally resolved video images obtained under the same operating conditions as the data in Fig. 2.15. Images **a**, **d** and **e** are from one cycle of a high-speed film. The LIF (**b**, **f**) and spectrally resolved video (**c**) images are shown using different rainbow false-color scales

obtained from the same engine cycle. Subsequent sections will discuss significant aspects of these processes in greater detail.

In order to create as stratified a mixture as possible and to optimize combustion phasing at light-load conditions such as TP1, fuel is injected as late in the compression process as is possible with the available air-assist injection pressure. The planar Mie-scattering image of Fig. 2.21a illustrates the fuel spray from the wide-spray injector under these conditions. As described earlier (Sect. 2.4.4), injectors of the type used here produce initially hollow-cone sprays that collapse and fill in with fine droplets and fuel vapor. For late injection, spray velocity, penetration and atomization typically deteriorate near the end of injection as the rising cylinder pressure approaches the injection pressure. The LIF image of Fig. 2.21b illustrates the resulting fuel distribution 1° before ignition, which is just at the end of the fuel-injection event in this test.

Heat-release analysis of cylinder-pressure data from stratified-charge SI engines generally suggests two distinct stages of combustion (see, e.g., [93, 94]). For this DI2S engine, in-cylinder combustion imaging suggests that the initial rapid-combustion peak in the plot of mass burning rate vs. crank angle (Fig. 2.20) corresponds to (i) (a) rapid propagation of a partially premixed flame through the region of the chamber where fuel-air-residual mixture exists within the flammability limits (illustrated by the spectrally resolved video image of CH luminosity in Fig. 2.21c), followed by (i) (b) burnout of small, locally rich parcels behind the flame front as remaining fuel and oxygen in the interior of the combustion chamber mix in the hot post-flame gases (illustrated by the high-speed film frame of Fig. 2.21d)[17]. This rapid initial stage is followed by (ii) (a) slow mixing-controlled combustion of fuel

---

[17] At least some of the fuel behind the flame front may have undergone thermal decomposition or partial oxidization.

that entered the chamber at reduced velocity near the end of injection and formed a large persistent pocket of rich mixture near the injector and along the slanted wall between the injector and the spark plug (Fig. 2.21e), perhaps accompanied by (ii) (b) continued slow, locally dilute premixed combustion at the periphery of the fuel cloud. Visible combustion luminosity ends around 40°–50° ATDC, but measurable heat release continues nearly until exhaust-port opening.

The onset of exhaust flow occurs a few degrees before the nominal exhaust-port opening time of 90° ATDC as the piston top ring land passes the top of the exhaust port. Until then, the fast FID samples essentially stagnant gas in the exhaust system, and the individual-cycle exhaust-hydrocarbon concentration trace in Fig. 2.20 is accordingly uneventful. The first exhaust gas to escape during the cylinder blow-down just after EPO is actually lower in HCs than the gas that had been trapped in the port. The exhaust-HC concentration then rises as the intake port is uncovered (120° ATDC) and fresh air enters the cylinder from the crankcase, mixing with and displacing some of the cylinder contents. Analysis of the fast-FID HC data to obtain individual-cycle exhaust-HC mass data is discussed in detail in [82] and will be summarized later (Sect. 2.5.7.1).

After TDC, as the cylinder pressure falls during expansion and blowdown, a plume of fuel (mostly vapor but also some liquid mist) emerges from the injector's nozzle-exit crevice, as shown by an LIF image in Fig. 2.21f. (The LIF intensities in Fig. 2.21b,f cannot be compared in terms of fuel concentration because the two images were recorded with different camera gains and in-cylinder conditions.) The fuel emerging from the nozzle-exit crevice remains unburned and largely unmixed until fresh air enters the cylinder after intake port opening, rapidly mixing the fresh gas with residual cylinder contents [80, 82].

2.5.5 Mixture Preparation

2.5.5.1 Spray Structure and Evolution

Average fuel distributions produced by the wide-spray injector at Test Point 3 are shown in Fig. 2.22 for several crank angles from just after the start of injection until 1° before ignition. For these measurements, the light sheet passed through the injector axis. Because the fluorescence intensities span a much wider dynamic range than can be reproduced adequately with printed gray-scale or false-color images, the results in Fig. 2.22 are shown side-by-side on two different intensity scales: on the left, a linear intensity scale with a varying upper limit is used; on the right, the images are shown on a fixed logarithmic intensity scale. All the images in this figure were recorded with fixed camera gain, so that the intensity values can be compared directly.

The fuel spray initially displays a hollow-cone structure, with the fuel emerging first on the left side of the injector and the LIF intensity remaining highest on the left side (although this is due in part to attenuation of the laser sheet by Mie scattering as it traverses the dense spray). As seen earlier for a different air-assist injector

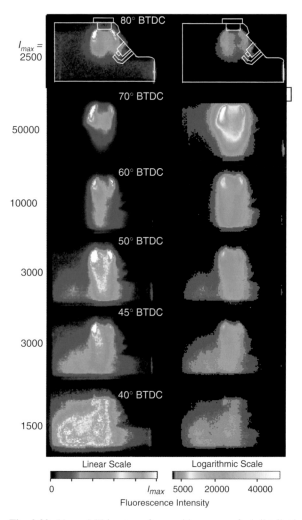

Fig. 2.22 Planar LIF images of ensemble-average fuel distribution during injection (wide-spray injector, Test Point 3)

using exciplex LIF, the spray collapses as injection proceeds, filling in appreciably as entrainment of air into the two-phase fuel-air jet draws small fuel droplets and vapor towards the spray axis. Although gasoline LIF does not distinguish spectrally between liquid- and vapor-phase fuel, comparison of Mie-scattering and LIF images (e.g., Fig. 2.17) shows that under these conditions, the LIF signal 1° before spark represents vapor except for a few-mm spot just at the injector nozzle exit, which represents the last liquid to leave the injector.

Immediately before ignition (Fig. 2.22, 40° BTDC), steep gradients in the fuel concentration (LIF intensity) remain, with the highest fuel concentrations (LIF intensities) existing along the injector axis and extending to the bottom of the field of view. Some fuel is present below the bottom of the quartz windows, and the high fuel concentration on the left side of the images represents fuel that has reentered the field of view after rebounding from the piston (see below).

Fuel distributions 1° before ignition at TP1 and TP3 are compared in Fig. 2.23 using LIF both from gasoline and from ketone-isooctane mixtures, which will be discussed in detail in Sect. 2.5.5.3. These images were obtained for a different wide-spray injector of the same model as that used elsewhere in this section. Comparison of the gasoline LIF images for TP3 in Fig. 2.23 with the 40° BTDC image in Fig. 2.22 indicates that there is some injector-to-injector variation in the average fuel distribution at ignition. This is typically reflected in slight (a few degrees crank angle) differences in optimum light-load injection and ignition timings between nominally identical injectors.

Figure 2.23 shows clearly that the fuel distribution just before ignition fills more of the combustion chamber at TP3 than it does at TP1. High-speed spray visualization shows that at TP1 the spray generally does not penetrate beyond the bottom of the quartz windows (cf. Fig. 2.15a), whereas at TP3 some fuel leaves the field

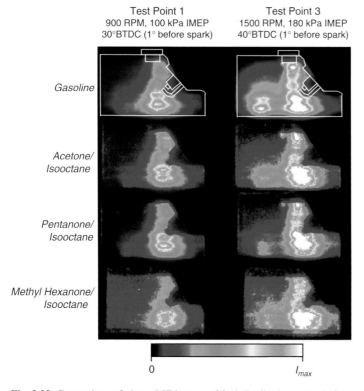

Fig. 2.23 Comparison of planar LIF images of fuel distributions at 1° before spark at Test Points 1 and 3 obtained with commercial gasoline and with high-purity isooctane doped with 20% of the indicated ketone as a fluorescent marker. Note that these images were obtained with a different (but nominally identical) wide-spray injector than the other LIF images in this section

of view and rebounds from the piston (which explains the characteristic J-shaped fuel distribution at ignition at TP3). Because injection begins earlier at TP3, the cylinder pressure and gas density are lower than at TP1 during most of the injection event. These conditions, together with the greater fuel and air masses injected, lead to greater spray penetration and spray-induced mixing at TP3.

At both TP1 and TP3, the rising cylinder pressure eventually reduces the injector-to-cylinder pressure difference and causes the last fuel to enter the chamber with reduced velocity, reduced penetration, and degraded atomization. Some large droplets (occasionally up to $\sim$0.25 mm) are typically seen near the end of injection in high-speed planar Mie-scattering films at both test conditions.

2.5.5.2 Tomographic 3-D Fuel Distributions

By repeating planar measurements such as the on-axis examples in Fig. 2.22 for a series of laser-sheet positions, mathematically stacking the resulting images, and interpolating between them, the average fuel distribution can be constructed in three dimensions. Experimentally, the different sheet positions are obtained by stepping the final laser-sheet turning mirror using a micrometer-driven translation stage (Fig. 2.16).

The resulting 3-D LIF intensity (fuel concentration) data set may be rendered in a variety of useful ways, e.g., as contour surfaces, as solids, as a series of planar contours in cutting planes of arbitrary orientation, etc. (see [79] for examples using the present data). In Fig. 2.24, 3-D LIF data for the wide-spray injector at TP3 are volume-rendered as cloudlike, partially transparent objects during mid-injection and at 1° before spark. Each volume element of the 3-D distribution has been assigned both a false color and an opacity that are related to the LIF intensity. Linear false-color and nonlinear opacity mappings are combined to show the periphery of the fuel cloud without totally obscuring the interior of the distribution.[18]

2.5.5.3 Fuel LIF Image Analysis and Interpretation

One of the major attractions of the fuel-LIF technique is that it offers some prospect for quantitative fuel-distribution measurements in fired engines. If the cylinder contents are optically thin at the excitation and fluorescence wavelengths and the fluorescence quantum yield and quenching are spatially uniform, then the LIF intensity incident on each detector pixel is proportional to the intensity of the laser light sheet and the number of fluorescing molecules (whether liquid or vapor) within the physical volume element corresponding to that pixel (see, e.g., [95, 96] and references therein).

Strictly speaking, our gasoline LIF provides a *qualitative* measure of the fuel distribution. For one thing, we have not corrected the images for pulse-to-pulse

---

[18] Clearly the actual fuel distribution at 40° BTDC extends beyond the front and rear laser-sheet positions used to obtain the series of LIF images for Fig. 2.24.

During Injection
(70° BTDC)

1° before Spark
(40° BTDC)

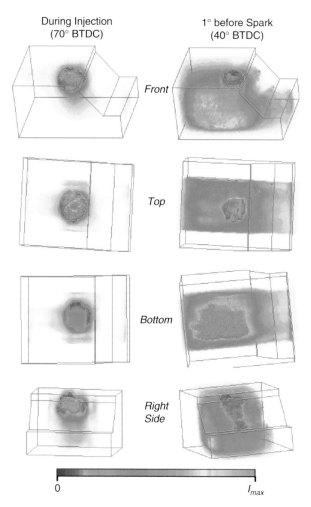

Front

Top

Bottom

Right
Side

0 I_{max}

Fig. 2.24 Three-dimensional reconstructions of ensemble-average fuel distributions produced by the wide spray injector at Test Point 3 during injection (70° BTDC) and just before ignition (40° BTDC). The fuel distributions are rendered using a combination of false color and opacity. A series of eleven positions of the UV laser sheet were used, with a spacing of 2.5 mm between adjacent positions

variations in overall laser power or for the spatial intensity distribution of the light sheet (recall, however, that the laser sheet is overexpanded so that only the central, approximately uniform-intensity portion illuminates the combustion chamber). More fundamentally, the temperature and pressure dependencies of the absorption, the rate of collisional quenching of the fluorescence, and the vapor pressure(s) of the unidentified fluorescing species in commercial gasoline are largely unknown. Toluene (111°C boiling point) is often present in commercial gasolines at the level of at least a few percent and yields strong LIF signals when excited at 266 nm (in fact, it has been used widely as an LIF dopant [96, 97]). Heavier multi-ring hydrocarbons will also contribute. LIF from toluene and multi-ring HCs is strongly quenched by collisions with O_2 [97].

To address these issues, Fig. 2.23 compares gasoline-LIF images recorded immediately before ignition at both TP1 and TP3 with corresponding LIF images obtained using fuels composed of 20% acetone, 3-pentanone or 5-methyl-2-hexanone doped into high-purity isooctane. Fluorescence from the three ketone dopants is much less affected by collisional quenching than fluorescence from typical aromatic hydrocarbons [96, 98, 99]. Furthermore, 3-pentanone has nearly the same boiling point (102°C) as isooctane (100°C), whereas acetone and 5-methyl-2-hexanone have significantly different boiling points (56°C and 134°C respectively), so that their LIF images offer information on the effects of fuel volatility on the fuel distribution at ignition. Although the results in Fig. 2.23 for gasoline and ketone-doped fuel LIF are not identical, they do agree in terms of their major features, particularly in the basic shape of the fuel distributions for the two operating conditions and in the presence of steep gradients near the spark gap. Note that the fuel distributions at ignition are consistently more compact at TP1 than at TP3.

With these observations in mind, we have used qualitative and semi-quantitative analysis of gasoline LIF images to provide useful insight into fuel sprays, fuel distributions at ignition, early stages of combustion, and HC emissions sources in this engine. In particular, although we have not performed a direct calibration of gasoline LIF intensity vs. absolute fuel concentration or fuel-air equivalence ratio, we have estimated a rough scale of fuel-air equivalence ratios from ensemble-average LIF images recorded by assuming that at 1° before ignition (1) the LIF intensity is proportional to the fuel concentration, (2) the residual gases and fresh scavenging air are well mixed,[19] and (3) the fuel-air equivalence ratio $\phi \approx 1$ near the spark gap for optimized engine operation [79].[20] Results for the wide-spray injector at Test Point 3 are illustrated in Fig. 2.25. The fuel distribution exhibits steep gradients and encompasses a wide range of equivalence ratios, with the estimated $\phi \approx 2.5$–3 in the rich core of the fuel cloud and ϕ approaching zero at the extremes of the fuel distribution (e.g., the upper left and lower right corners). In Fig. 2.26, the region occupied by ignitable mixture (estimated roughly as $0.6 \leq \phi \leq 1.6$) is rendered in 3-D along two orthogonal cutting planes. Just before spark, ignitable mixture is present only within a thin ($\approx$2–6 mm) shell near the periphery of the fuel cloud.

---

[19] This assumption is supported by fuel LIF in the residual gases, which shows very rapid mixing when the fresh scavenging air reaches the combustion chamber and a nearly uniform distribution of residual fuel in the combustion chamber immediately before the start of injection [80].

[20] More recent DI4S research indicates that in cases with large cycle-to-cycle variation in the equivalence ratio near the spark gap, the ensemble-mean equivalence ratio near the gap must be biased somewhat rich (e.g., $\langle \phi \rangle \approx 1.5$) in order to avoid lean misfires and partial burns [100].

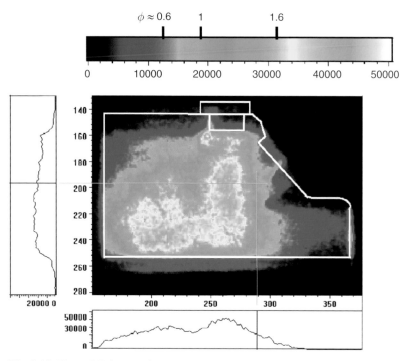

Fig. 2.25 Planar LIF image of average fuel distribution at 1° crank angle before ignition with rough scale of fuel-air equivalence ratio (Test Point 3, wide-spray injector). LIF intensities along vertical and horizontal cross sections through the spark gap (marked by the cross) are shown at the left and bottom of the image, respectively

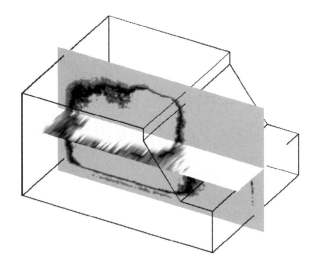

Fig. 2.26 3-D region of ignitable mixture ($0.6 \leq \phi \leq 1.6$) just before ignition at TP3 estimated roughly from LIF image analysis. Vertical and horizontal cutting planes (*light gray*) intersect just above the spark gap. Dark gray and black regions show the comparatively thin zone of ignitable mixture in each cutting plane

2.5.6 Combustion

The overview of combustion in Sect. 2.5.4 (Figs. 2.20 and 2.21) concentrated on TP1, whereas the discussion of mixture preparation in Sect. 2.5.5 focused primarily on fuel distributions at TP3. Here, additional heat-release (Fig. 2.27) and combustion-luminosity (Fig. 2.28) results that contrast stratified-charge combustion at these two speed-load conditions are presented.

Figure 2.27 plots individual-cycle heat-release data obtained with the wide-spray injector and optimum spark timing at TP1 and TP3, together with representative data for dilute, premixed-charge combustion in a four-stroke engine for comparison [101]. The data are conditional ensemble averages over all cycles that burned at least 80% of the total amount of fuel present in the cylinder (injected fuel plus unburned fuel retained from previous cycles). The mass-burning rate and the cumulative mass-burned fraction (which is used as a progress variable in Fig. 2.27b) are both normalized by the average mass of fuel injected per cycle.

Premixed-charge combustion typically yields nearly symmetric burning-rate curves, with the peak heat-release rate occurring at about 50% mass-burned fraction (Fig. 2.27b). In contrast, the asymmetric DI2S burning-rate curves exhibit long tails that are characteristic of stratified-charge combustion. Corresponding to the initial burning-rate peak (from ignition to about TDC at TP1 and to about $10°–15°$ ATDC at TP3 in Fig. 2.27a), combustion imaging shows that a flame propagates from the spark gap toward the periphery of the injected fuel cloud (Fig. 2.21c). In particular,

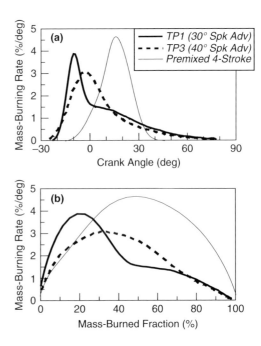

Fig. 2.27 Conditionally ensemble averaged heat-release data for wide-spray injector at TP1 and TP3

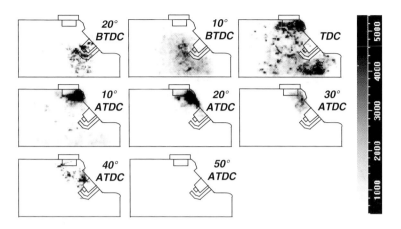

Fig. 2.28 Broadband combustion luminosity images at Test Point 1

single-shot fuel LIF and high-speed combustion visualization using planar Mie scattering from minute oil droplets seeded into the intake air both reveal the irregular contours and steep gradients between burned and unburned gases that are expected of a turbulent (partially) premixed flame [79].

At TP1, the initial rapid stage of combustion is richer overall than at TP3, and small regions of locally rich burning become evident as fuel and oxygen remaining behind the expanding flame mix in the hot post-flame gases. This gives the broadband combustion luminosity (Fig. 2.21d) a granular appearance at TDC, for example. At TP3, in contrast, the initial rapid stage of combustion is dominated by the near-UV (OH-radical) and deep-blue (CH-radical) emissions that indicate more nearly stoichiometric combustion. Longer wavelength C_2-radical and continuum emissions, which connote richer combustion, are prominent during the spark discharge but nearly disappear until the onset of rich burning near the fuel injector a few degrees before TDC. Apart from this, the prominent burnout of locally rich regions behind the flame front seen at TP1 is almost absent at TP3.

These observations imply that, despite the highly stratified fuel distributions observed just before ignition at both test conditions and the higher overall trapped-charge equivalence ratio at TP3, the expanding flame propagates through an appreciably leaner mixture at TP3 than at TP1, consistent with the lower peak burning rate (Fig. 2.27) at TP3. This reflects stronger mixing during and after injection at TP3, which is due in part to the greater spray penetration and spray-induced mixing that results from the lower cylinder pressure that prevails during most of the injection event at TP3 (Sect. 2.5.5.1). Also contributing to stronger mixing at TP3 are the stronger squish flow and squish-induced turbulence and the shorter real-time interval available between intake-port opening and injection for the scavenging flows and turbulence to decay in intensity.[21] The fraction of the fuel burned in the rapid initial stage (including the burnout of locally rich pockets behind the flame front) is about 40% at TP1 and 67% at TP3.

Another important distinction between early combustion at the two test conditions is that at TP1 the expanding flame generally fails to reach the side walls of the combustion chamber, as illustrated in Fig. 2.28. This is consistent with TP1's more compact fuel distribution (Fig. 2.23) and lower overall trapped-charge equivalence ratio (0.44 vs. 0.69 at TP3). Furthermore, because LIF reveals fuel at low concentration outside the maximum extent of the propagating flame, this is *strong evidence for the classic stratified-charge scenario of extinction as the expanding flame propagates into regions too lean to burn.*

At both test points, rich burning near the injector begins as combustion luminosity from the interior of the chamber disappears. For the optimized timings of Figs. 2.27 and 2.28, this occurs around TDC (corresponding to the beginning of the burning-rate plateau at TP1 in Fig. 2.27) and marks the transition from rapid combustion of the primary fuel cloud in the interior of the chamber to slow combustion of the fuel that entered the chamber at relatively low velocity near the end of injection as the rising cylinder pressure reduced the injector-to-cylinder pressure difference (Sect. 2.5.5.1). This last-injected fuel mixes slowly with the in-cylinder gases to form a persistent rich pocket along the slanted wall between the injector and the spark plug. At both test points, the last ∼30% of the fuel that is burned undergoes slow, mixing-controlled combustion (burning rate approximately proportional to mass of fuel remaining in the cylinder) in this region. For both test conditions and both injector types, the burning rate (Fig. 2.27) and the spectral distribution of the luminosity are almost identical during this final mixing-controlled stage, and are also relatively insensitive to spark advance [82].

2.5.7 Exhaust Emissions

A central problem in the development of automotive engines is meeting exhaust emissions requirements while simultaneously achieving performance and fuel-economy goals. Conventional premixed-charge four-stroke engines (see Chap. 1) typically operate at stoichiometric conditions over the entire speed-load range, permitting the optimization of three-way catalytic converters which can simultaneously reduce engine-out HC, CO and NOx emissions levels by factors of 100–1000 under steady-state, warmed-up conditions. Most vehicle or *tailpipe* (as opposed to engine-out) hydrocarbon emissions from such engines occur during cold starts before the catalyst reaches its operating temperature, while most tailpipe nitric oxide emissions occur during accelerations or decelerations where the engine control

---

[21] To first order, the impulsive scavenging flow scales with delivery ratio but not with engine speed in a crankcase-compression-scavenged engine (Sec. 2.3.4.1), whereas the piston-motion-driven squish flow does scale with engine speed.

system cannot maintain stoichiometric conditions accurately enough for optimum catalyst performance.

DI spark-ignition engines, both two- and four-stroke, offer new emissions challenges and opportunities (see also Chap. 3 and [37]). Direct fuel injection eliminates intake-port wall wetting and hence reduces HC emissions during cold starts and speed-load transients. However, part-load HC emissions from stratified-charge DISI engines can be significantly higher than from conventional premixed-charge SI engines, and the exhaust temperature is too low for good catalyst conversion efficiency. NOx production in stratified-charge SI engines (as mentioned in Sect. 2.1) can be significant because, even though the overall equivalence ratio may be very lean, some of the charge burns under locally stoichiometric conditions that maximize the local temperature and hence NO formation. Furthermore, conventional three-way catalysts are ineffective in reducing NOx when there is excess oxygen in the exhaust stream. Special lean-NOx, NOx trap and selective catalyst reduction (SCR) aftertreatment systems are therefore under development for diesel and stratified-charge gasoline engines [102]. Lean NOx aftertreatment demands are mitigated somewhat by the DI2S engine's intrinsically high residual fraction and the lighter *engine* load needed for a given *vehicle* load, which reduce in-cylinder temperatures and hence reduce NO formation [35]. At heavy load, DI2S and DI4S engines operate under homogeneous, stoichiometric conditions, with HC and NOx emissions mechanisms and catalyst operation similar to those of four-stroke engines. An additional challenge to DI gasoline engine control systems is presented by speed-load transients, which require changes in charge stratification (including shifts between stratified and homogeneous operation) while maintaining acceptable drivability and emissions.

2.5.7.1 Unburned Hydrocarbon Emissions

In stratified-charge SI engines (both two- and four-stroke), progressively-higher engine-out hydrocarbon emissions typically occur as the load is decreased and the degree of stratification increased. Based on conventional measurements (see, e.g., Fig. 2.19), the formation of overly dilute regions at the periphery of the injected fuel cloud has long been considered as a major HC source in such engines [44, 93, 94]. Sporadic misfires and partial burns due to slow combustion may also contribute significantly. In contrast, these mechanisms appear to be relatively unimportant to HC emissions from more conventional premixed-charge four-stroke SI engines (see Chap. 1), where the dominant HC source is generally considered to be the outgasing and incomplete combustion during expansion of fuel that is forced into crevices (in particular the piston top-ring-land crevice) during compression [30, 103].

Individual-cycle heat-release and fast-FID HC emissions measurements, combined with simultaneous high-speed spectrally resolved imaging and augmented by insight derived from (non-simultaneous) LIF imaging, have proven useful in understanding and quantifying HC emissions sources from this DI2S engine. The approach (already illustrated to some extent in Figs. 2.20 and 2.21) and salient results are summarized here, focusing on warmed-up operation at Test Point 1, which is representative of the light-load conditions that dominate HC emissions. For full details, see [82].

Interpreting measurements of individual-cycle HC concentration vs. crank angle (e.g., Fig. 2.20) in terms of individual-cycle HC *mass* emissions requires knowledge of the *individual-cycle* exhaust mass flow rate $dM_{exh}(\theta)/dt$ as a function of crank angle when there is significant cyclic variability in combustion. Here, $dM_{exh}(\theta)/dt$ has been evaluated from the measured cylinder and exhaust-port pressures $P_{cyl}(\theta)$ and $P_{exh}(\theta)$ as a function of crank angle θ by approximating the exhaust process as spatially uniform, quasi-steady flow through an orifice (the exhaust port) whose effective flow area $A_{eff}(\theta)$was determined in separate steady-flow measurements. The individual-cycle HC concentration [HC] and $dM_{exh}(\theta)/dt$ are then converted to HC mass flow rate $dM_{HC}(\theta)/dt$ using effective molecular weights of the fuel and exhaust gas (the latter evaluated from engine-out emissions data). Integrating $dM_{HC}(\theta)/dt$ from exhaust-port opening to exhaust-port closing yields the total exhaust-HC mass for each cycle.

As shown in Fig. 2.29, ensemble-averaged, mass-flow-weighted HC emissions evaluated thus from the fast-FID data agree well with simultaneously acquired time-averaged HC data from the conventional FID over a wide range of HC levels and operating conditions (0–30% misfires). This suggests that the fast-FID results are not significantly biased by any non-uniformity in the HC concentration across the exhaust port because the conventional FID (which is located downstream of the fast FID and which samples across most of the width of the exhaust runner) obtains a much better mixed exhaust sample than does the fast FID (see Fig. 2.14).

Heat-release and fast-HC sampling results for three consecutive engine cycles (numbers 14–16 of a 100-cycle ensemble) are shown in Fig. 2.30 for operation at TP1 with the narrow-spray injector. For normal cycles (e.g., Fig. 2.20b and cycle 14 in Fig. 2.30), the HC mass flow rate $dM_{HC}(\theta)/dt$ exhibits four important features: (1) the cylinder blowdown peak as the piston uncovers the exhaust port (90° ATDC); (2) a second, higher peak as the transfer (intake) ports are uncovered (120° ATDC) and fresh air enters the cylinder; (3) backflow from the exhaust into the cylinder

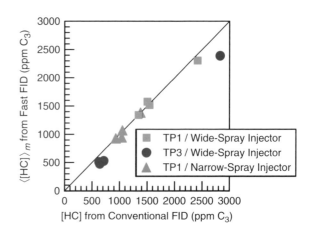

Fig. 2.29 Comparison of HC measurements by fast-FID and conventional FID

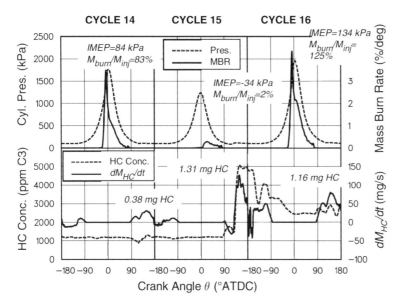

Fig. 2.30 Heat-release and fast-HC-sampling results for three consecutive engine cycles (narrow-spray injector, TP1). *Upper*: Cylinder pressure (*dashed line*) and mass burning rate (*solid line*). *Lower*: HC concentration (*dashed line*) and HC exhaust mass flow rate $dM_{HC}(\theta)/dt$ (*solid line*). Note that the HC concentration data have been shifted in crank angle to account for the fast-FID sampling delay. Also labeled for each engine cycle are the IMEP, the total mass of fuel burned relative to the average mass of fuel injected, and the total HC mass that leaves the cylinder during the exhaust event

around BDC; and (4) a final outflow peak as the rising piston covers first the transfer ports and finally the exhaust port ($120°$–$90°$ BTDC of the following cycle). Misfires (e.g., cycle 15) produce high HC concentrations despite the initial backflow that occurs instead of a normal cylinder blowdown.

High-speed (4000-frame/s) video images of spectrally resolved combustion luminosity at the end of the spark discharge and 1.5 ms later during early flame growth are shown in Fig. 2.31 for the same three cycles. Both the combustion luminosity here and the cylinder-pressure and heat-release data in Fig. 2.30 show that cycle

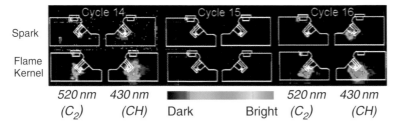

Fig. 2.31 Spectrally resolved dual-camera images at two times from the same three engine cycles as in Fig. 2.30 and Fig. 2.32

15 is a misfire. The spray images from these three cycles (not shown) are all fairly typical, but the near-total absence of 520-nm radiation (C_2 radical and continuum emission) during ignition suggests that cycle 15 is a lean misfire.

Because of the high average residual fraction ($\approx 65\%$) in this engine at light load, much of the unburned fuel from cycle 15 remains in the cylinder. Some of this retained fuel burns during cycle 16 to produce above-average combustion luminosity and heat release (about 25% more than the average mass of fuel injected is consumed). Much of the fuel retained after a misfire escapes combustion, however; note that the exhaust-HC mass from the cycle just after a misfire is comparable to the HC mass from the misfire cycle itself. Indeed, as Fig. 2.32a,b show, a misfire increases HCs appreciably for that cycle and the three succeeding cycles.

The full effect of a misfire on HC emissions should therefore be quantified by summing the exhaust HC mass from the misfire cycle itself and the next three cycles, and comparing the total to what would be emitted under the same operating conditions by four normal cycles. The *net excess HC mass* defined in this way is illustrated by the shaded area under the misfire spike in Fig. 2.32b and represents, on average, only about a third to a half of the average amount of fuel injected (4.5 mg) per cycle at TP1. Thus the engine's high residual fraction significantly mitigates the potential effect of sporadic misfires on HC emissions.

Figure 2.32c shows that the individual-cycle HC mass falls off linearly with the mass of fuel consumed, a trend due to the 60–70% of the HC mass that is exhausted during the main scavenging phase while the intake ports are open. The

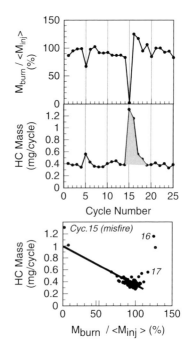

Fig. 2.32 Correlation between individual-cycle heat-release and exhaust-HC data. (**a**) Mass of fuel burned (normalized by average mass of fuel injected per cycle) and (**b**) HC mass for first 40 cycles of a 100-cycle ensemble. (**c**) HC mass vs. normalized mass burned for full ensemble. The points representing cycles 15 (misfire), 16 and 17 are labeled

linear trend includes the misfire cycles 15 and 34, but the HCs emitted one and two cycles after a misfire (i.e., cycles 16, 17, 35, and 36) are exceptions. Because of fuel retained after a misfire due to the high residual fraction, these exceptional cycles have high HCs and also consume more fuel than the amount injected per cycle.

Because of the late fuel-injection timing and the related spatial confinement of the fuel near the spark plug at light load, little fuel is expected to enter the piston top-ring-land crevice in this DI2S engine compared to a conventional premixed-charge four-stroke SI engine. The fast-FID concentration data directly confirm the diminished role of the top-ring-land crevice as a HC source here. By virtue of the exhaust port's location in the cylinder wall (see Fig. 2.14), the first gas to escape from the cylinder should come from the top-ring-land crevice. For normal-combustion cycles, as illustrated in Fig. 2.20, the measured HC *concentration* at exhaust-port opening is usually appreciably lower than during the main scavenging phase that begins at intake port opening.

Incomplete oxidation of fuel emerging from the injector nozzle-exit crevice during expansion (Fig. 2.21f; Sect. 2.5.4) may be another appreciable HC source. Additional evidence for this is provided by time-resolved sampling-valve measurements in the exhaust of a DI2S engine [104], which found that during the main scavenging event, the O_2 concentration increases, while the concentrations of the combustion products CO, CO_2 and NOx decrease, consistent with dilution of the exhaust stream by short-circuited fresh scavenging air. However, both the sampling-valve data of [104] and the present fast-FID data (e.g., Fig. 2.20a) show that the HC concentration *increases* after IPO, an effect that cannot be explained by dilution and that suggests an additional HC source such as the injector-tip crevice volume or vaporization of liquid fuel from combustion-chamber surfaces.

2.5.7.2 Nitric Oxide Emissions

Understanding in-cylinder NO formation and minimizing engine-out NOx emissions during stratified-charge operation motivated our efforts to measure NO concentrations in the optical two-stroke engine. Although sampling-valve probes and total-cylinder sampling have been used to measure in-cylinder NO [105], these techniques have limited spatial and temporal resolution and may suffer from systematic errors intrinsic to the sampling process. More recently, LIF has been used for point measurements and 2-D imaging of NO in flames [88, 89], in engine simulators [106, 107], in four-stroke engines [92, 108] (including a DI4S engine [109]), and, as described here, in a DI2S engine. NOx increases with engine load [35]; hence the experiments described here were all performed at the heavier-load Test Point 3, which contributes 18% of the total engine-out NOx on the six-point steady-state FTP simulation.

Figure 2.33 shows NO LIF images (averaged over 200 laser shots) taken in the optical DI2S engine under nearly-warmed-up conditions (fired 45 s prior to the start of imaging) at TP3, but with an increased delivery ratio of 0.35 and with isooctane fuel.[22] Compared to the fuel-LIF imaging discussed earlier (Sections 2.5.2.2, 2.5.5.4, and 2.5.5.5), the laser sheet here has been narrowed to a height of 12 mm and has been moved a few millimeters off the combustion-chamber center-line toward the camera in order to minimize laser scatter from the spark-plug electrodes and interfering LIF from deposits on the electrode surfaces.

It is useful to examine how the NO LIF images (Fig. 2.33) and the spatially integrated LIF intensities (Fig. 2.34) vary during the engine cycle. At 90° BTDC, the transfer ports are closed, the exhaust port has nearly closed, and the cylinder pressure is ~100 kPa. The LIF image at 90 BTDC indicates a nearly uniform NO distribution left over from the previous engine cycle. The ratio of integrated intensities at 90° BTDC to the peak integrated NO LIF intensity that occurs at 90° ATDC is ~0.6, consistent with the high residual fraction (0.54) at this operating condition. The NO LIF intensity level does not change much from 90° BTDC to 40° BTDC, suggesting that the isooctane fuel injected during this time does not absorb the laser

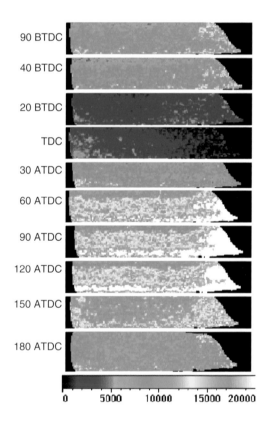

Fig. 2.33 Variation of in-cylinder NO LIF images with crank angle (Test Point 3, delivery ratio 0.35, isooctane fuel). Each image is an average over 200 laser shots

---

[22] To reduce window fouling and interferences due to combustion luminosity and LIF from combustion-chamber deposits, isooctane fuel was used for most of the LIF imaging.

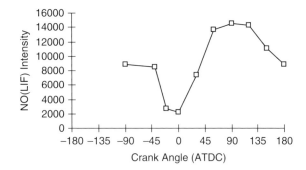

Fig. 2.34 Total NO LIF intensity integrated over each image in Fig. 2.33 plotted as a function of crank angle

or otherwise interfere with the NO LIF, and that the cylinder pressure increase to 500 kPa has little effect on the measured NO LIF intensity.[23]

Ignition starts at 39° BTDC, and detectable heat release lasts until ~75° ATDC. Although an increase in NO is expected as combustion proceeds, the integrated NO LIF intensity in Fig. 2.34 instead drops sharply at 20° BTDC, falling to a minimum at TDC which is some six times lower than the value at 90° BTDC. Unfortunately, this sharp drop in LIF intensity during combustion makes it impossible to determine anything about the locations and timing of NO formation in the DI2s engine studied here. Similar difficulties have also been encountered in NO LIF measurements in four-stroke SI and diesel engines [108, 110] and, according to more recent studies [111, 112, 113], are due to broadband laser absorption by hot CO_2 and transient hydrocarbon species.[24] This absorption becomes very strong at the short wavelengths (~193 nm or ~226 nm) needed to excite NO from the lowest vibrational level of the electronic ground state.[25] Curiously, LIF measurements in a heavy-duty diesel engine [114] using essentially the same ~226-nm laser wavelength as here did not show this strong laser attenuation during combustion; the results indicated that NO formation in diesels occurs predominately in the diffusion-flame regions.

After TDC, the NO LIF intensity (Fig. 2.34) increases rapidly, reaching post-combustion levels at 90° ATDC that are substantially above the initial values at 90° BTDC due to NO formation during combustion. The decrease in NO LIF intensity after exhaust-port opening (90° ATDC) is consistent with cessation of

---

[23] The NO number density and the NO collisional quenching rate both increase linearly with pressure, so the NO LIF intensity for a fixed NO concentration is reasonably independent of pressure, apart from pressure shifts and broadening effects (Sec. 2.5.2.2).

[24] Appreciable absorption by hot CO_2 as well as by transient hydrocarbon species is implied by our observation that the absorption begins earlier in the engine cycle (before ignition, in fact) and lasts longer when the engine is fueled with gasoline rather than with isooctane. Furthermore, the difference between the mass-averaged cylinder gas temperatures with the two fuels is fairly small (50 K in peak temperature).

[25] NO LIF excitation from the first vibrationally excited level of the electronic ground state $[X^2\prod(v''=1)]$ at 248-nm wavelength suffers much less from these absorption effects, but the LIF signal strengh is then proportional to the temperature-dependent population of the $v''=1$ level, which must be taken into account [112, 113].

NO formation in the cooler cylinder gases, escape of NO from the cylinder via the exhaust port, and (after intake-port opening at 120° ATDC) dilution of the cylinder contents by fresh scavenging air. At 180° ATDC (i.e., BDC), the NO LIF intensities are nearly equal to those measured at 90° BTDC, consistent with the strong mixing of fresh and residual gases after IPO that has been observed by fuel LIF imaging (Sect. 2.5.4; [80, 82]) and with mass-flow calculations that show that most of the scavenging mass transfer has been completed by BDC at this operating condition.

Measurements using a conventional chemiluminescence analyzer were also performed to examine the dependence of engine-out NO concentration on delivery ratio, fuel (commercial 91 RON gasoline or isooctane), and operating time after a warm start. (For these tests, the optical cylinder head was replaced by an all-metal head of exactly the same geometry.) As shown in Fig. 2.35, approximately two minutes of firing was required to obtain steady-state exhaust-NO levels for all conditions. Increasing delivery ratio greatly increased the measured steady-state NO concentration (from 20 ppm at DR = 0.25–480 ppm at DR = 0.35 for gasoline). This behavior is consistent with the fact that (as mentioned in Sect. 2.1) increasing delivery ratio increases the amount of fresh air and reduces the amount of residual gases in the cylinder. For fixed injection and spark timings, the higher oxygen concentration at higher DR leads to faster burning (shown by heat-release analysis), increased cylinder pressure (increasing the NO formation rate due to higher temperature) and an earlier crank angle of peak pressure (increasing the time available for NO formation). For given operating conditions, the measured exhaust NO concentrations for gasoline are higher than for isooctane, consistent with the faster

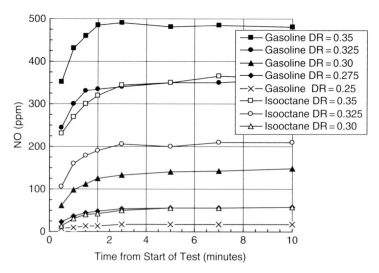

Fig. 2.35 Engine-out exhaust NO concentrations measured with a chemiluminescence analyzer as a function of time after a warm engine start for gasoline and isooctane fuels and for various delivery ratios

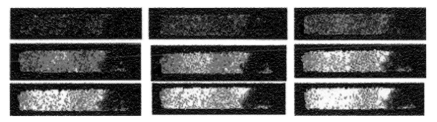

Fig. 2.36 Fifty-shot-average NO LIF images as a function of time after a warm engine start (TP3, DR = 0.33, isooctane fuel). The LIF images were recorded at crank angle 120° ATDC, after combustion had ceased. The upper-left image was recorded at the start of firing; the lower right image was taken after more than 1 min of firing

burning rate and slightly higher average cylinder temperature indicated for gasoline by the heat-release analysis.

All of the exhaust-NO trends in Fig. 2.35 were observed at least qualitatively in the in-cylinder NO LIF intensities. For example, Fig. 2.36 shows a series of NO LIF images recorded at 120° ATDC (TP3, DR = 0.33, isooctane fuel) as a function of engine running time after a warm start. Each image is a 50-laser-shot average, with the upper left image taken immediately after starting the engine and the last (lower right) taken after more than a minute of firing. Clearly, the overall NO LIF intensity increases rapidly with time, but the NO spatial distribution remains reasonably constant, with concentration minima occurring near the spray centerline and along the right-hand wall. In Fig. 2.37, the NO LIF intensities integrated over each image in Fig. 2.36 are plotted versus simultaneously acquired exhaust-NO concentration measurements. (Although not shown here, exhaust-NO data obtained with the optical and all-metal cylinder heads exhibit identical trends, but the optical-head data are typically ∼50 ppm lower.) The results in Fig. 2.37 demonstrate the excellent linearity between the NO LIF and exhaust-NO measurements. Similar linear

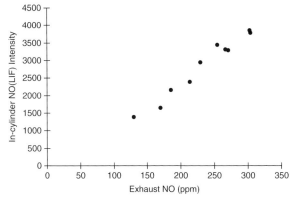

Fig. 2.37 Integrated NO LIF intensity at 120° ATDC vs. simultaneously acquired exhaust NO concentration measured by a chemiluminescence analyzer

results were obtained at other delivery ratios. However, substantial day-to-day and run-to-run variation in the intercepts of the curves results from long-term drifts in laser power, laser wavelength, and laser absorption (e.g., due to window fouling).

The sensitivity of NO formation to delivery ratio (illustrated in Fig. 2.35 and also observed by LIF imaging) suggests the primary strategy to minimize engine-out NO: operate with the minimum delivery ratio, i.e., with the maximum dilution by residual gases that is consistent with combustion stability. This is also true for DI4S engines (see Chap. 3 and [37]).

Although the chemiluminescence exhaust-NO measurements here have a minimum time resolution of a few seconds, cycle-to-cycle variations in in-cylinder NO concentrations can be characterized by single-laser-shot LIF images which have the time resolution of the laser pulse (<10 ns). This requires that shot-to-shot variations associated with the measurement technique be quantified, as illustrated (Fig. 2.38, top) by a histogram of total integrated LIF intensity from a set of 50 single-shot images of a uniform distribution of 240 ppm NO in N_2. The width of

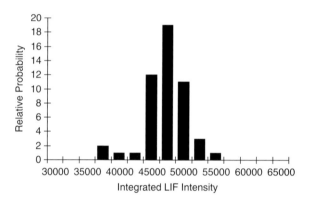

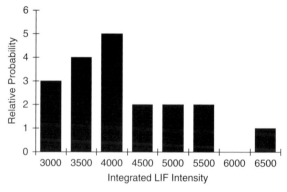

Fig. 2.38 Comparison of histograms of integrated NO LIF intensity. *Top*: Histogram of 50 images of a uniform distribution of 240 ppm NO in N_2. The histogram width is due to shot-to-shot variations in laser power. *Bottom*: Histogram of 20 images recorded in the firing engine. The histogram width is predominantly due to cyclic variations in NO concentration

the distribution (relative standard deviation of 7%) is due primarily to shot-to-shot variations in laser power because camera-related noise is not very significant for these relatively strong signal intensities.

To characterize relative cyclic variation in NO concentrations, a similar histogram evaluated from 20 single-laser shot LIF images in the firing engine (TP3, DR = 0.35, 120 ATDC) is shown at the bottom of Fig. 2.38. The average NO exhaust concentration (measured simultaneously) was 420 ppm, and the NO LIF signal intensities are ∼4000 counts/pixel, or ∼10 counts/pixel/ppm NO. The integrated LIF intensities from the firing engine vary by more than a factor of two, with a relative standard deviation of 21%. Clearly, the width of the NO LIF histogram from the firing engine is strongly dominated by cycle-to-cycle variation in NO concentrations. Furthermore, because about half the cylinder contents at this operating condition are retained from previous cycles, the factor-of-two overall spread in integrated NO LIF intensity implies a cycle-to-cycle variation range of about a factor of four in NO *production*.

2.5.8 Other Engine-Development Issues

2.5.8.1 Cyclic Variability

Cyclic variation in combustion is a significant practical problem in engine development that has not yet proven amenable to prediction by large-scale computational models. As suggested in Sect. 2.4.4, the steep gradients in fuel concentration that occur in late-injection stratified-charge engines are a likely source of combustion variability because, even when there is a stoichiometric fuel-air mixture at the spark gap *on average*, ignition problems (including misfires) can occur if the mixture is beyond the ignitability limits for any *individual* cycle.

In Fig. 2.39 a set of nine individual-cycle planar LIF images selected from a set of fifty recorded at Test Point 3 with the wide-spray injector are arranged so that the top row has the least fuel near the spark gap, the middle row has the average amount, and the bottom row has the most fuel at the spark gap. Although the overall shape of the fuel cloud and the total amount of fuel within the cloud are roughly the same in each case, the steep gradient in fuel concentration near the spark gap causes significant cycle-to-cycle variation, as quantified in Fig. 2.40 by the histogram of LIF intensity within a small region (0.75 × 1 mm image area × ∼0.5 mm laser-sheet thickness) just above the spark gap. Using the average over the fifty images in the set to relate LIF intensity roughly to equivalence ratio as described in Sect. 2.5.5.3 indicates that a few cycles were either too rich or too lean for successful ignition and rapid flame propagation, consistent with the incidence of misfires and partial burns determined from simultaneously acquired cylinder-pressure data [79].[26] Histograms

---

[26] This approach probably overestimates the incidence of misfires and partial burns because the spark can enflame gases from a somewhat larger region than the ∼0.4 mm^3 volume examined here, and the 1 ms spark duration is, moreover, sufficient for fuel-air mixture to reach the spark

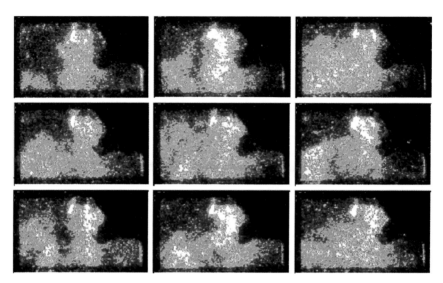

Fig. 2.39 Nine individual-cycle LIF images of fuel distribution just before ignition selected from a set of fifty images (TP3, wide-spray injector). The LIF intensities near the spark gap increase by row from top to bottom

from two other locations in the combustion chamber which have approximately the same average fuel LIF intensity are no narrower than the one in Fig. 2.40, suggesting that they would not have any advantage as alternate ignition locations.

2.5.8.2 Fuel Distribution During Engine Warmup

The full curve in Fig. 2.41 plots LIF fuel intensity near the spark gap (averaged over 50 images) acquired at uniform intervals during a six-minute period after a warm start of the optical engine (i.e., with the block and head heated to 90°C but with the engine not having been fired for at least five minutes). The labeled points show that indicated work (IMEP) and combustion stability [COV(IMEP)] both improved during the first three minutes after starting the engine, during which period the LIF intensity (and, by inference, fuel concentration) near the spark gap gradually increased by a factor of about 1.5. Assuming that $\phi \approx 1$ near the spark gap at this time implies $\phi \approx 0.65$ *on the average* near the spark gap just after starting the engine. Misfires or partial burns in some cycles would therefore be expected if there were even modest cyclic variability in the local equivalence ratio. As shown by the dashed curve in Fig. 2.41, however, a second injector produced a nearly constant fuel-LIF intensity near the spark gap over a three-minute period following a warm start, with indicated work and combustion stability reaching steady values within about 30 s.

---

gap from some distance away, depending on the local velocity field [100, 115]. Furthermore, the relatively long spark duration allows significant stretching of the discharge itself [116].

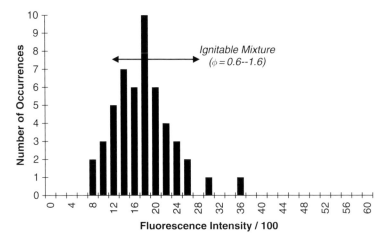

Fig. 2.40 Histogram of LIF intensity in a small region just above the spark gap determined from set of 50 individual-cycles images illustrated in Fig. 2.24, together with a rough scale of equivalence ratio determined using the procedure describe in Sect. 2.5.5.3)

Fig. 2.41 Fuel fluorescence intensity observed near the spark gap during engine warm-up with two injectors (A and B) at Test Point 3

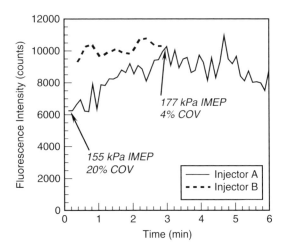

2.5.9 Fuel Economy and Emissions Control Considerations

The *indicated* specific fuel consumption (ISFC) of a DI2S engine is generally inferior to that of a comparable premixed-charge four-stroke SI engine because of the DI2S engine's lower compression ratio and its typically overadvanced combustion at light loads. In this study, for example, peak cylinder pressure typically occurs just a few degrees ATDC at light load, whereas for optimal work-conversion efficiency peak pressure should occur in the range of 12–15° ATDC. Stratified-charge combustion tends to be generically overadvanced due to the need to minimize overmixing by minimizing the time from injection to ignition. The modest pressure of the air-assist injection system used here worsens this situation by limiting how late the fuel can be injected before the rising cylinder pressure reduces the injector-to-cylinder pressure difference too much and degrades atomization unacceptably or halts injection entirely (Sections 2.5.4 and 2.5.5.1).

Nevertheless, the *brake* specific fuel consumption (BSFC) of the DI2S engine is typically ~10–30% better than that of a comparable four-stroke engine up to about 40% load due to the two-stroke engine's lower pumping losses at light load[27] as well as to its overall lower friction [35]. For a given vehicle, therefore, the relative fuel-economy benefit of the DI2S engine is greater for urban driving than for highway driving and for a higher-output engine than for a lower-output engine. The weighted-average engine load over the driving cycle also determines whether HCs or NOx is the leading-order engine-out emissions concern.

In marine applications, replacement of premixed-charge two-stroke engines by DI2S engines provides major reductions in both fuel consumption and HC emissions because short-circuiting of fuel to the exhaust is nearly eliminated. Overall fuel-economy improvements in the range of 35–40% have been reported (~80% for idle and trolling speeds and ~10% at top speed) [19].

The first-order control strategy for DI2S engines at part load is to operate at the lowest possible delivery ratio and with fuel injection and ignition as retarded as possible, consistent with acceptable combustion stability. Minimizing the duration between injection and ignition minimizes overmixing of fuel and air, while retarding ignition improves combustion phasing. Reducing delivery ratio increases the residual fraction, which serves multiple purposes: it reduces NO formation by decreasing peak cylinder pressure and temperature, it efficiently recycles unburned HCs for in-cylinder consumption during later cycles, it helps improve combustion phasing by slowing the rapid initial stage of the stratified combustion process, and it reduces pumping losses. Crankcase-compression-scavenging minimizes the need for external exhaust-gas recirculation and permits the residual fraction to be increased to over 60% merely by throttling the intake and so reducing delivery ratio.

2.6 Summary and Conclusions

This chapter has discussed key aspects of flow, mixture-preparation and combustion in DI2S engines with crankcase-compression scavenging, has described and illustrated the use of several advanced diagnostic techniques (mostly laser-based) to investigate these in-cylinder processes, and has indicated how these diagnostics can be used in a coordinated fashion both to increase knowledge and to address rather directly some practical questions that arise in engine development. The incorporation of the results into improved submodels for large-scale three-dimensional computational simulations and the use of the experimental data to test

---

[27] Pumping losses in crankcase-compression-scavenged two-stroke engines increase with load (strictly speaking, with delivered air), while they decrease with load for conventional four-stroke SI engines.

the predictions of such computations have not been discussed here, but have been described to some extent in the literature [56, 57, 58, 68].

Salient points of the chapter are summarized here. Recall that we have focused on part-load conditions where the engine operates in a stratified-charge mode and where exhaust-hydrocarbons are the major emissions concern. Overall, the two speed-load conditions examined here, together with a third condition intermediate between these two, account for about 55% of the fuel consumption, 80% of the engine-out HC emissions and 25% of the engine-out NOx emissions of a 3.0-L V6 DI2S engine in a mid-size sports sedan over a simulated US FTP driving cycle.

1. *Flow.* The scavenging process that removes combustion products from the cylinder and brings in fresh air – which is especially sensitive to the flow field in ported engines of this type because it is accomplished without the aid of overhead valves – was investigated using photon-correlation laser-Doppler velocimetry in an optical engine with a representative commercial crankcase and a simplified cylinder-head geometry.

   a. Fresh air enters the cylinder from the crankcase through the six transfer (intake) ports arrayed around the bottom of the cylinder in the form of impulsive, high-velocity (>200 m/s initially) jets that are fairly uniform circumferentially and that follow the intended mirror symmetry of the port layout closely but not perfectly.

   b. During the brief period of primary inflow to the cylinder (as the crankcase overpressure decreases towards the cylinder pressure), the port-efflux jets behave as highly transient but well behaved slot jets with relatively modest rms velocity-fluctuation intensities ($u' \lesssim 10\%$ of the local ensemble-mean velocity).

   c. The colliding-jet interaction of the port-efflux flows with each other and with the piston, cylinder and head surfaces produces an in-cylinder flow field with strong asymmetry, locally high mean velocities (~ 100 m/s) and very high rms velocities ($u'/\langle U \rangle \sim 30$–$40\%$).

   d. As the strong inflow diminishes, the intended scavenging-loop vortex becomes recognizable in the axial plane of nominal port symmetry, but the flow at the port-cylinder interface becomes complex and involves spatially non-uniform backflow into the transfer ports as the crankcase and cylinder pressures undergo damped oscillations toward equilibrium.

   e. The significant in-cylinder flow asymmetry, which includes unintended swirl, diminishes scavenging efficiency by increasing short-circuiting of fresh air to the exhaust and by increasing fresh-residual gas mixing.

   f. The observed in-cylinder *flow asymmetry*, the indicated *sensitivity to small geometric variations* (such as might easily occur in manufacturing), and the possibility of *large-scale cyclic flow-field variation* (expected for colliding-jet flows and implied by the observation of bimodal velocity PDFs at some locations) are all undesirable in emissions-regulated applications and, together with other experimental work and CFD computations, showed the *need for an improved port design* that would be less susceptible to these problems.

2. *Mixture preparation*. The spatial structure and temporal evolution of air-assist
   fuel sprays and the formation of a stratified fuel distribution at typical light-load
   ignition times was investigated first in a geometrically simplified, non-firing
   optical engine using planar exciplex LIF to obtain simultaneous but distinct
   single-cycle images of the liquid- and vapor-phase fuel. Later experiments examined fuel sprays, mixture formation and combustion in a continuously fired
   optical engine that closely simulated the geometry and operation of an experimental V6 DI2S engine. These measurements employed continuous, high-speed
   planar Mie scattering imaging of liquid fuel and single-cycle and multi-cycle-
   averaged LIF of commercial gasoline and of ketone-doped isooctane in both
   liquid and vapor phases. LIF images from a set of several measurement planes
   were also used to construct 3-D renderings of the fuel distribution in the firing
   engine.

   a. The *initially hollow-cone fuel sprays collapse and fill in* with liquid and vapor
      fuel due to entrainment of air.
   b. For the late-injection conditions needed to stratify the charge, *atomization
      and spray penetration typically deteriorate near the end of injection* as the
      rising cylinder pressure approaches and exceeds the injection air pressure.
      This effect, which also reduces spray-induced mixing during and after injec-
      tion, is most severe at the lightest loads, for which injection durations are
      shortest and the start of injection is most retarded.
   c. With the air-assist injectors used here, *evaporation is very rapid*. For typical
      ignition timings (just at or after the end of injection), nearly all of the fuel
      within the viewable region has evaporated, as revealed both by the exciplex
      results (which were carried out at reduced compression ratio and with dopants
      significantly less volatile than most components of commercial gasoline) and
      by comparison of gasoline LIF with Mie scattering in a realistic firing optical
      engine.
   d. For ignition at the end of the nominal injection period, the *charge is highly
      stratified* around the spark plug, with a core that is too rich to burn and outer
      regions that are too dilute to ignite and burn well. *Readily-ignitable mixture is
      present only in a thin (few-mm) shell around the periphery of the fuel cloud*,
      where the average fuel distribution displays a steep gradient.
   e. Significant *cycle-to-cycle and injector-to-injector variation* in the fuel sprays
      and in the fuel distribution at ignition was quantified.

3. *Combustion*. Combustion of the stratified charge was investigated under demon-
   strably realistic conditions using optical imaging techniques (planar Mie scattering, fuel LIF, and spectrally resolved flame luminosity) combined with simultaneous cylinder-pressure measurement.

   a. *Stratified-charge combustion occurs in two principal stages*:

      (i) (a) *Rapid propagation of a partially premixed flame* through regions
              where the fuel-air-residual distribution has a high laminar burning velocity,

followed by (b) *rapid burnout of locally rich regions behind the flame front* as remaining fuel and oxygen mix in the hot post-flame gases.

(ii) (a) *Slow mixing-controlled combustion of large, rich fuel pockets, a principal source of which is the last-injected fuel*, which enters the chamber at low velocity and with degraded atomization to form a persistent rich pocket between the injector and the spark plug. This stage of slow mixing-controlled combustion is probably accompanied by (b) *continued slow premixed flame propagation into highly dilute zones* at the periphery of the fuel cloud.

b. *At the lightest loads, the expanding flame never fills the combustion chamber*, despite the existence of fuel at low concentration outside the flame's maximum extent.

c. As injection timing is advanced with increasing load, combustion becomes more premixed in character.

d. Observable combustion luminosity ends around 40°–50° ATDC, but *measurable heat release continues until the onset of exhaust flow.*

4. *Hydrocarbon emissions.* Simultaneous heat-release measurements, time-resolved exhaust-hydrocarbon sampling with a fast-flame-ionization detector, and high-speed spectrally resolved imaging have been combined with insight from fuel LIF imaging to identify and, to some extent, quantify sources of HC emissions. Individual-cycle exhaust-HC mass has been evaluated from a calculation of the individual-cycle exhaust mass flow rate based on measured cylinder and exhaust-port pressures.

a. The fast-FID HC concentration data directly confirm that *fuel trapped in the piston top ring-land-crevice is a comparatively unimportant light-load HC source* in this engine, as expected from the late injection timing and the observed spatial confinement of the fuel near the spark plug.

b. Sporadic *misfires and partial burns are an appreciable HC source*, although their effect on total HC emissions is mitigated significantly by the high residual fraction, which allows more than half the injected fuel from a misfire cycle to remain in the cylinder and be consumed over the next three cycles.

c. *Bulk quenching is a major HC source*, contributing in two distinct ways:

i. Extinction occurs as combustion propagates into regions that are too dilute to burn, as shown conclusively by comparison of fuel-LIF and combustion-luminosity images at the lightest loads.

ii. Some rich or combustible mixture fails to burn during the progressively slower mixing-controlled stage of combustion.

d. The release of fuel (mostly vapor but also some liquid mist) trapped in the nozzle-exit crevice has been observed by LIF and Mie scattering, and may be another appreciable HC source.

5. *Nitric oxide emissions.* In-cylinder NO imaging using LIF has been performed simultaneously with exhaust-NO concentration measurements using a standard

chemiluminescence analyzer for DI2S engine operation on gasoline and isooctane fuels over a range of conditions.

  a. Images of NO fluorescence have been obtained successfully prior to ignition and after combustion is complete, but measured NO LIF intensities are strongly reduced during combustion due to strong absorption of the 226-nm laser sheet. The locations and sources of NO formation could therefore not be determined.

  b. Prior to ignition, uniformly distributed NO concentrations are observed, consistent with the strong mixing between fresh charge and residual gases observed in this engine by fuel LIF imaging.

  c. Late in the engine cycle, the integrated NO LIF intensities are linearly correlated with the exhaust NO concentration. *In-cylinder NO levels corresponding to exhaust concentrations as low as 10 ppm can readily be detected in the firing engine using LIF imaging.*

  d. Consistent increases in NO concentration with increasing delivery ratio and with time after the start of firing are observed with both in-cylinder LIF and conventional exhaust-gas measurements.

  e. *Single-laser-shot LIF imaging reveals substantial cycle-to-cycle variation in in-cylinder NO levels.*

6. *Application to engine development problems.* In addition to providing the detailed insight just outlined about HCs (which constitute the major development problem with direct-injection gasoline engines at light load), the techniques here have proven useful in addressing other practical issues.

  a. *High-speed planar Mie scattering has proven to be an effective and comparatively rapid "injector du jour" development tool* for comparing different types of injectors in terms of spray structure and penetration, identifying such malfunctions as secondary injections or fuel dribble, and assessing cycle-to-cycle and injector-to-injector repeatability. Even in a motored optical engine, realistic conditions of in-cylinder flow and gas density can be replicated readily.

  b. Single-cycle fuel-LIF imaging coordinated with cylinder-pressure measurement has established the important role of *cycle-to-cycle fluctuations in the fuel concentration near the spark plug as a primary source of misfires and partial burns* and has pointed out the need for a more repeatable fuel distribution at the time of ignition.

  c. Fuel LIF imaging has also revealed important differences in the fuel distributions produced by two nominally identical injectors during engine warmup over a period of six minutes.

7. *Relation to four-stroke DISI engines.* DI2S engines continue to be of interest for marine, motorcycle, scooter and other light-vehicle applications, although their development as automobile powerplants has given way to four-stroke DISI engines, which are discussed in the next chapter. DI4S engines encounter most of the mixture-preparation, combustion and emissions-formation issues discussed

in this chapter, and application of many of the same diagnostic techniques, together with concurrent advances in fuel-injection equipment, catalyst technology and engine control systems, have proven decisive in the introduction of commercial DI4S engines (see Chap. 3 and [37, 38].

Acknowledgments D.T. French collaborated in all the experimental work described in this chapter. We are also grateful to E.D. Klomp for valuable advice on the individual-cycle exhaust mass flow calculations, to P. Meernik and G. Lalonde for advice on installing and calibrating the fast FID, and to P.M. Najt and A.S.P. Solomon for providing the heat-release code as well as tutelage in its use. E.G. Groff and P.E. Reinke, who (together with R.A. Bolton) led the GM two-stroke-engine project at the operational level, provided much helpful information on engine-system and vehicle issues. In addition, the following colleagues generously shared data and insight obtained through their extensive experimental work on two- and four-stroke DI stratified-charge engines: W.C. Albertson, H.E. Evans, R.M Frank, M.R. Galasso, R.M. Otto, K.B. Rober, A.J. Shearer, and L.H. Weinand.

References

1. D. Scott and J. Yamaguchi, "Pneumatic Fuel Injection Spurs Two-Stroke Revival," *Automotive Engineering*, **94**, 74, 1986.
2. P. Duret, A. Ecomard, and M. Audinet, "A New Two-Stroke Engine with Compressed-Air Fuel Injection for High Efficiency Low Emissions Applications," SAE Paper 880176, 1988.
3. K. Schlunke, "Der Orbital Verbrennungsprozess des Zweitaktmotors," 10th International Motor Symposium, Vienna, VDI No. 122, April 1989, pp. 63–78.
4. D.A. Smith and S.R. Ahern, "The Orbital Ultra Low Emissions and Fuel Economy Engine," 14th Vienna Motorsymposium, VDI No. 182, pp. 203–209.
5. L. Brooke and P.J. Mullins, "To Stroke . . . Or Not Two Stroke," *Automotive Industries*, May 1988.
6. B. Brooks, "Clean Air Fuels Engine Evolution, Not Revolution," *Ward's Auto World*, June 1991, p. 33.
7. F.A. Wyczalek, "Two-Stroke Engine Technology for Passenger Car Automobiles," SAE Paper 912163, 1991.
8. L. Brooke, "Two-Stroke Showdown," *Automotive Industries*, April 1993, pp. 36–43.
9. B. Brooks, "GM Winding Down 2-Stroke Engine Program," *Ward's Engine & Vehicle Technology Update*, April 1994, pp. 1–2.
10. B. Brooks, "Ford Straying from Orbital's 2-Stroke Technology," *Ward's Engine & Vehicle Technology Update*, July 1994, p. 11.
11. K. Buchholz, "Chrysler Updates Two-Stroke Engine Progress," *Automotive Engineering,* January 1997, p. 84.
12. B. Brooks, "EU Invests in 2-Stroke Technology," *Ward's Engine & Vehicle Technology Update*, July 1998, p. 8.
13. D. Scott, "Euro 2-Stroke Project Packs Novel Technology," *Ward's Engine & Vehicle Technology Update*, Sep. 1998, p. 3.
14. J. Personnaz and C. Stan, "Car Hybrid Propulsion Strategy Using an Ultra-Light GDI Two-Stroke Engine," SAE Paper 1999-01-2940.
15. P. Duret et al., "The Air Assisted Direct Injection ELEVATE Automotive Engine Combustion System," SAE Paper 2000-01-1899.
16. D. Shawcross, C. Pumphrey, and D. Arnall, "A Five-Million Kilometre, 100-Vehicle Fleet Trial of an Air-Assist Direct Fuel Injected Automotive 2-Stroke Engine," SAE Paper 2000-01-0898.
17. B. Brooks, "Mercury/Orbital Team Ready with Strat-Charge Outboard," *Ward's Engine & Vehicle Technology Update*, Mar. 1996, p.3.

18. B. Brooks, "Outboard Marine Introduces First DI Outboards," *Ward's Engine & Vehicle Technology Update*, July 1996, p. 4.
19. B. Brooks, "Low Emissions Outboards: Big Success," *Ward's Engine & Vehicle Technology Update*, Sep. 1998, p. 5.
20. S. Strauss, Y. Zeng, and D.T. Montgomery, "Optimization of the E-TEC^{TM} Combustion System for Direct-Injected, Two-Stroke Engines Toward 3-Star Emissions," SAE Paper 2003-32-0007.
21. *Popular Science*, **252**, 18, 1998.
22. D.E. Johnson and H.-C. Wong, "Electronic Direct Fuel Injection System Applied to an 1100-cc Two-Stroke Personal Watercraft Engine," SAE Paper 980756, 1998.
23. "Kawasaki Signs Agreement with Outboard Marine Corporation to Equip Personal Watercraft with FICHT Fuel Injection," *PR Newswire*, April 1998.
24. L. Arnone et al., "Development of a Direct Injection Two-Stroke Engine for Scooters," SAE Paper 2001-01-1782.
25. N. Bradbury et al., "University of Idaho's Clean Snowmobile Design Using a Direct-Injection Two-Stroke," SAE Paper 2005-01-3680.
26. N. Lorenz, T. Bauer, and B. Willson, "Design of a Direct Injection Retrofit Kit for Small Two-Stroke Engines," JSAE Paper 20056601/SAE Paper 2005-32-0095.
27. J.C. Dabadie et al., "DI Two-Stroke Engine Catalyst Development for 2 Wheelers Application," SAE Paper 2001-01-1847.
28. S. Strauss and Y. Zeng, "The Effect of Fuel Spray Momentum on Performance and Emissions of Direct-Injected, Two-Stroke Engines," SAE Paper 2004-32-0013.
29. J. Cromas and J.B. Ghandhi, "Particulate Emissions from a Direct-Injection, Spark-Ignition Engine," SAE Paper 2005-01-0103.
30. J.B. Heywood, *Internal Combustion Fundamentals*, McGraw Hill, Inc., New York, 1988.
31. C. Ferguson, *Internal Combustion Engines: Applied Thermosciences*, Wiley, New York, 1986.
32. G.P. Blair, *The Basic Design of Two-Stroke Engines*, SAE, Warrendale, PA, 1990.
33. G.P. Blair, *Design and Simulation of Two-Stroke Engines,* SAE, Warrendale, PA, 1996.
34. J.B. Heywood and E. Sher, *The Two-Stroke Cycle Engine: Its Development, Operation and Design,* Taylor & Francis, Philadelphia, 1999.
35. P. Duret and J.-F. Moreau, "Reduction of Pollutant Emissions of the IAPAC Two-Stroke Engine with Compressed Air Assisted Fuel Injection, SAE Paper 900801, 1990.
36. H.R. Ricardo, "Recent Research Work on the Internal Combustion Engine," *SAE Transactions*, **14**, 30–32, 1922.
37. F.-Q. Zhao, D.L. Harrington, and M.-C. Lai, *Automotive Gasoline Direct-Injection Engines*, SAE, Warrendale, PA, 2002.
38. M.C. Drake and D.C. Haworth, "Advanced Gasoline Engine Development Using Optical Diagnostics and Numerical Modeling," *Proc. Combust. Inst.*, **31**, 99–124, 2007.
39. Y. Iwamoto, K. Noma, O. Nakayama, T. Yamauchi, and H. Ando, "Development of Gasoline Direct Injection Engine," SAE Paper 970541, 1997.
40. T. Kume, Y. Iwamoto, K. Iida, N. Murakami, K. Akishino, and H. Ando, "Combustion Control Technologies for Direct Injection SI Engines," SAE Paper 960600, 1996.
41. T. Tomoda, S. Sasaki, D. Sawada, A. Saito, and H. Sami, "Development of Direct Injection Gasoline Engine – Study of Stratified Mixture Formation," SAE Paper 970539, 1997.
42. J. Harada, T. Tomita, H. Mizuno, Z. Mashiki, and Y. Ito, "Development of Direct-Injection Gasoline Engine," SAE Paper 970540, 1997.
43. M.C. Drake, T.D. Fansler, A.S. Solomon, and G.A. Szekely, "Piston Fuel Films as a Source of Smoke and Hydrocarbon Emissions from a Wall-Controlled SIDI Engine," SAE Paper 2003-01-0547.
44. C.D. Wood, "Unthrottled Open-Chamber Stratified-Charge Engines," SAE Paper 780341, 1978.

45. M. Alperstein, G. Schafer, and F. Villforth, "Texaco's Stratified Charge Engine: Multifuel, Efficient, Clean and Practical," SAE Paper 740563, 1974
46. A.J. Scussel, A.O. Simko, and W.R. Wade, "The Ford PROCO Engine Update," SAE Paper 780699, 1978.
47. D.R. Lancaster, "Diagnostic Investigation of Hydrocarbon Emissions from a Direct-Injection Stratified-Charge Engine with Early Injection," I. Mech. E. Paper C397-80, 1980.
48. H. Schäpertöns, K.-D. Emmenthal, H.-J. Grabe, and W. Oppermann, "VW's Gasoline Direct Injection (GDI) Research Engine," SAE Paper 910054, 1991.
49. A. Waltner, et al., "Die Zukunftstechnologie des Ottomotors: strahlgeführte Direktein-spritzung mit Piezo-Injektor," 27, Internationales Wienermotoren Symposium, 2006.
50. F. Altenschmidt, et al., "The analysis of the ignition process on SI-engines with direct injection in stratified mode," 7th Intl. Symp. Internal Combustion Diagnostics, Baden-Baden, 2006.
51. J. Fisher, et al., "Methods for the Development of the Spray Guided BMW DI Combustion System," 7th Intl. Symp. Internal Combustion Diagnostics, Baden-Baden, 2006.
52. P. Langen, et al., "Neue BMW Sechs- und Vierzylinder-Ottomotoren mit High Precision Injec-tion und Schichtbrennverfahren," 28. Internationales Wienermotoren Symposium, 2007.
53. A. Jante, "Scavenging and Other Problems of Two-Stroke-Cycle Spark-Ignition Engines," SAE Trans Vol. 77, SAE Paper 680468, 1968.
54. T.D. Fansler and D.T. French, "The Scavenging Flow Field in a Crankcase-Compression Two-Stroke Engine – A Three-Dimensional Laser-Velocimetry Survey," SAE Paper 920417, 1992.
55. T.D. Fansler and D.T. French, "High-Speed Flow Measurements in a Two-Stroke Engine by Photon-Correlation Laser Velocimetry," *Applied Optics*, **32**, 3846–3854, 1993.
56. A.A. Amsden et al., "Comparison of Computed and Measured Three-Dimensional Velocity Fields in a Motored Two-Stroke Engine," SAE Paper 920418, 1992.
57. D.C. Haworth, M.S. Huebler, S.H. El Tahry, and W.R. Matthes, "Multidimensional Calcula-tions for a Two-Stroke-Cycle Engine: A Detailed Scavenging Model Validation," SAE Paper 932712, 1993.
58. T.W. Kuo and R.D. Reitz, "Three-Dimensional Computations of Combustion in Premixed-Charge and Fuel-Injected Two-Stroke Engines," SAE Paper 920425, 1992.
59. T.D. Fansler, "Photon-Correlation Laser Velocimetry in Reciprocating Engine Research," in *Photon-Correlation Techniques and Applications*, J.B. Abbiss and A.E. Smart, eds., Vol. 1 of OSA Conf. Proc. Series (Optical Soc. Amer., Washington, DC, 1988), pp. 54–77.
60. T.E. Hepner, "State-of-the-Art Laser Doppler Velocimeter Processors: Calibration and Evalu-ation," AIAA Paper 94-0042, 1992.
61. Y. Ikeda and T. Nakajima, "Burst Digital Correlator as Laser-Doppler Velocimetry Signal Processor," *Applied Optics*, **35**, 3243–3249, 1996.
62. J.B. Abbiss, "The Structure of the Doppler-Difference Signal and the Analysis of Its Autocor-relation Function," *Physica Scripta*, **19**, 399–395, 1979.
63. C. Arcoumanis and J.H. Whitelaw, "Fluid Mechanics of Internal Combustion Engines – A Review," *Proc. I. Mech. E.*, **201**, 57–74, 1987.
64. L.A. Melton, "Spectrally Separated Fluorescence Emissions for Diesel Fuel Droplets and Vapor," *Applied Optics*, **22**, 2224–2226, 1983.
65. L.A. Melton and J.F. Verdieck, "Vapor/Liquid Visualization for Fuel Sprays," *Combust. Sci. & Tech.*, **42**, 217–222, 1985.
66. M.E.A. Bardsley, P.G. Felton, and F.V. Bracco, "2-D Visualization of Liquid and Vapor Fuel in an I.C. Engine," SAE Paper 880521, 1988.
67. M.E.A. Bardsley, P.G. Felton, and F.V. Bracco, "2-D Visualization of a Hollow-Cone Spray in a Cup-in-Head, Ported, I.C. Engine," SAE Paper 890315, 1989.

68. R. Diwakar et al., "Liquid and Vapor Fuel Distributions from an Air-Assist Injector – An Experimental and Computational Study," SAE Paper 920422, 1992.
69. P.G. Felton, F.V. Bracco, and M.E.A. Bardsley, "On the Quantitative Application of Exciplex Fluorescence to Engine Sprays," SAE Paper 930870, 1993.
70. J.M. Desantes, J.V. Pastor, J.M. Pastor, and J.E. Julia, "Limitations on the Use of the Planar Laser Induced Exciplex Fluorescence Technique in Diesel Sprays," *Fuel*, **84**, 2301–2315, 2005.
71. L.A. Melton, "Exciplex-Based Vapor/Liquid Visualization Systems Appropriate for Automotive Gasolines," *Appl. Spectrosc.*, **47**, 782, 1993.
72. J.B. Ghandhi, P.G. Felton, B. Gajdezcko, and F.V. Bracco, "Investigation of the Fuel Distribution in a Two-Stroke Engine with an Air-Assisted Injector," SAE Paper 940394, 1994.
73. A.P. Fröba et al., "Mixture of Triethylamine (TEA) and Benzene as a New Seeding Material for the Quantitative Two-Dimensional Laser-Induced Exciplex Fluorescence Imaging of Vapor and Liquid Fuel Inside SI Engines," *Combust. Flame*, **112**, 199–209, 1998.
74. A.A. Rotunno, M. Winter, G.M. Dobbs, and L.A. Melton, "Direct Calibrations Procedures for Exciplex-Based Vapor/Liquid Visualization of Fuel Sprays," *Combust. Sci. & Tech.*, **71**, 247–261, 1990.
75. C.-N. Yeh, T. Kamimoto, H. Kosaka, and S. Kobori, "Quantitative Measurement of 2-D Fuel Vapor Concentration in a Transient Spray via Laser-Induced Fluorescence Technique," SAE Paper 941953, 1994.
76. R.S. Schafer, "The Development of an Improved Quantitative Calibration for an Exciplex Liquid/Vapor Visualization System," in *Laser Applications in Combustion and Combustion Diagnostics II*, SPIE Vol. 2122, p. 61, 1994.
77. P. Wieske, S. Wissel, G. Grünefeld, and S. Pischinger, "Improvement of LIEF by Wavelength-Resolved Acquisition of Multiple Images Using a Single CCD Detector – Simultaneous 2D Measurement of Air/Fuel Ratio, Temperature Distribution of the Liquid Phase and Qualitative Distribution of the Liquid Phase with the Multi-2D Technique," *Appl. Phys. B*, **83**, 323–329, 2006.
78. E. Giffen and A. Muraszew, *The Atomisation of Liquid Fuels*, Wiley, New York, 1953.
79. T.D. Fansler, D.T. French, and M.C. Drake, "Fuel Distributions in a Firing Direct-Injection Spark-Ignition Engine Using Laser-Induced Fluorescence Imaging," SAE Paper 950110, 1995.
80. M.C. Drake, T.D. Fansler, and D.T. French, "Crevice Flow and Combustion Visualization in a Direct-Injection Spark-Ignition Engine using Laser Imaging Techniques," SAE Paper 952454, 1995.
81. M.C. Drake, D.T. French, and T.D. Fansler, "Advanced Diagnostics for Minimizing Hydrocarbon Emissions from a Direct-Injection Gasoline Engine," *Proc. Combust. Inst.*, **26**, 2581–2587, 1996.
82. T.D. Fansler, D.T. French, and M.C. Drake, "Individual-Cycle Measurements of Exhaust-Hydrocarbon Mass from a Direct-Injection Two-Stroke Engine," SAE Paper 980758, 1998.
83. J.A. Gatowski, E.N. Balles, K.M. Chun, F.E. Nelson, J.A. Ekchian, and J.B. Heywood, "Heat Release Analysis of Engine Pressure Data," SAE Paper 841359, 1984.
84. C.A. Poynton, *A Technical Introduction to Digital Video*, Wiley, New York, 1996.
85. T.A. Baritaud and R.M. Green, "A 2-D Flame Visualization Technique Applied to the I.C. Engine," SAE Paper 860025, 1986.
86. A.O. zur Loye and F.V. Bracco, "Two-Dimensional Visualization of Premixed-Charge Combustion Flame Structure in an IC Engine," SAE Paper 870454, 1987.
87. G.F.W. Ziegler, A. Zettlitz, P. Meinhardt, R. Herweg, R. Maly, and W. Pfister, "Cycle-Resolved Two-Dimensional Flame Visualization in a Spark-Ignition Engine," SAE Paper 881634, 1988.

88. M.D. DiRosa, K.G. Klavuhn, and R.K Hanson, "LIF Spectroscopy of NO and O_2 in High-Pressure Flames," *Combust. Sci. Tech.*, **118**, 257–283, 1996.
89. M.C. Drake and J.W. Ratcliffe, "High-Temperature Quenching Cross Sections for Nitric Oxide Laser-Induced Fluorescence Measurements," *J. Chem. Phys.*, **98**, 3850, 1993.
90. T.B. Settersten, B.D. Patterson, and J.A. Gray, "Temperature- and Species-Dependent Quenching of NO $A^2\sum(v'=0)$ Probed by Two-Photon Laser-Induced Fluorescence Using a Picosecond Laser," *J. Chem. Phys.*, **124**, 234308, 2006
91. J.R. Reisel, C.D. Carter, N.M. Laurendeau, and M.C. Drake, "Laser-Induced Fluorescence Measurements in Laminar, Flat $C_2H_6/O_2/N_2$ Flames at Atmospheric Pressure," *Combust. Sci. Tech.*, **91**, 271, 1993.
92. F. Hildenbrand, C. Schulz, V. Sick, G. Josefsson, I. Magnusson, Ö. Andersson, and M. Aldén, "Laser-Spectroscopic Investigation of Flow Fields and NO Formation in a Realistic SI Engine," SAE Paper 980148, 1998.
93. A.J. Giovanetti et al., "Analysis of Hydrocarbon Emissions in a Direct-Injection Spark-Ignition Engine," SAE Paper 830587, 1983.
94. R.M. Frank and J.B. Heywood, "Combustion Characterization in a Direct-Injection Stratified-Charge Engine and Implications on Hydrocarbon Emissions," SAE Paper 892058, 1989.
95. E.W. Rothe and P. Andresen, "Application of Tunable Excimer Lasers to Combustion Diagnostics: A Review," *Applied Optics*, **36**, 3971–4033, 1997.
96. C. Schulz and V. Sick, "Tracer LIF Diagnostics: Quantitative Measurement of Fuel Concentration, Temperature and Fuel/Air Ratio in Practical Combustion Systems," *Progr. Energy Combust. Sci.*, **31**, 75–121, 2005.
97. J. Reboux, D. Peuchberty, and F. Dionnet, "A New Approach of Planar Laser Induced Fluorescence Applied to Fuel/Air Ratio Measurement in the Compression Stroke of an Optical S.I. Engine," SAE Paper 941988, 1994.
98. F. Grossmann, P.B. Monkhouse, M. Ridder, V. Sick, and J. Wolfrum, "Temperature and Pressure Dependences of the Laser-Induced Fluorescence of Gas-Phase Acetone and 3-pentanone," *Appl. Phys. B*, **62**, 249–253, 1996.
99. J.B. Ghandhi and P.G. Felton, "On the Fluorescent Behavior of Ketones at High Temperatures," *Expts. in Fluids*, **21**, 143–144, 1996.
100. T.D. Fansler, M.C. Drake, B. Stojkovic, and M. Rosalik, "Local Fuel Concentration, Ignition and Combustion in a Stratified Charge Spark Ignited Direct Injection Engine: Spectroscopic, Imaging and Pressure-Based Measurements," *Intl. J. Engine Res.*, **4**, 61–86, 2003.
101. J.A. Eng, W.R. Leppard, P.M. Najt, and F.L. Dryer, "The Interaction Between Nitric Oxide and Hydrocarbon Oxidation Chemistry in a Spark-Ignition Engine, SAE Paper 972889, 1997.
102. Z. Liu and S.I. Woo, "Recent Advances in Catalytic DeNOx Science and Technology," *Catalysis Reviews*, **48**, 43–89, 2006.
103. A.C. Alkidas, "Combustion-Chamber Crevices: The Major Source of Engine-Out Hydrocarbon Emissions Under Fully Warmed Conditions," *Progr. Energy Combust. Sci.*, **25**, 253–273, 1999.
104. E. Hudak and J.B. Ghandhi, "Time-Resolved Exhaust Sampling Measurements of a Two-Stroke Direct-Injection Engine," SAE Paper 1999-01-3309.
105. X. Liu and D.B. Kittelson, "Total Cylinder Sampling from a Diesel Engine (Part II)," SAE Paper 820360, 1982.
106. A. Bräumer, V. Sick, J. Wolfrum, V. Drewes, M. Zahn, and R. Maly, "Quantitative Two-Dimensional Measurements of Nitric Oxide and Temperature Distributions in a Transparent Square-Piston SI Engine," SAE Paper 952462, 1995.
107. C. Schulz, V. Sick, J. Wolfrum, V. Drewes, M. Zahn, and R. Maly, "Quantitative 2D Single-Shot Imaging of NO Concentrations and Temperatures in a Transparent SI Engine," *Proc. Combust. Inst.*, **26**, 2597–2604, 1996.
108. P. Andresen, G. Meier, H. Schluter, H. Voges, A. Koch, W. Hentschel, W. Operman, and E. Rothe, "Fluorescence Imaging Inside an Internal Combustion Engine Using Tunable Excimer Lasers," *Applied Optics*, **29**, 2393, 1990.

109. W.G. Bessler et al., "Quantitative In-cylinder NO-LIF Imaging in a Realistic Gasoline Engine with Spray-Guided Direct Injection," *Proc. Combust. Inst.*, **30**, 2667–2674, 2005.
110. B. Alatas, J.A. Pinson, T.A. Litzinger, and D.A. Santavicca, "A Study of NO and Soot Evolution in a DI Diesel Engine via Planar Imaging," SAE Paper 930973, 1993.
111. F. Hildenbrand, C. Schulz, E. Wagner, and V. Sick, "Investigation of Spatially Resolved Light Absorption in a Spark-Ignition Engine Fueled with Propane/Air," *Applied Optics*, **38**, 1452–1458, 1999.
112. C. Schulz, V. Sick, U.E. Meier, J. Heinze, and W. Stricker, "Quantification of NO *A–X*(0,2) Laser-Induced Fluorescence: Investigation of Calibration and Collisional Influences in High-Pressure Flames," *Applied Optics*, **38**, 1434–1443, 1999.
113. W.G. Bessler et al., "Strategies for Laser-Induced Fluorescence Detection of Nitric Oxide in High-Pressure Flames. III. Comparison of A–X Excitation Schemes," *Applied Optics*, **42**, 4922–4936, 2003.
114. J.E. Dec and R.E. Canaan, "PLIF Imaging of NO Formation in a DI Diesel Engine," SAE Paper 980147, 1998.
115. J.B. Ghandhi and F.V. Bracco, "Fuel Distribution Effects on the Combustion of a Direct-Injection Stratified-Charge Engine," SAE Paper 950460, 1995.
116. T.D. Fansler, M.C. Drake, I. Düwel, and F. Zimmermann, "Fuel-Spray and Spark-Plug Interactions in a Spray-Guided Direct-Injection Gasoline Engine," *Proc. 7th Intl. Symp. on Internal Combustion Diagnostics*, Baden-Baden, Germany, May 18–19, 2006.

Chapter 3
Flow, Mixture Preparation and Combustion in Four-Stroke Direct-Injection Gasoline Engines

Hiromitsu Ando and Constantine (Dinos) Arcoumanis

3.1 Introduction

Since the launch of the first direct-injection (DI) gasoline engine in 1996, many Japanese and European car manufacturers have introduced this technology into the market with the expectation that it would spread worldwide. There was a commonly-shared view that it represented one of the most effective fuel economy improvement technologies for CO_2 reduction. Unfortunately, this technology hasn't achieved significant penetration into the mass production market even after 10 years from its launch, despite its obvious scientific and technological advantages.

In this chapter, the basic features of direct injection gasoline engines and the technologies adopted to realize them will be described. Then the technological evolutions over the last 10 years, through worldwide efforts by researchers in both industry and academia, will be reviewed and the prospects for the future of the direct-injection gasoline engine technology summarized.

Finally, the latest developments, i.e. the spray-guided combustion system representing the most promising concept for meeting the future CO_2 targets, are presented and examples are provided of the new generation of fuel injection systems that may allow the mass production of stratified direct injection gasoline engines for automotive applications.

3.2 Significance and Limits of the Direct-Injection Gasoline Engine Technology

It is generally accepted that the fuel economy improvements of DI gasoline engines are due to their higher compression ratio, higher specific heat ratio, pumping loss reduction (lean burn, EGR) and cooling loss reduction. The effects of these factors

Hiromitsu Ando
University of Fukui, Fukui, Japan, e-mail:

Constantine (Dinos) Arcoumanis
City University London, London, UK, e-mail: c.arcoumanis@city.ac.uk

C. Arcoumanis, T. Kamimoto (eds.), *Flow and Combustion in Reciprocating Engines*,   137
DOI: 10.1007/978-3-540-68901-0_3, © Springer-Verlag Berlin Heidelberg 2009

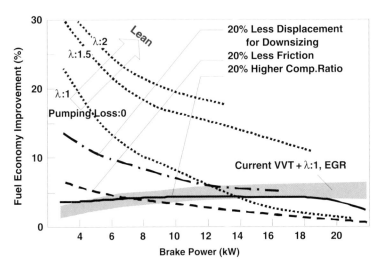

Fig. 3.1 Factors of fuel economy improvements

on fuel economy improvement are shown schematically in Fig. 3.1 and can be summarized as follows:

- Although there are large effects in the low load region, their influence diminishes at higher loads
- The effect of lean burn is large mainly due to the higher specific heat ratio rather than reduction of the pumping loss.
- The effect of higher specific heat ratio is maintained even at higher loads.
- The reason the DI gasoline engine technology is so efficient lies in its ability to take advantage of all the factors shown in Fig. 3.1, that is, the higher specific heat ratio realized by the stable lean burn, the higher compression ratio permitted by various anti-knock measures, the pumping loss reduction through lean burn and the cooling loss reduction through lowering of the burned gas temperature and mixture stratification effects.

Although the fuel economy benefits of direct-injection gasoline engines compared with their conventional port fuel injection counterparts were large at the timing of their launch in 1996, they have gradually lost some of their advantages over the last 10 years or so. The main factors are the following:

- Port fuel injection engines have improved significantly
- Breakthrough transmission technologies have emerged
- Operating conditions for the lean operation of DI gasoline engines could not be extended in order to meet the increasingly more stringent emission control requirements.

In addition, diesel engines have entered the European market with superior fuel efficiency and outstanding performance which reduced the momentum for research into DI gasoline engines.

The most distinctive evolution of port fuel injection engines has been the introduction of Variable Valve Timing (VVT) technologies which are now quite standard. Through the VVT, pumping loss reduction can be realized by using the Atkinson cycle effect where valve overlap is reduced under low load conditions to reduce the amount of residual gases. As a result of the improved combustion stability, engine idle speed can be reduced to the same level as in DI gasoline engines. In addition, by adjusting the timing of the valve overlap, hot exhaust gas is evacuated selectively which results in a slight improvement of the anti-knock characteristics; further improvement can be achieved through optimization of the thickness of the coolant flow wall between the combustion chamber and the water gallery. This can be realized through advances in the casting techniques which also contributes to the improvement of the anti-knock characteristics.

Almost all of the fuel economy improvement technologies are aiming at the low load conditions where thermal efficiency is rather poor. New transmissions such as CVT, multi-stage AT, multi-stage MT and AMT (Automated Manual Transmission) are technologies able to improve the powertrain efficiency by minimizing low load engine operation. The end result has been that the significant effects on fuel consumption reduction introduced by the DI gasoline engine have been gradually depreciated by the introduction of these transmission technologies.

Finally, although the evolution of the lean NOx catalyst technology has been remarkable, as will be described later, and its efficiency almost exceeds 90% under relatively wide ranging conditions, the efficiency of its competitor, the three-way catalyst used by port fuel injection engines, frequently exceeds 99%. It can therefore be argued that the efficiency of DI gasoline engines has been sacrificed in order to achieve emission levels comparable to those of port fuel injection engines. However, at present fuel consumption is attracting more attention than local exhaust emissions due to the emphasis on global warming, which is responsible for the renewed interest on DI Diesels and spray-guided stratified gasoline engines.

3.3 Technologies for Direct-Injection Gasoline Engines

It is well known that there are three basic concepts of direct injection gasoline engines: wall-guided, air-guided and spray-guided (Table 3.1). The concept investigated since the 1950s had been the spray-guided concept in a configuration known as narrow/close spacing which locates closely the fuel spray and the spark plug in the central region of the cylinder head. [1, 2, 3, 4, 5, 6, 7]. In this concept, the spark plug is projected into a specific region of the fuel spray in order to achieve stable spark ignition and, as such, is it exposed to fuel droplets and/or a rich fuel-air mixture which encourages soot formation and spark plug fouling leading to misfires. It is only recently that the first encouraging signs have emerged for a successful

Table 3.1 Characteristics of gasoline direct-injection concepts

	Wall-guided	Air-guided	Spray-guided
Fuel economy	0	+	++
HC	0	+	++
Smoke	0	+	++
Power	0	−	0 to +
Robustness	0	−	−
Mixture preparation			

implementation of the spray-guided concept in a production engine; more details are provided later.

In 1996 a new concept called wide spacing was proposed [8]. This concept locates the fuel spray and the spark plug at some distance from each other. The fuel is injected towards the hot piston surface and the formed fuel vapour is directed towards the spark plug through a combination of piston-bowl geometry and strong in-cylinder bulk flow. The main advantage of this concept is that the spark plug fouling problem is resolved since the fuel approaches the spark plug after vaporization and mixing with air. The wall-guided engine concept used the following four main technologies:

- An in-cylinder flow called the reverse tumble generated by the upright straight intake ports.
- A swirl injector nozzle preparing well dispersed and atomized fuel spray at relatively low injection pressures.
- A piston with a cavity designed to optimize fuel spray impingement and re-direction towards the spark plug
- Mixing control mechanism combining a homogeneous mixture realized by injection during the early stages of the intake stroke and a stratified mixture realized by injection during the later stages of the compression stroke.

After the launch of the first mass production DI gasoline engine in 1996, many alternative systems have been introduced into the market, employing swirl or tumble in place of the reverse tumble, injection systems using a swirl nozzle with an asymmetric spray shape named casting net spray and achieved through a novel nozzle hole design, and a fan-spray nozzle producing a two-dimensional thin shaped spray through a slit located at the outlet of the nozzle hole [9, 10, 11, 12]. Both systems are classified into the wide spacing concept, and have the common feature of using an intense in-cylinder flow and a purpose-designed piston cavity. Although some systems have been referred to as wall-guided and others as flow-guided, this classification seems to have little meaning since all first-generation DI gasoline engines in production have employed guidance by flow and wall simultaneously and experienced increased heat losses as a result of the complicated piston cavity with its larger surface area. Flow-guided is probably a term introduced to highlight the necessity for simplification of the piston crown geometry which more recently led to the introduction of the air-guided concept. Existing wall-guided combustion systems in production today are illustrated in Fig. 3.2 in terms of their corresponding piston and spray configurations.

The spray-guided concept under intense development today offers the best potential partly because of the possibility of using a simpler piston geometry. In order to realize the spray-guided concept, it is important to prevent impingement of fuel droplets onto the spark plug. To achieve this, many ideas have been proposed such as using an air-assisted injector, a spray collision injector, an outwards opening nozzle and a multi-hole nozzle, the latter two combined with higher pressure injection [13, 14, 15, 16, 17]. The targets are to dilute the fuel through air entrainment and

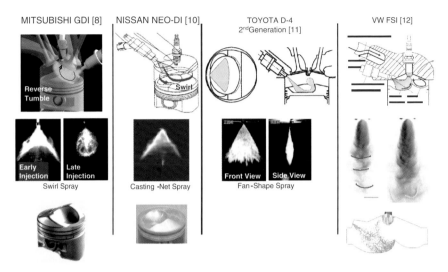

Fig. 3.2 Wall-guided gasoline direct-injection engines

improved atomization and to prevent the concentration of large fuel droplets and
rich fuel vapour around the spark plug at the time of ignition through control of the
spray development as a function of the thermodynamic in-cylinder conditions. Out
of the emerging advanced fuel injection technologies, the multi-hole and outward-
opening nozzles shown in Figs. 3.3 and 3.4, have real potential for becoming a

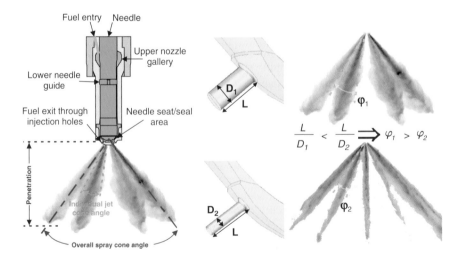

Fig. 3.3 Multi-hole injector for spray-guided gasoline direct injection engines and its key
geometric parameters (spray and jet cone angle, hole diameter D and length L)

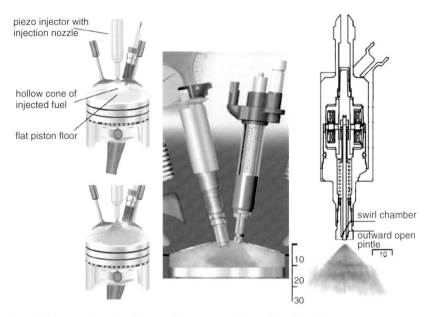

piezo injector with
injection nozzle

hollow cone of
injected fuel

flat piston floor

swirl chamber

outward open
pintle

Fig. 3.4 Outward-opening injectors for spray-guided gasoline direct injection engines

mass production technology. In case of the multi-hole nozzle, the spark plug can be located between two of the sprays or over one of the sprays with a small aperture.

On the other hand, an outward-opening nozzle allows intense interaction between fuel droplets and the surrounding air, with atomized fuel concentrated on the outer periphery of the spray and the spark plug located in the recirculation zone formed at the leading edge of the spray. Both nozzle types are able to prepare a fuel-air mixture with near stoichiometric mixture strength around the spark plug while avoiding direct fuel spray impingement on the spark electrodes. More details about these two emerging fuel injection technologies and the generated sprays are given in subsequent sections. Overall, it can be argued that the potential of the spray-guided concept is much higher than the other two, despite its questionable robustness relative to the wall-guided concept (Table 3.1) due to the fact that its combustion development is controlled exclusively by the stability of the fuel spray characteristics.

3.4 General Features of Direct-Injection Gasoline Engines

The distinctive features of the DI gasoline engine are centred on its potential to achieve stable lean burn, to reduce and even eliminate engine-out unburned hydrocarbon emissions, to improve the anti-knock characteristics and to control torque precisely and quickly.

3.4.1 Lean Burn

Combustion of a lean fuel/air mixture should be discussed from two viewpoints: thermal efficiency and NOx formation. With respect to thermal efficiency, reduction of pumping losses, reduction of heat losses and higher specific heat ratio are the main advantages of lean burn. These are realized by increasing the total amount of working fluid, lowering the mean burned-gas temperature and increasing the percentage of diatomic molecules in the working fluid. All these effects can be achieved when the fuel-air mixture is overall lean. The DI gasoline engine which can maintain stable combustion under overall lean mixture conditions has the best potential to fully explore the thermodynamic advantages of lean burning.

With respect to NOx reduction, it is necessary to control the local equivalence ratio to remain always lean and to suppress flame propagation. The first requirement is to lower the local temperature or oxygen concentration in the burned gas region to prevent or reduce thermal NOx formation. The local temperature of the burned gas continues to increase by means of adiabatic compression realized in the region ahead of the propagating flame. In particular, when combustion is controlled by flame propagation, a non-uniform temperature distribution takes place in the cylinder which results in substantial NOx formation in the high temperature region. This is the reason for the second prerequisite for NOx reduction which is the elimination of flame propagation. The developing consensus on this point is that the ultimate solution for satisfying the requirement for NOx reduction is HCCI combustion, where simultaneous ignition of the mixture takes place across the cylinder volume and any temperature increases maintain mixture homogeneity.

It can be argued from the above that, from the viewpoint of NOx reduction, the direct-injection gasoline engine is not the optimal solution since the local combustion field is generated by spark ignition and the resulting flame traverses the combustion chamber tracing the fuel/air mixture with moderate local equivalence ratio. The temperature of the burned gas zone continues to rise and generates large amounts of thermal NOx. Therefore, the engine-out NOx emissions of direct injection gasoline engines operating under lean overall conditions remain at relatively high levels of about 0.2–0.35 of those in a conventional port fuel injection engine operating at stoichiometric conditions which necessitates a sophisticated aftertreatment system for reducing NOx.

When high output is desirable, it is necessary to operate the engine under stoichiometric or slightly rich conditions, because it is the fuel quantity that determines the engine output under conditions of fixed air mass. Therefore, lean engine operation with higher efficiency is quite restricted under realistic driving conditions. This represents the essential limit to the advantages that can be offered by direct injection gasoline engines.

3.4.2 Minimization of Fuel Escaping Power Generating Combustion

In port fuel injection engines, a lot of fuel is captured in the oil film formed on the cylinder liner surface and/or in the piston crevices. This fuel is scraped and released from the walls and cavities by the ascending piston during the exhaust stroke and is burned rather quickly in the cylinder when the intake valves open, due to the intense turbulence generated and the large amount of fresh air introduced into the cylinder during the opening period of the intake valves. Although this fuel is not strictly counted as unburned hydrocarbons, the fuel burned during the exhaust and intake strokes which does not contribute to the work output of the engine should be treated as escaped fuel. It is sometimes overlooked that escaping fuel is one of the most important factors penalizing the thermal efficiency of port fuel injection engines. It is thus very fortunate that DI gasoline engines, where the air around the cylinder liner does not contain fuel, offer the possibility of minimizing the fuel escape and enhancing further their thermal efficiency.

3.4.3 Improvement of Anti-Knock Characteristics

At higher loads, the DI gasoline engine operates by adopting the early injection strategy where the combustion characteristics are identical to those of the premixed port injection engine. However, improved engine performance can be realized by the higher volumetric efficiency and the improved anti-knock characteristics.

In the case of port fuel injection engines, the latent heat of evaporation is supplied by the surface of the intake port(s), the intake valves or the cylinder liner. In the case of early direct injection, the fuel spray follows the piston and the impingement of the liquid fuel on its crown can be carefully minimized. Therefore, the latent heat is supplied to the fuel by the intake air and this causes efficient charge air cooling; it is estimated that the charge air is cooled by about 15 K. This implies that the gas temperature at the end of the compression stroke can be reduced by about 30 K, which enhances the volumetric efficiency and suppresses knocking.

In the case of port fuel injection engines, significant transient knock can take place during several cycles at the start of vehicle acceleration. Transient knock is caused by selective transport of low boiling point gasoline components with their lower octane numbers [18]. The direct injection engine, however, is not affected by such transient knock because all of the gasoline components are transported into the cylinder. A knock suppression period, caused by delay in the surface heating of the combustion chamber, follows the transient knock period. As a result, in the case of direct-injection gasoline engines that are not sensitive to transient knock, ignition timing can be advanced by 10 s or more at the start of the acceleration period which, in general, is sufficient to complete the vehicle acceleration.

For further improvements of the full-load performance, an alternative knock suppression method named 'two-stage mixing' has been proposed [19]. Since there is consensus that the most distinctive feature of a direct injection engine is the 'freedom of mixing', 'two-stage mixing' takes advantage of this freedom for knock suppression; Fig. 3.5 illustrates the relevant procedure. Fuel is injected into the engine cylinder twice; the first injection takes place during the early stages of the intake stroke to prepare a premixed very lean mixture while the second injection takes place during the later stages of the compression stroke to prepare a stratified mixture for ignition.

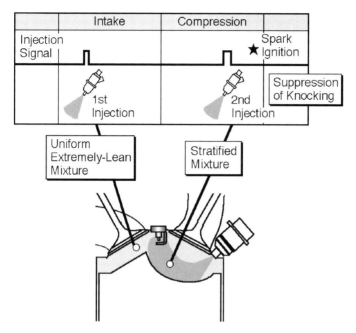

Fig. 3.5 Characteristics of two-stage mixing [18]

This knock-prevention process can be explained as follows:

- The premixed mixture prepared by the first injection is too lean to cause knock.
- The stratified mixture prepared by the second injection may form a stoichiometric mixture at some locations, however, the life time of that mixture prior to ignition is not long enough for knock precursor reactions to proceed.

Although in the case of a single late injection, a large amount of soot can be emitted when the average mixture strength becomes rich, in the case of two-stage mixing soot emission is not a problem except at very low engine speeds when the average air/fuel ratio is 12. The process of soot formation and oxidation can be explained as follows:

- When spark ignition takes place, only the stratified charge is ignited because the surrounding premixed mixture is too lean to be ignited.
- Combustion products generated in the rich stratified mixture containing CO and soot are transported by means of flame propagation into the surrounding lean mixture.
- Soot acts as ignition site and ignites the very lean mixture beyond its flammability limit; in this process CO may assist ignition and soot is oxidized in the combustion zone of the lean mixture.

3.4.4 Precise and Rapid Torque Management

In terms of engine management, the direct injection gasoline engine technology offers another advantage, the 'freedom of control', that is the potential of precise and quick management of torque with the best possible robustness.

In the case of port fuel injection engines, engine torque is controlled by the intake-air volume and, thus, is dependent on throttle valve operation. Even with an electronic throttling system, a change in air intake is somewhat delayed because the intake manifold with its large volume is located downstream of the throttle valve. In the case of direct-injection gasoline engines, however, torque can be easily controlled by just changing the injected fuel quantity while retaining the air intake constant; to be more specific, the torque in each cylinder and at each engine revolution can be freely controlled. This freedom can be used in the idle stop application [20] of a hybrid vehicle with a small capacity motor and in the advanced engine-transmission integrated control system. To gain customer acceptance, the idle-stop system needs to be able to achieve fast start and acceleration of the vehicle under the standstill mode.

Figure 3.6 compares the starting process in a port fuel injection engine with that of a direct injection engine. In the port fuel injection engine, the procedure for the earliest re-starting process is as follows: (i) detect the cylinder under exhaust stroke at the engine cranking, (ii) inject fuel into the port of the cylinder, (iii) the injected fuel is carried into the cylinder during the intake stroke and is compressed and burned. In the case of the direct injection gasoline engine, the cylinder under compression is detected at the beginning of the engine cranking and fuel is injected

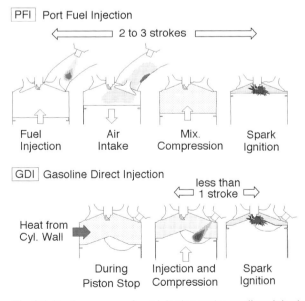

Fig. 3.6 Starting process of port injection engine vs. direct injection engine [19]

into this cylinder. The air temperature in that cylinder is already high enough due to the heated cylinder walls even at the suspended engine operation, and the fuel injected at the later stage of the compression stroke becomes ready for ignition, with complete combustion taking place once the mixture is ignited. Consequently, the delay in torque generation is minimized to one-fifth or one-tenth of that in the port fuel injection engine. Accordingly, the vehicle can execute a desirable idle-stop without giving the driver any uncomfortable feeling. Recently, the possibility to start the engine only by fuel injection, that is, without using the starter motor has been considered. The applicability of this idea will depend on the simplicity of the device to control the piston phase at the starting point of the engine.

The concept of 'freedom of control' can be applied to hybrid electric vehicles which require starting the engine when the vehicle is driven by the motor. Smooth and gradual torque increase from zero is desirable to suppress the shock caused by the torque gap. In the case of port fuel injection engines, however, it is difficult to maintain stable combustion under extremely low air and fuel conditions and some torque gap is inevitable. On the contrary, in a direct injection gasoline engine where stable combustion can be realized by a very small fuel quantity while keeping an overall very lean mixture, very low torque can be generated and the shock associated with engine starting can be eliminated completely. This effect is quite distinctive in a hybrid system using a small motor where frequent switching between motor and engine takes place.

The advantage of the DI gasoline engine in controlling the cycle- and cylinder-resolved torque by tuning only the injected fuel quantity is used in the integrated control of engine and transmission. In the case of CVT, for example, it is used to compensate the harsh and tough trade-offs of the efficiency, drivability and durability. It is also used to solve the problem of the tensional resonance caused by the connection of the shaft with poor rigidity to the pulleys with large inertia through rapid torque management synchronized with the natural vibration of the system. It can thus be argued that the strategies for resolving the trade-off problems using the 'freedom of torque' management approach may allow advances in various fields where technological breakthroughs are urgently needed.

3.5 Possibility for Further Reduction in NOx Emissions

The worldwide enforcement of exhaust emission reductions and the trend towards more stringent targets is creating difficulties for production direct injection gasoline engines to meet future regulations. Although it cannot be denied that they are handicapped relative to the mature three-way catalyst technology for port fuel injection engines, the catalyst and the catalyst reaction control technologies for direct injection gasoline engines have also advanced. At present, the developed exhaust after treatment system has the potential to meet the most stringent worldwide emission regulations of Japan, Europe, U.S.A. and California, in particular.

3.5.1 Advances in Lean NOx Catalyst Technologies

The first production direct injection gasoline engine adopted for NOx emission control a selective reduction catalyst which is less sensitive to sulphur poisoning. This has later been replaced by a NOx trap catalyst of higher performance as a result of the worldwide trend in gasoline sulphur reduction. A NOx trap catalyst operates by capturing NOx under lean operating conditions and converting it into nitrate of alkali metals and alkaline-earth metals. When the engine operating condition changes from lean to rich during short acceleration periods, the nitrates are converted into carbonates by means of the CO contained in the exhaust gas and the adsorbed NOx is released.

The first NOx trap used Ba as a NOx adsorbing metal. Although it was known that K shows better performance than Ba, K is lower in electronegativity and is thermally unstable. It is, therefore, for the following reasons that the K adsorbent has been unpopular in its use for automotive NOx control:

- The higher the catalyst temperature is, the more K adsorbent outpours from the wash coat.
- K penetrates into the catalyst substrate.

Despite all its problems, recent advances of NOx trap technologies have been remarkable and some solutions were found [21, 22]. The first solution is to mix K adsorbent with zeolite for K stabilization, while the second solution is to coat the substrate surface with Si which, due to its stronger chemical affinity to K, prevents K penetration into the substrate. It was thus found that, by adding silica compound to the wash coat, it is possible to have high-storage performance up to 550°C and extended lean-operating region, while the heat resistance can be improved to 850°C.

Overall, progress of NOx trap technologies has steadily continued. New methods, such as a method to stabilize K by adding MgO, a method to use Na in place of K and a pore distribution optimization method [23, 24], were proposed. The most effective breakthrough technology has been the introduction of metal substrates. Because the problem of the K penetration can now be resolved, higher efficiency can be achieved by the increase of doped K. Besides these NOx trap technologies, new direct NOx decomposition catalysts using new materials such as La-Ba-Mn-In-Cu perovskite or the oxides of Ba and CO were also proposed [25, 26, 27]. It is expected that one of these advanced concepts for lean NOx catalysts will reach sooner or later mass production.

3.5.2 Catalyst Reaction Control by Mixture Preparation and Combustion Control

Various catalyst reaction control technologies that make use of the distinctive feature of direct injection gasoline engines, the 'freedom of mixing', have been proposed [28, 29]. One of them is the method of 'two-stage combustion', illustrated in

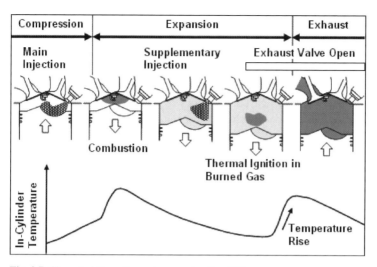

Fig. 3.7 Characteristics of two-stage combustion [28]

Fig. 3.7. The main fuel injection takes place at the later stages of the compression stroke for achieving stratified charge combustion and combustion is completed at an early stage of the expansion stroke; a second injection of a smaller amount of fuel takes place at the later stage of the expansion stroke. Since a large amount of excess air remains in the burned gas and its temperature is sufficiently high, this supplementary fuel is easily burned and the exhaust gas temperature rises rapidly; the end result is that the catalyst is warmed up quickly. Unfortunately, two-stage combustion consumes approximately twice the fuel of conventional combustion because this supplementary fuel is converted to heat rather than to work. Therefore, increased fuel consumption has accompanied the concept of 'two stage combustion'.

To solve this problem, an additional method named 'Stratified Slightly Lean Combustion' was proposed where the fuel is injected at an early stage of the compression stroke to realize a stratified slightly lean mixture. Figure 3.8 illustrates the situation of the exhaust gas which contains substantial amounts of CO generated in the slightly rich zone and O_2 remaining in the slightly lean zone; CO and O_2 are carried over the catalyst surface where the light-off temperature of CO is as low as 150°C. The catalyst is heated for several seconds up to this temperature in order to start the catalytic oxidation of CO on its surface by the two-stage combustion. Thereafter, the combustion mode is switched over to the stratified slightly lean combustion where CO oxidation instantly begins on the catalyst surface and HC oxidation is also induced because the catalyst surface is heated selectively to a temperature higher than the light-off temperature of HC. In such a manner, the stratified slightly lean combustion substantially reduces the time duration of the two-stage combustion. The end result is that better fuel economy and faster catalyst light-off can be simultaneously achieved.

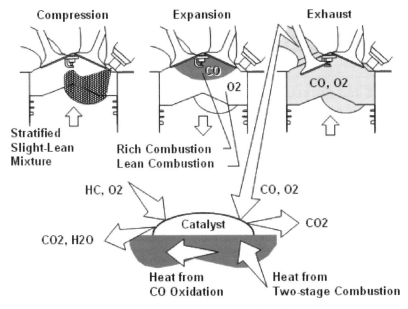

Fig. 3.8 Characteristics of stratified slight-lean combustion [28]

The NOx reduction method using a NOx-trap catalyst is basically a reversible reaction of nitrate and carbonate. NOx is adsorbed by the formulation of nitrate and desorbed by the formulation of carbonate. The desorbed NOx is then converted into O_2 and N_2 in the three-way catalyst located downstream of the NOx trap by a reducing agent such as the CO present in the exhaust gas. During this process, part of the CO in the exhaust is consumed for carbonate formulation which leads to a shortage of the reducing agent to purify the desorbed NOx; the result is that substantial NOx is temporarily emitted from the tail-pipe. The increase of engine-out CO leads to more residual CO in the exhaust gas downstream of the NOx trap catalyst, but it is not sufficient to reduce subsequent NOx levels since the desorbed NOx also increases with CO emissions. Thus the catalyst needs to be provided with a reducing agent that is less effective on NOx desorption. For such purpose, a method of introducing an additional fuel injection for a very limited period during the expansion stroke, to supply hydrocarbons to the three-way catalyst, was invented. A large amount of K loaded onto the NOx-trap catalyst as adsorbent interacts with π electrons in the HC, and then the HC adsorption into the noble metal is disturbed. Consequently, the HC emissions are not reduced by the NOx trap catalyst and flow into the three-way catalyst downstream; this HC is decomposed into CO and H_2 which achieve the reduction of the desorbed NOx. The end result of this control technology is that NOx emissions at NOx purging are minimized.

In summary, these are the methods realized by the inherent advantage of direct injection gasoline engines, the 'freedom of mixing'. In the relevant research of catalyst reactions, it is important to remember this important advantage and pursue any innovative ideas in an effort to solve the remaining problem of the quite complicated NOx trap catalyst system which remains a prerequisite for the future market success of direct injection gasoline engines.

3.6 Present and Future Prospects

More than 10 years have already elapsed since the launch of the first mass production direct injection gasoline engine. In the meantime, some non-engine breakthrough technologies have emerged, other combustion concepts such as the Homogeneous Charge Compression Ignition (HCCI) have received worldwide attention, diesel engines are now well established and recently the spray-guided concept using either an outward opening nozzle or a multi-hole nozzle is emerging as the most serious technology offering hope for the mass production of the direct-ignition gasoline engine. Needless to say that continuation of the evolution of catalyst technologies and catalyst reaction control technologies is also very important to the success of the spray-guided concept. It can also be argued that the outstanding torque control capability of DI engines will be very useful in various applications and may contribute to their popularity.

3.6.1 Stoichiometric DI Gasoline Engines

As research into the spray-guided concept has intensified and the expectations from the wall-guided concept have diminished, a compromise mid-way solution, the stoichiometric direct injection gasoline engine has started attracting a lot of attention [30, 31, 32]. This is the concept which aims to achieve relatively modest improvements in thermal efficiency through a higher compression ratio. In particular, the fuel is injected into the cylinder during the induction stroke and a homogeneous and stoichiometric mixture is available for ignition in a similar way to PFI engines. This approach has the advantage of simultaneously reducing hydrocarbon emissions, improving power and fuel economy while taking advantage of the 99% or so efficiency of catalytic converters. These simultaneous benefits are the result of:

- High pressure (5–20–MPa) fuel injection and careful timing of injection which prevent impingement of the fuel on the piston and cylinder walls leading to low HC emissions.
- Charge cooling by the evaporating spray ($\sim$15 K) which allows higher compression ratios ($\sim$12:1) to be used, leading to increased power (up to $\sim$15%) and a very modest fuel economy advantage (3–5%) relative to PFI engines [33].

Further benefits can be achieved with cylinder deactivation or stratified-charge-start options [34]. Although the overall fuel economy improvement is not significant when it is applied to naturally aspirated engines, as shown in Table 3.2, the homogeneous and stoichiometric concept offers higher potential when it is applied

Table 3.2 Feasibility of gasoline direct-injection applications

Features	Sources	Stoichiometric		Stoichiometric and lean	
		NA	TC	NA	TC
Fuel economy	Lean burn	None	None	++	+++
	High EGR tolerance	+	+	+	+
	High copm. ratio	+	++	+	++
Power	Charge air cooling	+	++	+	++
	2-Stage mixing	+	+++	+	+++
Exhaust gas	Catalyst warm-up by 2-stage combustion	+	++	+	++
	Catalyst warm-up by stratified slight-lean	+	++	+	++
	Necessity of NOx trap	None	None	–	—
Response	Torque control by fuel quantity	None	None	++	++
	Quick start-up	+ for HEV or idle stop application			

to turbocharged and/or supercharged engines. It has been argued for quite some time that engine downsizing with turbocharging is one of the candidates for fuel economy improvements, but the idea has not spread widely. To increase the knock limited torque, it is required to lower the compression ratio resulting in lower thermal efficiency. The improved anti-knock characteristics of the direct injection gasoline engine minimize the requirement for compression ratio reduction. Therefore, the combination of turbocharging and direct injection gasoline engine technology makes a lot of scientific sense since it takes advantage of all the positive aspects of turbocharging but with no negatives as direct injection counterbalances these. When combined with the downsizing concept, it offers the possibility that a certain percentage of the thermal efficiency loss caused by 'giving up' lean burn may be recovered. In the case of a direct injection gasoline, engine turbo-lag which represents one of the generic weaknesses of turbocharging, can be reduced. Because the engine is operated in the lean mode before the acceleration, the mass of exhaust gas flowing into the turbine is larger than that of a port fuel injection turbocharged engine. The rotational speed of the turbine and compressor before acceleration is maintained at two or three times higher level than that of a port fuel injection engine, thus reducing turbo-lag. Inferior low-end torque has been considered to be another disadvantage of turbocharging; however, this problem has been improved by advanced transmission. Through the rapid shift of the gear position or the pulley ratio to switch the engine speed to high, the requirement for low-end torque has been reduced. What remains an issue is the customers' degree of acceptability of the turbocharging concept. However, through turbocharging the driver and passengers can experience the superb acceleration feeling during overtaking from high speed cruising since, at the same gear position, the vehicle speed increases linearly with increasing boost pressure. Although this is a big advantage for European drivers, it is less so for Japanese drivers who cannot experience a similar feeling since the speed limit in Japanese highways is only 100 km/h.

3.6.2 Stratified Spray-Guided DI Gasoline Engines

As already mentioned earlier, more than 10 years after the introduction into the mass market of the GDI Mitsubishi engine the expectations are that the wall-guided concept (1st generation DI engine) will soon be replaced by the more promising spray-guided concept where the injector and the spark plug are closely spaced in the central part of the cylinder head (Fig. 3.9). The main reason for the introduction of the spray-guided concept, despite its obvious technical difficulties, is that it is capable of substantially expanding the speed-load range of the stratified-charge operation, thus taking better advantage of its superior fuel economy relative to the first generation DI engines. The expansion of the stratified-charge operating range is due to the different mechanism for achieving fuel stratification which does not depend on either the shape of the piston cavity or the in-cylinder flow. The advantages of the spray-guided concept in terms of fuel-economy, HC and smoke emissions are summarized in Fig. 3.2 relative to the two alternative configurations of the wall-guided and air-guided systems. The close proximity of the injector to the spark-plug results in short time separation between injection and ignition which necessitates very accurate positioning of a vaporized fuel cloud of the right mixture stoichiometry at the spark plug gap at the time of ignition. Although conceptually this seems rather straightforward, in practice it is very difficult to have high

Fig. 3.9 Cylinder head configuration for spray-guided concept [35]

repeatability in the spray structure, mixture formation and ignition for every injection that takes place in the engine cylinder. It should be stressed that the local conditions are quite unfavourable for robust ignition due to:

- high gas-and liquid-phase velocities in the vicinity of the spark gap
- the presence occasionally of large droplets
- fluctuations in the spray cone angle
- variations in the spark duration and its stretching during ignition
- variations in the local pre-injection velocity field and air/fuel ratio.

It can thus be argued that the characteristics of the fuel injection system play the dominant role in the repeatability of the combustion process which will determine the chances of success of the spray-guided stratified DI gasoline engine concept as a mass-production technology for automotive applications.

Two types of fuel injectors have been developed over the last 2 years specifically for DI gasoline engines: the solenoid-actuated multi-hole injectors and the piezoelectric outward-opening pintle injectors. Both offer significant advantages relative to the swirl injectors, used extensively in the first-generation DI gasoline engines based on the wall-guided concept (Table 3.3), and deserve further attention and discussion.

3.6.2.1 Piezoelectric Outward-Opening Pintle Injectors

Contrary to the multi-hole injectors which show close similarities with their diesel counterparts, the outward-opening pintle injector represents a relatively new design purpose-built for the second-generation DI gasoline engines. Figure 3.10 gives detailed information about their standard design as well as some of its numerous variants which have been tested as means of understanding the all-important link between needle internal design and spray characteristics. The fuel from the rail and high-pressure connecting pipe is entering into the nozzle gallery and is directed towards the nozzle exit through three or four flow passages located at the space between the lower needle guide and the nozzle body. Below the lower guide there is a dead volume where the four high speed jets entering from the flow passages are mixing before the fuel exits the injector in the form of a hollow-cone spray; the direction of the spray is mainly determined by the seat angle of the pintle-type needle. More details about the internal flow and its link with the spray characteristics are provided in [37, 38, 39, 40] representing experiments and CFD calculations in both real-size and large scale models of outward-opening pintle injectors. High speed spray images taken just at the nozzle exit at atmospheric back pressure (Fig. 3.11) have revealed the formation of strings/ligaments originating at the annular area of the opening pintle whose location is not fixed during the injection period and their spacing varies as a function of the fuel flow rate. Although the origin of these ligaments/strings has been attributed to the formation of a two-phase flow inside the nozzle, associated with either cavitation or air entrainment, there are still uncertainties in the cause-and-effect relationship between internal flow and spray

Table 3.3 Comparison of the three dominant high-pressure injector nozzles [36]

Nozzle configurations	Swirl	Outward opening	Multi-hole
Spray stability/tolerance	+	++	+
Flexibility of spray pattern	+	0	++
Resistance against backpressure influence	–	++	++
Multi-injection capability	0	+	0
Costs	0	–	+
Robustness against fouling	+	+	+

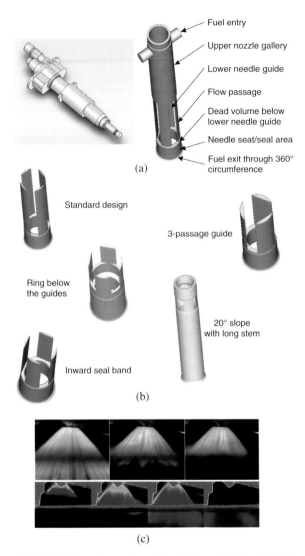

Fuel entry

Upper nozzle gallery

Lower needle guide

Flow passage

Dead volume below
lower needle guide

Needle seat/seal area

Fuel exit through 360°
circumference

(a)

Standard design

3-passage guide

Ring below
the guides

20° slope
with long stem

Inward seal band

(b)

(c)

Fig. 3.10 Outward-opening pintle injector. (**a**) Standard design details. (**b**) Nozzle alternative designs (**c**) in-cylinder Mie and LIF images

structure. Independent of the mechanism of formation of the hollow-cone spray and its emergence as clearly separated liquid ligaments rather than as a continuous film, the penetration of this structure into the engine cylinder is a function of the prevailing thermodynamic conditions (pressure/density/temperature). As Fig. 3.12 shows, the 'streaky' structure of the spray under nearly atmospheric conditions gradually diminishes with increasing back pressure, giving rise to a more compact fuel cloud with clearly identifiable leading edge vortices: Furthermore, the position of

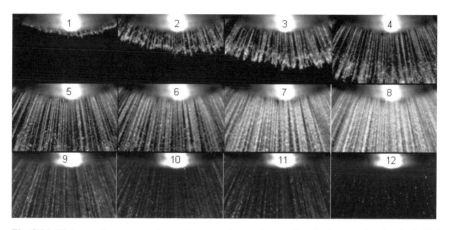

Fig. 3.11 High-speed sequence of spray images taken at the nozzle exit vicinity showing the initial development of the strings (1–4), during the main injection period (5–8) and at the end of injection (9–12) [57]

the leading edge vortex varies with cylinder pressure and, fortunately, higher back pressure forces this vortex to remain close to the spark plug [41]; injection timing thus offers some mechanism for controlling the position of this critical for ignition vortex.

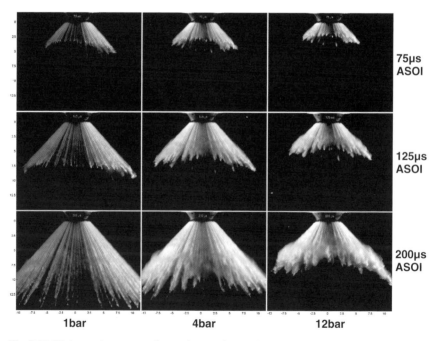

Fig. 3.12 High-speed sequence of spray images from a single injection event, showing the spatial distribution of fuel at different chamber pressures [40]

The creation of vortices at the spray edge and their stability in space relative to the spark plug represent jointly the key to the success of the spray-guided concept. In addition, the leading vortex grows in size as more fuel is injected [42], thus offering some flexibility in the positioning of the spark plug for both stratified and stoichiometric operating conditions. Due to the hollow spray configuration, air is entrained from underneath, which creates an upwards air motion and sets up a second, clockwise rotating vortex that balances the main anti-clockwise leading edge vortex formed earlier (Fig. 3.13a). The two vortices together are converting the kinetic energy of the spray into rotational motion, thus effectively reducing the spray penetration [42] and containing it near the spark plug. Interestingly enough, if the same amount of fuel is divided into two separate injections separated by a small dwell time, a third vortex is formed at the outer edge of the spray, which assists in bringing the fuel cloud closer to the spark plug (Fig. 3.13b). Confirmation of the degree of atomization and the 'ignitability' of the spray has been provided by PDA measurements of the droplets' size and velocity near the spark plug (Fig. 3.14) which confirmed that the droplet velocities are lower and their size smaller in the recirculation zone near the spark plug than in the leading edge zone. However, in the case of split injection the opposite trend has been observed [41].

The relative positioning of the spray's 'ignitable' recirculation zone with respect to the spark gap is illustrated in Fig. 3.15 for two extreme cases representing unsuccessful ignition and associated with large injection-to-injection, variations of the spray cone angle. In one case (Fig. 3.15a) the spray impinges on the electrodes as a result of air trapped at the needle head which gives rise to a spray cone angle larger than the nominal and, in the other (Fig. 3.15b) the cone angle is smaller than the nominal, representing the well-known phenomenon of hydraulic flip where air is trapped this time on the cartridge side.

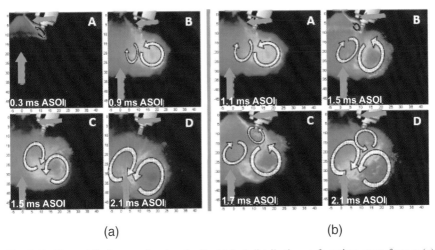

(a)   (b)

Fig. 3.13 Planar Mie images showing the liquid fuel distribution at four time steps from a (**a**) single and (**b**) split injection event [42]

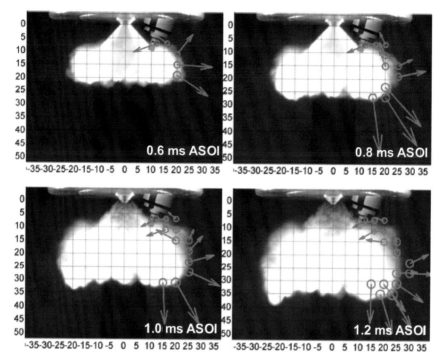

Fig. 3.14 High-speed images taken at four time steps during a single injection event with superimposed averaged PDA data (D_{10} droplet diameters and velocities); fuel temperature is at 90°C [42]

The combustion stability and robustness of an outward-opening pintle injector were examined in an optical engine under stratified charge conditions [42]. A stable operating window for misfire-free combustion was identified, extending over 4–6 CAD, with partial burns and misfires increasing as the rich and the lean burn zones were approached (Fig. 3.16a); this operating window of ±2.5 CAD is defined in terms of phase shifting the ignition from the end of injection at the given operating conditions (Fig. 3.16b).

Finally, the effect of flash-boiling on the spray generated by an outward-opening pintle injector has been investigated in a spray chamber [43]; flash-boiling occurs when the fuel temperature is high and the cylinder pressure is low as in the case of a warmed-up engine starting with nearly closed throttle. The results shown in Fig. 3.17 have revealed that at low fuel temperatures and back-pressures (non-flashing) the recirculation zone at the spray edge almost diminishes but increasing fuel temperature initiates flash boiling (superheated state) which increases the spray frontal area and restores the leading edge vortex. Nevertheless, the spray formed by the outward-opening pintle injector maintained its characteristic spray throughout these fuel temperature variations, with a small reduction in spray penetration occurring at flash boiling.

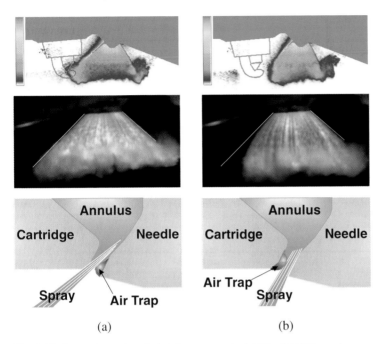

Fig. 3.15 Outward-opening pintle injector spray instability. (**a**) Wide angle spray and (**b**) narrow angle spray [57]

In concluding, it has to be stressed that, in addition to the previously identified advantages of the sprays formed by pintle injectors, the piezoelectric actuation of the needle available in all production outward-opening pintle injectors adds significantly to its 'market value'. Despite their much higher cost than multi-hole injectors, the emerging piezoelectric outward-opening pintle injectors allow very rapid opening and closing times and, therefore, short injection durations, multiple injections per engine cycle, and control of needle lift through voltage adjustment (thus the ability to change the fuel mass flow independent of injection duration). Furthermore, there is consensus that these injectors exhibit reduced tendency for fouling and spark-plug wetting. It is, therefore, not surprising that Mercedes-Benz and BMW have introduced into the market the first spray-guided stratified DI gasoline engines for medium-size passenger cars equipped with piezoelectric outward-opening pintle injectors [35, 44, 45].

3.6.2.2 Solenoid-Driven Multi-Hole Injectors

This type of injector is geometrically very familiar to the automotive industry due to its popularity or, better, dominance of the direct-injection diesel engine market. It has many advantages, as already indicated in Table 3.3, which have encouraged its early use in research and development of various spray-guided configurations and

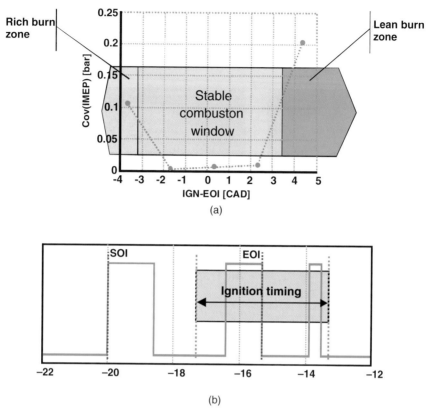

(a)

(b)

Fig. 3.16 (**a**) Stable combustion window for stratified charge operation. (**b**) Example of injection timing and the stable combustion window (shaded area)

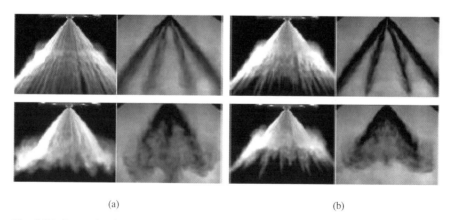

(a)   (b)

Fig. 3.17 Spray chamber measurements of multi-hole (*black background*) and pintle outward opening nozzle (*white background*) sprays in non-flashing (*top row*) and flashing (*bottom row*) conditions at chamber pressures of (**a**) 0.3 bar and (**b**) 0.8 bar [42]

concepts, but also some disadvantages such as a tendency for injector fouling and spark-plug wetting.

The main advantage of multi-hole injectors has been the enhanced flexibility in the geometry and spatial orientation of the nozzle holes (see Fig. 3.18) which

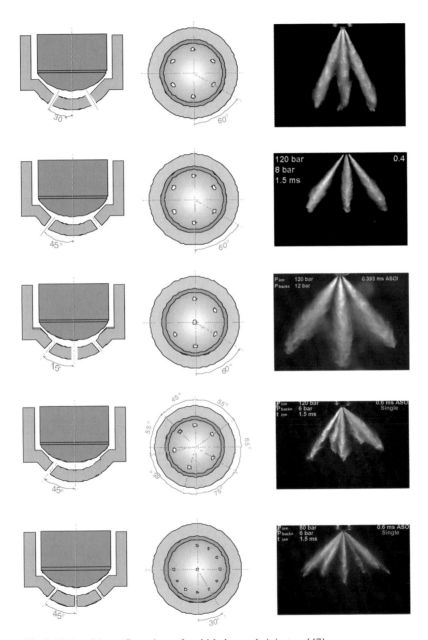

Fig. 3.18 Possible configurations of multi-hole nozzle injectors [47]

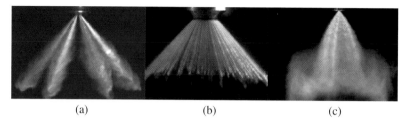

(a)   (b)   (c)

Fig. 3.19 Comparison of sprays generated by (**a**) multi-hole, (**b**) outward opening and (**c**) swirl pressure atomizers

allows a wide variety of hole numbers (6–12), hole sizes, hole pattern (symmetric and asymmetric) and length-to-diameter ratios (L/D). The generated sprays from each of these holes have very different structure than those generated by pintle and swirl-type injectors, as shown in Fig. 3.19; their fishbone structure, which has been for many years the subject of intense investigations by researchers in industry and academia involved with diesel engines, is closely linked not only with the internal nozzle flow but also with the air-entrainment pattern (Fig. 3.20a). Sprays from multi-hole nozzles are characterized in global terms by the overall spray cone angle, the individual jet cone angle and its tip penetration (Fig. 3.20b and also Fig. 3.3). Concerning the local spray characteristics, these are quantified by phase Doppler anemometry (PDA) in terms of droplet velocities and sizes (diameter) visually represented as in Fig. 3.21 by spheres of various size (μm) and vectors of various length (m/s). Under atmospheric conditions and typical injection pressures, sprays from multi-hole injectors [46, 47, 48] can exhibit typical droplet sizes in the range 10–20 μm and droplet velocities well exceeding 100 m/s which leads to steep velocity gradients near the spark plug. According to [49], an increase of L/D, through reduction of the hole diameter, gives rise to smaller droplet sizes while the spray angle increases with decreasing L/D.

As mentioned earlier, multi-hole injectors are considered to be more susceptible to fouling (Fig. 3.22) due to their proximity to the spark plug which facilitates

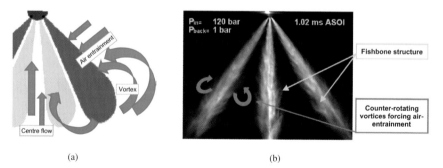

(a)   (b)

Fig. 3.20 Example of air-entrainment in multi-hole nozzle sprays. (**a**) Schematic representation [42], (**b**) Mie spray image [58]

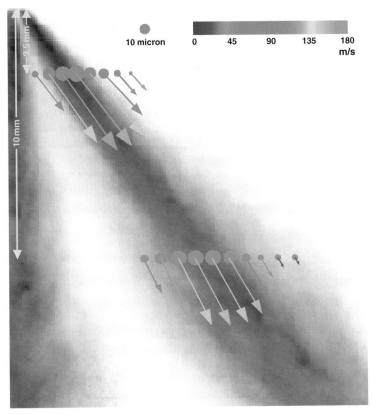

Fig. 3.21 Droplet sizes and velocities from a 6-hole nozzle superimposed on a shadowgraphy spray image [58]

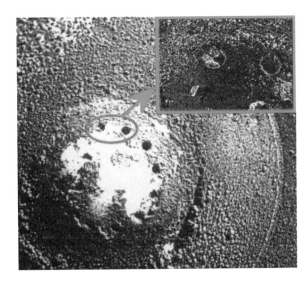

Fig. 3.22 Multi-hole injector nozzle fouling due to carbon deposits

contact of the propagating flame with the external nozzle surface in a high temperature gas environment. At the same time, the need for spark ignition of the mixture in the recirculation zone at the spray edges, in a region of steep velocity gradients, can easily lead to spark electrode wetting. As Fig. 3.23 shows, one of the jets/sprays from the multi-hole injector strikes the ground electrode producing a stream of large droplets; this could be the result of a small variation in the individual spray cone angle leading to poor spray targeting of the spark gap area. As discussed in [50], concerning the same figure, although the orientation of the ground electrode facing the oncoming spray shields the spark plug from the spray plume and reduces the local fuel velocities at the time of spark, the formed wake behind the electrode causes strong fluctuations and degrades ignition stability. In the worst case, misfire takes place which is totally unacceptable in production engines: conditions which are unfavourable to robust ignition and early flame-kernel growth include [33]:

- high gas-and-liquid-phase velocities
- large number of droplets
- steep temporal gradients in air/fuel ratio
- poor design and orientation of spark plug electrodes.

Another factor which has recently attracted attention is nozzle cavitation as a mechanism that contributes to spray instability and can cause significant cone angle fluctuations. Although most of the understanding about the onset and development of cavitation comes from diesel injectors [51, 52, 53], confirmation about the various forms of in-nozzle cavitation has been provided by experiments, supported in some cases by CFD calculations, in real-size and enlarged multi-hole injectors [38, 54]; further evidence was provided by experiments in a real-size, single-hole injector [55]. In particular, there are at least three types of cavitation in the nozzle of multi-hole gasoline injectors:

- geometric cavitation in the holes
- string or vortex-type cavitation in the sac volume
- needle cavitation originating in the vicinity of the needle and extending to the opposite hole.

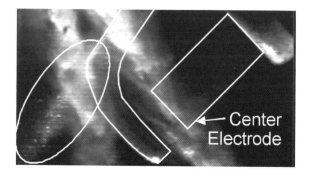

Fig. 3.23 Example of spray/spark plug interaction in a firing direct-injection spark-ignition engine [50]

It is generally accepted that all these cavitation types are affected and controlled by needle lift, radius of curvature of hole inlet, sac volume and hole orientation and geometry. In addition, string cavitation is the most likely cause of spray cone fluctuations but the detailed link with the spray structure is still relatively unknown in multi-hole injectors.

Another thermodynamic effect which can change significantly the fuel distribution at the nozzle exit and the resulting spray structure is flash boiling, a phenomenon occurring when the fuel temperature is high and already discussed in connection with the sprays generated by pintle injectors. As Fig. 3.24 clearly shows, at high fuel temperatures and low ambient pressures, the individual sprays are rapidly atomized to such a degree that it becomes impossible to distinguish the spray plumes; although flash boiling is highly desirable from the atomization point of view, it can only take place within a very narrow window of engine operating conditions and it is not considered important as an atomization mechanism. However, in relative terms, it can be argued that the effect of flash boiling is much more pronounced in the sprays generated by multi-hole rather than pintle injectors.

Overall, it can be argued that the solenoid-driven multi-hole injector is a simpler and less expensive system than piezoelectric outward-opening pintle injectors offering significant advantages in terms of the flexibility for adjusting the spray configuration to the engine geometry, the narrow cone angle of individual sprays, and the control of tip penetration and atomization through injection pressure and timing of injection (back pressure). Although experiments in single and multi-cylinder engines employing multi-hole injectors have been promising, it is widely accepted that further refinement of their design is needed to ensure stable ignition and combustion with acceptable gaseous and particulate emissions. Faster opening

Fig. 3.24 The effect of pressure/temperature on the flash boiling of gasoline sprays from multi-hole nozzle injectors [56]

and closing times combined with much higher than today's injection pressures may make multi-hole injectors directly comparable to the piezoelectric outward-opening pintle-type injectors that, at present, seem to be the preferred fuel injection system.

3.7 Conclusions

Following nearly 10 years of refinement of the first-generation direct-injection gasoline engines, based on the wall-guided concept, the second-generation of stratified DI engines are now entering the mass market equipped with the highly sophisticated piezoelectric outward-opening pintle injectors. In the meantime, stoichiometric DI engines have reached production, taking advantage of the outstanding efficiency of three-way catalysts. All three initiatives highlight in a very consistent way that the concept of direct in-cylinder fuel injection is here to stay due to its unquestionable advantages relative to port fuel injection. Further developments in the technology of NOx catalysts and high pressure multi-hole injection systems are expected to assist the spray-guided combustion mode in establishing itself as the preferred gasoline engine technology concept of the future. Nevertheless, it can be argued that unless the DI gasoline engine is either turbocharged and/or supercharged, it won't achieve its fuel potential and may never reach 'mass production' status at a time of intense competition from advanced diesel and hybrid gasoline/diesel engine systems. It thus remains to be seen whether the many years of research into two-and-four stroke direct-injection gasoline engines will have the desired outcome for the global automotive industry in producing a second-generation supercharged/turbocharged DI gasoline engine which will compete directly with the most advanced diesels in terms of fuel efficiency, performance and exhaust emissions and be superior to the most advanced port-injection engines.

Acknowledgments The authors would like to acknowledge the valuable support of Dr N. Mitroglou to the editing of this chapter and the sponsors of the various research programmes.

References

1. Bishop, I.N., Simko, A., A New Concept of Stratified Charge Combustion – The Ford Combustion Process (FCP), SAE Transactions, Paper 680041, Vol.77, pp. 93–117, 1968.
2. Mitchell, E., Cobb, J.M., Frost, R.A., Design and Evaluation of a Stratified Charge Multi-fuel Military Engine, SAE Transactions, Paper 680042, Vol.77, pp.118–128, 1968.
3. Alperstein, M., Schafer, G.H., Villforth, F.J., Texaco's Stratified Charge Engine – Multi-fuel, Efficient, Clean, and Practical, SAE Paper 740563, 1974.
4. Urlaub, A.G., Chmela, F.G., High-Speed, Multi-fuel Engine: L9204 FMV, SAE Paper 740122, 1974.
5. Scussel, A.J., Simko, A.O., Wade, W.R., The Ford PROCO Engine Update, SAE Transactions, Paper 780699, Vol. 87, pp. 2706–2725, 1978.
6. Wood, C.D., Unthrottled Open-Chamber Stratified Charge Engines, SAE Paper 780341, 1978.

7. Schapertons, H., Emmenthal, K.-D., Grabe, H.-J., Opperman, W., VW's Gasoline Direct Injection (GDI) Research Engine, SAE Paper 910054, 1991.
8. Iwamoto, Y., Noma, K., Nakayama, O., Yamauchi, T., Ando, H., Development of Gasoline Direct Injection Engine, SAE Transaction, Paper 970541, 1997.
9. Harada, J., Tomita, T., Mizuno, H., Mashiki, Z., Ito, Y., Development of Direct Injection Gasoline Engine, SAE Paper 970540, 1997.
10. Tagaki, Y., Itoh, T., Muranaka, S., Iiyama, A., Iwakiri, Y., Urushihara, T., Naitoh, K., Simultaneous Attainment of Low Fuel Consumption, High Output Power and Low Exhaust Emissions in Direct Injection SI Engines, SAE Paper 980149, 1998.
11. Kanda, M., Baika, T., Kato, S., Iwamuro, M., Koike, M., Sito, A., Application of a New Combustion Concept to direct injection gasoline engine, SAE Paper 2000-01-0531, 2000.
12. Hentschel, W., Block, B., Hovestadt, T., Meyer, H., Ohmstede, G., Richter, V., Stiebels, B., Winkler, A., Optical Diagnostics and CFD-Simulations to Support the Combustion Process Development of the Volkswagen FSI Direct-Injection Gasoline Engine, SAE Paper 2001-01-3648, 2001.
13. Cathcart, G., Railton, D., Improving Robustness of Spray Guided DI Combustion Systems: The Air-Assisted Approach, JSAE Spring Convention 20015360, 2001.
14. Befrui, B., Kneer, R., Breuer, S., Reckers, W., Robart, D., Wanlin, H., Weiten, C., Investigation of a DISI Fuel Injector for a Close-Arranged Spray-Guided Combustion System, SAE Paper 2002-01-1133, 2002.
15. Ortmann, R., Arndt, S., Raimann, J., Wuerfel, G., Grzeszik, R., Methods and Analysis of Fuel Injection, Mixture Preparation and Charge Stratification in Different Direct-Injection SI Engines, SAE Paper 2001-01-0970, 2001.
16. Achleitner, E., Berger, S., Frenzel, H., Klepatsch, M., Warnecke, V., Gasoline Direct Injection System with Piezo Injectors for Spray-Guided Combustion Processes, MTZ worldwide 5/2004, Vol. 65, p. 2, 2004.
17. Green Car Congress Mercedes-Benz Premiers New Gasoline Direct Injection System for More Power and Lower Fuel Consumption, http://www.greencarcongress.com/2006/02/mercedesbenz_pr.html.
18. Ando. H., Takemura, J., Koujina, E., A Knock Anticipating Strategy Basing on the Real-Time Combustion Mode Analysis, SAE Paper.890882, 1989.
19. Kuwahara, K., Ueda, K., Ando, H., Mixing Control Strategy for Engine Performance Improvement in a Gasoline Direct Injection Engine, SAE Paper 980158, 1998.
20. Ueda, K., Kaiahara, K., Krose, K., Ando, H., Idling Stop System Coupled With Quick Start Features of Gasoline Direct Injection, SAE Paper 2001-01-0545, 2001.
21. Hori, M., Taniguch, S., Noda, N., Abe, F., Iwachido, K., Tanada, H., Watanabe, T., Yamada, N., Ando, H., Development of the NOx Adsorber Catalyst for Use with High-Temperature Condition, SAE Paper 2001-01-1298, 2001.
22. Tamura, Y., Kikuchi, S., Okada, K., Koga, K., Dogahara, T., Nakayama, O., Ando, H., Development of Advanced Emission-Control Technologies for Gasoline Direct-Injection Engines, SAE Paper 2001-01-0254, 2001.
23. Kojima, S., Jimbo, T., Katoh, K., Miyashita, S., Watanabe, M., Analysis and Simplification of Thermal Endurance Tests of NOx Storage-Reduction Catalysts, SAE Paper 2004-01-1496, 2004.
24. Iizuka, H., Kaneeda, M., Higashiyama, K., Kuroda, O., Shinotsuka, N., Watanabe, H., Improvement of Heat Resistance for Lean NOx Catalyst, SAE Paper 2004-01-1495, 2004.
25. Iwamoto, S., Kouno, Y., Inoue, M., Direct Decomposition of NO by Ba-Loaded Catalyst", 90th CATSJ A-Page 239, 2002.
26. Hamada, H., Evolution of NOx removal catalyst 90th CATSJ A-Page 244, 2002.
27. Ishihara, T., Ando, M., Sada, K., Takiishi, K., Nishiguchi, N., Takita, Y., Direct Decomposition of NO into N2 and O2 over La(Ba)Mn(In)O3 Perovskite Oxide, J. Catal., 220, 104–114, 2003.

28. Noma, K., Iwamoto, Y., Murakami, N., Iida, N., Nakayama, O., Ando, H., Optimized Gasoline Direct-Injection Engine for the European Market, SAE Paper 980150, 1998.

29. Yamamoto, S., Tanaka, D., Takemura, J., Ando, H., Mixing Control and Combustion in Gasoline Direct Injection Engines for Reducing Cold-Start Emissions, SAE Paper 2001-01-0550, 2001.

30. Sadakane, S., Sugiyama, M., Kishi, H., Abe, S. Harada, J., Sonoda, Y., Development of a New V-6 High Performance Stoichiometric Direct Injection Gasoline Engine, SAE Paper 2005-01-1152, 2005.

31. Nishida, M., Isobe, R., Goto, T., Hanzawa, H., Aiga, S., The New 2.3L Direct Injection Turbo Gasoline Engine from Mazda, 14th Aachen Automotive Colloquium, Aachen, pp. 939–960, 2005.

32. Middendorf, H., Krebs, R., Szengel, R., Pott, E., Fleis, M., Hagelstein, D., Volkswagen Introduces the Worlds First Double Charge Air Direct Injection Petrol Engine, 14th Aachen Automotive Colloquium, Aachen, pp. 961–986, 2005.

33. Drake, M.C., Haworth, D.C., Advanced Gasoline Engine Development using Optical Diagnostics and Numerical Modeling, Invited Plenary Lecture, 31st International Symposium on Combustion, Combustion Institution, Heidelberg, Germany, August 6–11, 2006.

34. Ladenfield, T., Kufferath, A., Gerhardt, J., Gasoline Direct Injection – SULEV Emission Concept, SAE Paper 2004-01-0041, 2004.

35. Altenschmidt, F., Bertsch, D., Bezner, M., Laudenbach, N., Zahn, M., Schaupp, U., Kaden, A., The Analysis of the Ignition Process on SI-Engines with Direct Injection in Stratified Mode, International Symposium on Internal Combustion Diagnostics, Baden-Baden, Germany, pp. 395–411, 2006.

36. Stach, T., Schlerfer, J., Vorbach, M., New Generation Multi-Hole Full Injector for Direct-Injection SI Engines – Optimization of Spray Characteristics by Means of Adapted Injector Layout and Multiple Injection", SAE Paper 2007-01-1404, 2007.

37. Marchi, A., Nouri, J.M., Yan, Y., Arcoumanis, C., Internal Flow and Spray Characteristics of Pintle-type Outwards Opening Piezo Injectors for Gasoline Direct-Injection Engines, SAE Paper 2007-01-1406, 2007.

38. Papoulias, D., Giannadakis, E., Mitroglou, N., Gavaises, M., Cavitation in Fuel Injection Systems for Spray-Guided Direct Injection Gasoline Engines, SAE Paper 2007-01-1418, 2007.

39. Nouri, J.M., Hamid, M.A., Abo-Serie, E., Marchi, A., Mitroglou, N., Arcoumanis, C., Internal and Near Nozzle Flow Characteristics in an Enlarged Model of an Outwards Opening Pintle-Type Gasoline Injector, Proceedings of Third International Conference on Optical and Laser Diagnostics (ICOLAD 2007), City University, London, 22–25 May 2007, 2007.

40. Nouri, J.M., Hamid, M.A., Yan, Y., Arcoumanis, C., Spray Characterization of a Piezo Pintle-Type Injector for Gasoline Direct Injection Engines, Proceedings of Third International Conference on Optical and Laser Diagnostics (ICOLAD 2007), City University, London, 22–25 May 2007, 2007.

41. Skogsberg, M., Dahlander, P., Denbratt, I., Spray Shape and Atomization Quality of an Outward-Opening Piezo Gasoline DI Injector, SAE Paper 2007-01-1409, 2007.

42. Skogsberg, M., A Study on Spray-Guided Stratified Charge Systems for Gasoline DI Engines, PhD Thesis, Department of Applied Mechanics, Chalmers University of Technology, Gothenburg, Sweden, 2007.

43. Dahlander, P., Lindgren, R., Denbratt, I., High-Speed Photography and Phase Doppler Anemometry Measurements of Flash-Boiling Multihole Injector Sprays for Spray-Guided Gasoline Direct Injection, Paper ICLASS06-0112, ICLASS, Kyoto, Japan, 2006.

44. Schwarz, C., Schünemann, E., Durst, D., Fischer, J., Witt, A., Potentials of the Spray-Guided BMW DI Combustion System, SAE Paper 2006-01-1265, 2006.

45. Fischer, J., Kern, W., Unterweger, G., Witt, A., Durst, B., Schünemann, E., Schwarz, C., Methods for the Development of the Spray-Guided BMW DI Combustion System, International Symposium on Internal Combustion Diagnostics, Baden-Baden, pp. 413–423, 2006.

46. Yan, Y., Gashi, S., Nouri, J.M., Lockett, R.D., Arcoumanis, C., Investigation of Spray Char-
    acteristics in a Spray-Guided DISI Engine Using PLIF and LDV, Proceedings of Third
    International Conference on Optical and Laser Diagnostics (ICOLAD 2007), City University,
    London, 22–25 May 2007, 2007.
47. Mitroglou, N., Nouri, J.M., Yan, Y., Gavaises, M., Arcoumanis, C., Spray Structure Generated
    by Multi-Hole Injectors for Gasoline Direct-Injection Engines, SAE Technical Paper Series,
    2007-01-1417, 2007.
48. Mitroglou, N., Nouri, J.M., Gavaises, M., Arcoumanis, C., Spray Characteristics of a Multi-
    Hole Injector for Direct-Injection Gasoline Engines, Int. J. Eng. Res., 7, No.3, 255–270, 2006.
49. Skogsberg, M., et al., Effects of Injector Parameters on Mixture Formation for Multi-Hole
    Nozzles in a Spray-Guided Gasoline DI Engine, SAE Technical Papers, 2005-01-0097, 2005.
50. Fansler, T.D., Drake, M.C., Düwel, I., Zimmerman, F.P., Fuel-Spray and Spark-Plug Inter-
    actions in a Spray-Guided Direct-Injection Gasoline Engine, International Symposium on
    Internal Combustion Diagnostics, Baden-Baden, pp. 81–97, 2006.
51. Arcoumanis, C., Gavaises, M., Nouri, J.M., Abdul-Wahab, E. Horrocks, R., Analysis of the
    Flow in the Nozzle of a Vertical Multi-Hole Diesel Engine Injector, SAE Paper 980811, 1998.
52. Arcoumanis, C., Badani, M., Flora, H. Gavaises, M., Cavitation in Real-Size Multi-Hole
    Diesel Injector Nozzles, SAE Paper 2000-01-1249, 2000.
53. Roth, H., Gavaises, M., Arcoumanis, C., Cavitation Initiative, its Development and Link with
    Flow Turbulence in Diesel Injector Nozzles, SAE Paper 2002-01-0214, 2002.
54. Nouri, J.M., Mitroglou, N., Yan, Y., Arcoumanis, C., Internal Flow and Cavitation in a Multi-
    Hole Injector for Gasoline Direct-Injection Engines, SAE Technical Paper Series, 2007-01-
    1405, 2007.
55. Birth, I.G., Recks, M., Spicher, U., Experimental Investigation of the In-Nozzle Flow of
    Valve Covered Orifice Nozzle for Gasoline Direct Injection Engines, International Sympo-
    sium Internal Combustion Diagnostics, Baden-Baden, pp. 59–78, 2006.
56. Dahlander, P., Annual Report, Combustion Engine Research Centre, Chalmers University of
    Technology, Gothenburg, 2006.
57. Marchi, A., Internal Flow and Spray Characteristics of the Pintle Type Piezo Injector, PhD
    Thesis in preparation, School of Engineering and Mathematical Sciences, City University,
    London, 2008.
58. Mitroglou, N., Multihole Injectors for Direct-Injection Gasoline Engines, PhD Thesis, School
    of Engineering and Mathematical Sciences, City University, London, 2006.

Chapter 4
Turbulent Flow Structure in Direct-Injection, Swirl-Supported Diesel Engines

Paul C. Miles

4.1 Introduction

The in-cylinder turbulent flow structure—encompassing both mean flow and turbulence—plays a pivotal role in the mixture preparation, combustion, and pollutant formation and destruction processes in direct-injection diesel engines. Prior to TDC, mean flow structures and turbulence influence both the bulk fuel and residual gas stratification as well as the small-scale, molecular level mixing in pre-mixed, 'homogeneous' charge combustion systems. Pre-TDC structures also exert a prominent influence on the transport and turbulent diffusion of heat and species generated from pilot fuel injection events in more traditional diesel combustion systems. Later in the cycle the bulk flow structures are equally important. During the 'diffusion' phase of diesel combustion, the oxidation of unburned fuel and particulates is limited by the rate at which fuel or particulate matter can be mixed with oxidants. In this phase, bulk flow structures are responsible not only for ensuring that fuel and fresh oxidants are transported to the same location within the cylinder, but also for generating the intense turbulence required to bring about small-scale mixing before falling temperatures due to cylinder volume expansion slow chemical reactions excessively. As low-emission combustion systems that employ high intake charge dilution levels or retarded injection timings become more prevalent, the need to enhance these mixing processes becomes more urgent. A clear understanding of and an ability to predict the effects of engine operating parameters and geometry on bulk flow structures and turbulence generation is therefore necessary for the development of diesel engines with optimal fuel economy and pollutant emissions.

A comprehensive overview of the measurement, analysis, and modeling of in-cylinder flows, and a description of the flow structures generated during the induction and compression processes, has been provided by Gosman [46]. In the two decades since publication of that work, there have been numerous additional studies pertinent to the understanding and modeling of diesel engine flows. Accordingly, one objective of this work is to provide a review of the recent literature,

Paul C. Miles
Sandia National Laboratories, Livermore, CA, USA

C. Arcoumanis, T. Kamimoto (eds.), *Flow and Combustion in Reciprocating Engines*,
DOI: 10.1007/978-3-540-68901-0_4, © Springer-Verlag Berlin Heidelberg 2009

with an emphasis on highlighting the new physical understanding that has been gained regarding the development of these flows. This review is supplemented with a summary of recent measurements made in the author's laboratory, closely coupled with numerical simulations performed by researchers at the University of Wisconsin Engine Research Center, which aims to explicitly identify the dominant sources of flow turbulence and the mean flow structures conducive to turbulence generation in swirl-supported diesel engines. Subsequently, the effects of fuel injection and combustion on the in-cylinder flow structures are described. At the time of Gosman's writing, little work had been performed in this area—and few additional studies have since been published. As will be seen below, the fuel injection event profoundly alters the in-cylinder flow structures, and its interaction with the flow swirl can generate significant levels of flow turbulence that persist well into the expansion stroke.

Additionally, although numerical modeling of the in-cylinder flow structures can often reproduce the mean flow with impressive accuracy, the prediction of the flow turbulence, both before and after injection, requires further refinement. To this end, efforts to understand and to identify the source of the discrepancies between measurements and model predictions, based on detailed characterization of the mean flow velocity gradients, the turbulent stresses, and the turbulent length and time scales, are reviewed. This detailed assessment of the turbulence modeling is essential to guide and improve the modeling of engine turbulence, particularly in direct-injection diesel engines that incorporate a high level of flow swirl.

The extraction of physical understanding, both from the literature and from the recent results summarized herein, relies on a firm grasp of the behavior dictated by the governing Reynolds-averaged Navier-Stokes (RANS) equations. Similarly, examination of the turbulence modeling, and identification of the sources of discrepancies between the measurements and predictions, can be best achieved by contrast of the RANS equations with the simpler model equations. A review of the relevant theoretical concepts is thus provided to assist with this process.

Finally, there are many aspects of the fluid mechanics of the fuel injection event in which there have been significant contributions in recent years, but which are not discussed here. For example, several groups have clarified the detailed flow field surrounding a free diesel fuel jet [83, 85] or a jet impacting a combustion chamber wall [73]. Similarly, the effect of strong in-cylinder flows on the fuel jet structure (e.g., [76, 105]) has also been considered in numerous studies. These aspects are outside of the scope of the present writing, in which attention is focused primarily on the influence of the fuel injection event on formation of large-scale bulk flow structures, and on the turbulence that these structures generate.

4.2 Theoretical Background

General overviews of both measurement and modeling issues relevant to engine flows have been provided previously [45, 46, 82]. Similarly, many texts are available that describe the Reynolds-averaged equations governing turbulent flows and modeling approaches which are employed to 'close' these equations [e.g., 51, 78, 101]. To facilitate discussion of both the flow physics and the modeling concepts that follow, however, the basic equations and relevant modeling concepts are reviewed briefly in this section.

4.2.1 Mean Flow Equations

In writing the equations governing the mean flow velocities, the density and fluid properties are assumed to be spatially uniform, but to change in time due to the bulk compression and expansion processes. Away from the combustion chamber walls, this assumption is appropriate for 'motored' engine flows in which fuel injection and combustion do not occur. This approach allows the important flow physics to be illustrated without unduly complicating the equations. Additionally, a polar-cylindrical coordinate system is employed—which is a natural choice for flows in modern, axisymmetric diesel engines. Finally, ensemble (or phase) averages are used throughout this work. The choice of this averaging procedure is discussed in greater detail below. Ensemble mean quantities are denoted by angle brackets, e.g., $\langle U_z \rangle$, while fluctuations from the mean are denoted with a 'prime', e.g., u'_z.

With the above assumption of a spatially-uniform density ρ, conservation of mass is given by

$$(\nabla \cdot \langle \mathbf{U} \rangle) + \frac{1}{\rho} \frac{\partial \rho}{\partial t} = 0 \tag{4.1}$$

Expanding Eq. (4.1) in polar-cylindrical coordinates, and introducing the cylinder volume V

$$\frac{1}{r} \frac{\partial (r \langle U_r \rangle)}{\partial r} + \frac{1}{r} \frac{\partial \langle U_\theta \rangle}{\partial \theta} + \frac{\partial \langle U_z \rangle}{\partial z} = \frac{1}{V} \frac{\partial V}{\partial t} \tag{4.2}$$

Thus, at all locations in the cylinder $(\nabla \cdot \langle \mathbf{U} \rangle)$ takes on a uniform, non-zero value determined by the rate-of-change of the cylinder volume—which is negative during compression and positive during expansion. In writing Eq. (4.2), blowby and crevice flows have been neglected.

Since gradients in $(\nabla \cdot \mathbf{U})$ are zero, the momentum equations are identical to those that apply to incompressible flow. Further, because turbulent Reynolds numbers near TDC in engines are typically greater than 500, the viscous terms can be neglected in the equations governing the mean flow momentum:

$$\frac{\bar{D} \langle U_r \rangle}{\bar{D}t} - \frac{\langle U_\theta \rangle^2}{r} = -\frac{1}{\rho} \frac{\partial \langle P \rangle}{\partial r} - \left(\frac{1}{r} \frac{\partial \left(r \langle u'^2_r \rangle \right)}{\partial r} + \frac{1}{r} \frac{\partial \langle u'_r u'_\theta \rangle}{\partial \theta} + \frac{\partial \langle u'_r u'_z \rangle}{\partial z} - \frac{\langle u'^2_\theta \rangle}{r} \right) \tag{4.3}$$

$$\frac{\bar{D}\langle U_\theta\rangle}{\bar{D}t} + \frac{\langle U_r\rangle\langle U_\theta\rangle}{r} = -\frac{1}{\rho}\frac{1}{r}\frac{\partial\langle P\rangle}{\partial\theta} - \left(\frac{1}{r}\frac{\partial\left(r\langle u_r'u_\theta'\rangle\right)}{\partial r} + \frac{1}{r}\frac{\partial\langle u_\theta'^2\rangle}{\partial\theta} + \frac{\partial\langle u_\theta'u_z'\rangle}{\partial z} + \frac{\langle u_r'u_\theta'\rangle}{r}\right)$$

$$(4.4)$$

$$\frac{\bar{D}\langle rU_\theta\rangle}{\bar{D}t} = -\frac{1}{\rho}\frac{\partial\langle P\rangle}{\partial\theta} - \left(\frac{1}{r}\frac{\partial\left(r^2\langle u_r'u_\theta'\rangle\right)}{\partial r} + \frac{1}{r}\frac{\partial\left(r\langle u_\theta'^2\rangle\right)}{\partial\theta} + \frac{\partial\left(r\langle u_\theta'u_z'\rangle\right)}{\partial z}\right) \quad (4.4a)$$

$$\frac{\bar{D}\langle U_z\rangle}{\bar{D}t} = -\frac{1}{\rho}\frac{\partial\langle P\rangle}{\partial z} - \left(\frac{1}{r}\frac{\partial\left(r\langle u_r'u_z'\rangle\right)}{\partial r} + \frac{1}{r}\frac{\partial\langle u_\theta'u_z'\rangle}{\partial\theta} + \frac{\partial\langle u_z'^2\rangle}{\partial z}\right) \quad (4.5)$$

The substantive derivative operator is defined by

$$\frac{\bar{D}}{\bar{D}t} = \frac{\partial}{\partial t} + \langle U_r\rangle\frac{\partial}{\partial r} + \frac{\langle U_\theta\rangle}{r}\frac{\partial}{\partial\theta} + \langle U_z\rangle\frac{\partial}{\partial z} \quad (4.6)$$

Several observations regarding Eqs. (4.3)–(4.5) are appropriate:

1. The radial momentum equation, Eq. (4.3), has an additional mean flow acceleration term on the left-hand-side (LHS) due to the centripetal acceleration of a fluid element. For a flow with circular streamlines, in the absence of turbulent stresses, this acceleration is caused by the differential pressure forces associated with the radial pressure gradient. An alternate view is that the apparent outward centrifugal force experienced by a fluid element is balanced by a net inward pressure force.

2. The first three turbulent stress terms in the radial momentum equation are clearly associated with the gradients in the mean turbulent radial momentum flux. The last term, however, is better identified as a component of the mean centripetal acceleration, $\langle U_\theta^2\rangle = \langle U_\theta\rangle^2 + \langle u_\theta'^2\rangle$. In this light, the radial momentum equation simply states that the differential radial pressure force and the turbulent flux of radial momentum balance the total radial mean flow acceleration.

3. The tangential momentum equation, usually written in a form similar to Eq. (4.4), also appears to have additional mean acceleration and turbulent flux terms. However, multiplying by r and re-arranging, Eq. (4.4a) is obtained. In this simpler form, it is clear that the tangential momentum equation is an expression of the conservation of mean flow angular momentum, which is influenced by the net turbulent flux of angular momentum and the mean tangential pressure gradient. It is important to recognize that for axisymmetric flows, in the absence of significant turbulent transport, the angular momentum of a fluid element is conserved. Although perhaps trivial, this observation is key to understanding the development of swirling, bowl-in-piston diesel engine flows.

4. Items 1 and 3 jointly describe a stability condition imposed on the mean flow by the balance between the net pressure force and the apparent centrifugal force

on a fluid particle (e.g., [17, 94, 95]). If a fluid element is displaced outward, conserving angular momentum, it will, on average, have an angular momentum (and $\langle U_\theta \rangle$) deficit compared to the surrounding fluid—provided the mean flow angular momentum gradient is greater than zero. Accordingly, the centrifugal forces acting on the fluid element are insufficient to oppose the net pressure force imposed by the mean radial pressure gradient, and the element is forced back inward. Conversely, if the mean angular momentum gradient is negative, the pressure gradient will be insufficient to restrain the excessive centrifugal forces and the fluid particle will continue its outward migration. These ideas are illustrated in Fig. 4.1.

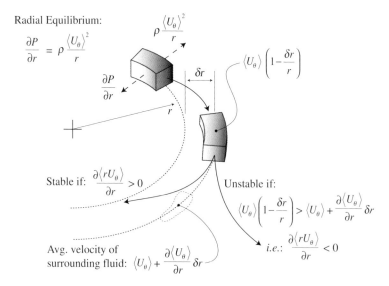

Fig. 4.1 Illustration of the stability of a rotating flow when a fluid element is displaced radially. Adapted from [17]

5. As will be discussed below, if the turbulence is isotropic then all turbulent shear stresses are zero. Similarly, gradients in the normal stresses must be zero. In isotropic turbulence, therefore, the turbulent stresses in Eqs. (4.3)–(4.5) vanish, and the turbulence has no effect on the mean flow development.

4.2.2 Turbulent Stresses and Kinetic Energy

The central task in the modeling and prediction of turbulent flows is the determination of the turbulent stresses in Eqs. (4.3)–(4.5). Several texts are available which describe the various methods employed to model and predict these stresses, e.g., [78, 101], and a review of these methods is outside the scope of this writing. However, the mechanisms by which turbulent stresses are generated in engines are explored in detail below, as is the modeling of these stresses within the ubiquitous

k-ε turbulence model—which is the industry standard for computing turbulent engine flows. Accordingly, the necessary background material is introduced briefly here.

Before discussing the evolution and modeling of the turbulent stresses, it will be useful to describe some of their fundamental properties. The six stresses in Eqs. (4.3)–(4.5) make up the elements of a symmetric second-order tensor. Students of solid mechanics will recall that the stress tensor (or any symmetric second-order tensor) can be diagonalized through choice of an appropriate coordinate system— i.e., the principal axes. Thus there is no inherent distinction between the normal stresses and the shear stresses. The inherent distinction is between the isotropic and anisotropic part of the stress tensor. Defining the isotropic part as the sum of the diagonal elements—twice the turbulent kinetic energy k—the turbulent stress tensor can be written as

$$\left\langle u_i' u_j' \right\rangle = \underbrace{\left\langle u_i' u_j' \right\rangle - \tfrac{2}{3} k \delta_{ij}}_{\text{Anisotropic Part}} + \underbrace{\tfrac{2}{3} k \delta_{ij}}_{\text{Isotropic Part}} \tag{4.7}$$

The momentum equations can be manipulated to obtain six additional conservation equations, one for each of the turbulent stresses. Summing the equations for each of the three normal stresses, an equation for the conservation of the kinetic energy of the turbulence—the k-equation—is obtained:

$$\frac{\bar{D}k}{\bar{D}t} = P - \varepsilon - (\nabla \cdot \mathbf{T}) \tag{4.8}$$

In the k-equation, P denotes the production of turbulent kinetic energy, ε the viscous dissipation, and $\mathbf{T}$ the turbulent transport. The turbulent transport, while not negligible, is not a dominant source of turbulence energy and its modeling is not controversial. It will not be considered further.

The turbulence production, under the assumption of spatially-uniform density, is

$$P = - \left\langle u_i' u_j' \right\rangle \left\langle S_{ij} \right\rangle \tag{4.9}$$

$\left\langle S_{ij} \right\rangle$ denotes the mean rate-of-strain tensor. Note that P is independent of rotation of the mean flow. Mean flow rotation serves to re-distribute energy among the different normal stresses, but results in no net increase in k. It is convenient to break out the contributions to P from the isotropic and the anisotropic portions of the stress tensor:

$$P = - \left(\left\langle u_i' u_j' \right\rangle - \frac{2}{3} k \delta_{ij} \right) \left\langle S_{ij} \right\rangle - \frac{2}{3} k \left(\nabla \cdot \left\langle \mathbf{U} \right\rangle \right) \tag{4.10}$$

In a constant density flow, $\nabla \cdot \left\langle \mathbf{U} \right\rangle = 0$, and only the anisotropic portion of the stress tensor contributes to production. In engines flows this is not true, due to the changing gas density during the compression and expansion processes. As will be seen below, production associated with the isotropic stresses is often the dominant term contributing to k. During compression, $\nabla \cdot \langle \mathbf{U} \rangle < 0$, and the production by the isotropic stresses is positive. On the other hand, during expansion the isotropic stresses will reduce k, at a rate that is proportional to k. Clearly, it is difficult to maintain high levels of turbulence during the expansion stroke.

Equation (4.10) can be written in polar-cylindrical coordinates as

$$
P = -\left(\langle u_r'^2 \rangle - \frac{2}{3}k \right) \frac{\partial \langle U_r \rangle}{\partial r} - \underbrace{\langle u_r' u_\theta' \rangle \left(\frac{1}{r} \frac{\partial \langle U_r \rangle}{\partial \theta} - \frac{2 \langle U_\theta \rangle}{r} \right) - \langle u_r' u_z' \rangle \frac{\partial \langle U_r \rangle}{\partial z}}_{\text{Terms contributing to } \langle u_r'^2 \rangle}
$$

$$
- \langle u_r' u_\theta' \rangle \left(\frac{1}{r} \frac{\partial \left(r \langle U_\theta \rangle \right)}{\partial r} \right) - \underbrace{\left(\langle u_\theta'^2 \rangle - \frac{2}{3}k \right) \left(\frac{1}{r} \frac{\partial \langle U_\theta \rangle}{\partial \theta} - \frac{\langle U_r \rangle}{r} \right) - \langle u_z' u_\theta' \rangle \frac{\partial \langle U_\theta \rangle}{\partial z}}_{\text{Terms contributing to } \langle u_\theta'^2 \rangle}
$$

$$
- \underbrace{\langle u_r' u_z' \rangle \frac{\partial \langle U_z \rangle}{\partial r} - \langle u_z' u_\theta' \rangle \left(\frac{1}{r} \frac{\partial \langle U_z \rangle}{\partial \theta} \right) - \left(\langle u_z'^2 \rangle - \frac{2}{3}k \right) \frac{\partial \langle U_z \rangle}{\partial z}}_{\text{Terms contributing to } \langle u_z'^2 \rangle}
$$

$$
\underbrace{-\frac{2}{3} \frac{k}{V} \frac{dV}{dt}}_{\text{Production by isotropic stresses}}
$$

$$(4.11)$$

Anisotropy in the normal stresses is largely expected to result from differences in the production of each normal stress. The measured anisotropy can thus be used to help identify or confirm the dominant sources of turbulence production. To aid in this process, the re-distributive terms that arise when the normal stress convection terms are transformed into a cylindrical coordinate system are included in Eq. (4.11), where the sources of the individual normal stresses have been identified. While these re-distributive terms influence the magnitude of the individual normal stresses, their overall contribution to k and P is zero.

Equation (4.11) has also been written in a form that emphasizes production associated with the isotropic stresses. When considering the anisotropy of the normal stresses, however, it should be recalled that the isotropic production does not contribute equally to each stress component. From Eqs. (4.2) and (4.11) it is clear that

$$
-\frac{2}{3} \frac{k}{V} \frac{dV}{dt} = -\frac{2}{3}k \left(\underbrace{\frac{\partial \langle U_r \rangle}{\partial r}}_{\langle u_r'^2 \rangle} + \underbrace{\frac{1}{r} \frac{\partial \langle U_\theta \rangle}{\partial \theta} + \frac{\langle U_r \rangle}{r}}_{\langle u_\theta'^2 \rangle} + \underbrace{\frac{\partial \langle U_z \rangle}{\partial z}}_{\langle u_z'^2 \rangle} \right)
\tag{4.12}
$$

Thus production by the isotropic part of the normal stresses influences the individual normal stresses (component energies) differently, depending on the details of the mean flow structure.

With the exception of the radial (squish) velocity near the bowl lip during the peak squish/reverse-squish periods, the tangential (swirl) velocity is often the dominant velocity component in direct-injection diesel engines. Turbulence production directly associated with the swirl velocity thus merits closer examination. In an axisymmetric flow, the production terms associated with the swirl velocity are

$$P_{\text{swirl}} = -\langle u'_r u'_\theta \rangle \left(\frac{1}{r} \frac{\partial (r \langle U_\theta \rangle)}{\partial r} - \frac{2 \langle U_\theta \rangle}{r} \right) - \langle u'_z u'_\theta \rangle \frac{\partial \langle U_\theta \rangle}{\partial z} \qquad (4.13)$$

The last term on the right-hand-side (RHS) of Eq. (4.13) is self-explanatory and will be considered no further. The first two terms are related to the radial distribution of the swirl velocity, and may be more readily recognized as a portion of the r-θ element of the mean rate-of-strain tensor when re-written as

$$\frac{1}{r} \frac{\partial (r \langle U_\theta \rangle)}{\partial r} - \frac{2 \langle U_\theta \rangle}{r} = r \frac{\partial}{\partial r} \left(\frac{\langle U_\theta \rangle}{r} \right) \qquad (4.14)$$

From the form of Eq. (4.14) it can be seen immediately that if $\langle U_\theta \rangle$ is locally proportional to r (i.e., solid-body-like), then the net production associated with these terms is zero. However, for non-zero $\langle u'_r u'_\theta \rangle$, Eq. (4.11) indicates that there will be a net production of one normal stress ($\langle u'^2_r \rangle$ or $\langle u'^2_\theta \rangle$)—with a corresponding reduction in the remaining stress. Thus, although a locally solid-body-like swirl profile is not a net source of turbulent kinetic energy, it generally will re-distribute turbulence energy between $\langle u'^2_r \rangle$ and $\langle u'^2_\theta \rangle$.

Due to the potential of the large swirl velocities for turbulence generation, it is worthwhile examining the expected sign and magnitude of $\langle u'_r u'_\theta \rangle$ in greater detail. Algebraic approximations to the equations governing the evolution of the turbulence stresses [78] suggest that the magnitude of each stress component is proportional to the magnitude of its production rate. In our polar-cylindrical coordinate system, we therefore anticipate

$$\langle u'_r u'_\theta \rangle \propto - \langle u'^2_r \rangle \frac{1}{r} \frac{\partial (r \langle U_\theta \rangle)}{\partial r} - \langle u'^2_\theta \rangle \left(\frac{1}{r} \frac{\partial \langle U_r \rangle}{\partial \theta} - \frac{2 \langle U_\theta \rangle}{r} \right) - \langle u'_r u'_\theta \rangle (\langle S_{rr} \rangle + \langle S_{\theta\theta} \rangle)$$
$$- \langle u'_r u'_z \rangle \frac{\partial \langle U_\theta \rangle}{\partial z} - \langle u'_z u'_\theta \rangle \frac{\partial \langle U_r \rangle}{\partial z}$$
$$(4.15)$$

Following the practice employed in writing Eq. (4.11), sources of $\langle u'_r u'_\theta \rangle$ associated with transformation of the convective terms into a polar-cylindrical coordinate system are included with the production terms in Eq. (4.15).

It will again be convenient below to refer to the role of the isotropic and the anisotropic normal stresses in generating the shear stress. Accordingly, Eq. (4.15) can be re-written as:

$$\langle u'_r u'_\theta \rangle \propto -\frac{2}{3} k \langle S_{r\theta} \rangle$$

$$- \left(\langle u'^2_r \rangle - \frac{2}{3} k \right) \frac{1}{r} \frac{\partial (r \langle U_\theta \rangle)}{\partial r} - \left(\langle u'^2_\theta \rangle - \frac{2}{3} k \right) \left(\frac{1}{r} \frac{\partial \langle U_r \rangle}{\partial \theta} - \frac{2 \langle U_\theta \rangle}{r} \right) \qquad (4.16)$$

$$- \langle u'_r u'_\theta \rangle (\langle S_{rr} \rangle + \langle S_{\theta\theta} \rangle) - \langle u'_r u'_z \rangle \frac{\partial \langle U_\theta \rangle}{\partial z} - \langle u'_z u'_\theta \rangle \frac{\partial \langle U_r \rangle}{\partial z}$$

Before further simplifying Eqs. (4.15) and (4.16), it should be noted that during compression, the magnitude of $\langle u'_r u'_\theta \rangle$, whatever its sign, can be expected to be enhanced due to an anticipated negative average dilatation in the r-θ plane ($(\langle S_{rr} \rangle + \langle S_{\theta\theta} \rangle)$) as the flow is compressed into the bowl. During expansion, as the flow spreads into the squish volume, the opposite is anticipated. This observation further highlights the difficulty of maintaining high levels of turbulence production during the expansion stroke.

For an axially-uniform, axisymmetric flow on circular streamlines, Eq. (4.15) can be simplified to read

$$\langle u'_r u'_\theta \rangle \propto - \langle u'^2_r \rangle \frac{1}{r} \frac{\partial (r \langle U_\theta \rangle)}{\partial r} + \langle u'^2_\theta \rangle \frac{2 \langle U_\theta \rangle}{r} \qquad (4.17)$$

Several observations can be made regarding Eq. (4.17):

1. For most engine flows, in which the radial gradient in mean angular momentum is positive, there are two competing mechanisms that influence the shear stress $\langle u'_r u'_\theta \rangle$. The first mechanism, discussed by Lumley [62], is associated with displacement of fluid by the radial velocity fluctuations. On average, a fluid particle displaced in the positive radial direction (by a positive u'_r) will have an angular momentum deficit (and, hence, a negative u'_θ) if the radial gradient in the mean flow angular momentum is positive. This mechanism thus tends to make $\langle u'_r u'_\theta \rangle$ negative. The second mechanism is associated with the increased apparent centrifugal force on a fluid element experiencing a positive u'_θ. The increased centrifugal force cannot be balanced by the mean radial pressure gradient (v. Eq. (4.3) and the discussion of the mean flow stability), and a net positive u'_r results—thereby producing a positive $\langle u'_r u'_\theta \rangle$.
2. For a solid-body-like mean flow structure and equal normal stresses, these mechanisms exactly balance each other and the resulting shear stress is expected to be zero. If the mean flow is solid-body-like and $\langle u'^2_r \rangle > \langle u'^2_\theta \rangle$, then $\langle u'_r u'_\theta \rangle < 0$ and, according to Eq. (4.11), energy will be abstracted from the radial stresses and transferred to the tangential stresses. Conversely, if $\langle u'^2_r \rangle < \langle u'^2_\theta \rangle$, then $\langle u'_r u'_\theta \rangle > 0$ and energy will be transferred from the tangential stresses to the radial stresses. In a solid-body-like turbulent flow, there is evidently a self-stabilizing process that attempts to equalize the normal stresses and reduce the shear stress; i.e., a tendency toward isotropy.

3. Generalizing to a non-solid-body-like mean flow structure, if the radial gradient in mean angular momentum becomes sufficiently small, but remains positive, the shear stress $\langle u'_r u'_\theta \rangle$ will be positive. Under these circumstances the radial fluctuations will be enhanced, but a negative production of $\langle u'^2_\theta \rangle$ will occur. Consequently, the shear stress will decrease—followed by reduced net production of k. Conversely, for a sufficiently large angular momentum gradient, $\langle u'_r u'_\theta \rangle$ will be negative. Accordingly, $\langle u'^2_\theta \rangle$ will increase, but $\langle u'^2_r \rangle$ will decrease. The net effect will be to increase the shear stress (decrease its magnitude), again reducing the production of k. Thus for a non-solid-body-like flow structure, there is a tendency for reduced shear stresses and reduced production of k analogous to the tendency toward isotropy discussed above.

4. If the mean angular momentum gradient is negative, then both the radial and the tangential fluctuations will lead to increased $\langle u'_r u'_\theta \rangle$. Similarly, production of both $\langle u'^2_r \rangle$ and $\langle u'^2_\theta \rangle$ will be positive, leading to still further increased $\langle u'_r u'_\theta \rangle$. Plainly, a large increase in turbulence energy can be expected under these circumstances. Recall that, as discussed in the context of Eqs. (4.3)–(4.5) and Fig. 4.1, a negative angular momentum gradient defines an unstable mean flow condition.

Although the above observations are made for a simplified flow, and are based on approximate equations for the Reynolds stresses, it is expected that they will provide guidance regarding the behavior of more realistic engine flows. As will be seen in greater detail below, this expectation is largely justified.

4.2.3 Turbulence Modeling

The modeling of the turbulent stresses is vital to not only the correct prediction of the mean flow development given by Eqs. (4.3)–(4.5), but also to the prediction of the production of k. The production figures not only in the conservation equation for k, but also in the model equation governing the dissipation ε. Within the k-ε turbulence model, the turbulent stresses are modeled in terms of the anisotropic mean rate-of-strain tensor $\langle S^*_{ij} \rangle$ and an isotropic eddy viscosity ν_T:

$$\langle u'_i u'_j \rangle - \frac{2}{3} k \delta_{ij} = -2\nu_T \langle S^*_{ij} \rangle \tag{4.18}$$

where

$$\langle S^*_{ij} \rangle = \langle S_{ij} \rangle - \frac{1}{3} (\nabla \cdot \langle \mathbf{U} \rangle) \delta_{ij} \tag{4.19}$$

and

$$\nu_T = C_\mu \frac{k^2}{\varepsilon} \tag{4.20}$$

Several points should be noted from Eqs. (4.18)–(4.20):

1. The last term on the RHS of Eq. (4.19) is required in variable density flows to ensure that the modeled $\langle u_i' u_i' \rangle = 2k$. This term is absent in descriptions of the k-ε turbulence modeling of constant density flows.
2. There is no assumption embodied in Eqs. (4.18)–(4.20) that the turbulent stresses are isotropic. In fact, it is only the anisotropic portion of the stress tensor that is modeled in terms of $\langle S_{ij} \rangle$. Only the eddy viscosity ν_T is assumed isotropic—that is, independent of the component of $\langle u_i' u_i' \rangle$ being modeled. This scalar isotropic eddy viscosity is thus modeled only in terms of other scalars: k, ε, and the model constant C_μ.
3. Equation (4.18) is commonly regarded as a postulated relationship. However, as will be seen below, it is the leading term in a more general constitutive relation that can be derived directly from the Reynolds-averaged equations [42, 77, 87]. This may also be inferred by comparison of Eqs. (4.16) and (4.18).

If the adequacy of Eqs. (4.18)–(4.20) in modeling the turbulent stresses is to be assessed experimentally, both the appropriate mean flow gradients and turbulence quantities must be obtained. Although k can be obtained from measurement of the normal stresses, the dissipation ε is less accessible. It can be estimated, however, by the relation

$$\varepsilon = A \frac{\left(\frac{2}{3}k\right)^{3/2}}{\ell} \tag{4.21}$$

Equation (4.21) implies that the turbulence is in equilibrium—the production of turbulence energy, its rate-of-transfer to smaller scales, and its viscous dissipation are all approximately in balance. In engine flows, where the characteristic time scale of the turbulence can be considerably greater than the time scales describing the mean flow evolution, this condition may not be fulfilled. An alternative perspective on this issue may be gained by considering Eq. (4.18) as an expression of a mixing length hypothesis, which can be obtained by substitution of Eq. (4.21) into Eqs. (4.18) and (4.20):

$$\langle u_i' u_j' \rangle - \frac{2}{3} k \delta_{ij} = -2Cu'\ell \langle S_{ij}^* \rangle \tag{4.22}$$

where C is a constant and $u' \propto k^{1/2}$. From this viewpoint, the quantity $u'\ell \propto \frac{k^2}{\varepsilon}$ is directly related to the transport of momentum by a turbulent fluctuation with velocity u' over a distance ℓ, and use of Eq. (4.21) seems justified. However, employing a value for the constant A obtained in equilibrium turbulence may not be appropriate. Nevertheless, in the results presented below we adopt the relation $A = 0.55$, obtained from grid turbulence measurements [25] when the integral scale employed is the transverse scale ℓ_g, as defined by Hinze [51].

To improve the prediction of the normal stress anisotropy [88], and the prediction of the stresses in swirling flows [21, 28, 87] alternatives to the linear relation between the turbulent stresses and the mean strain rate embodied in Eq. (4.18) have been explored. These relations (retaining terms only up to the third order) take the form

$$
\frac{\left\langle u'_i u'_j \right\rangle - \frac{2}{3} k \delta_{ij}}{2k} = -\alpha_1 \left(\frac{k}{\varepsilon}\right) S^*_{ij} + \alpha_2 \left(\frac{k}{\varepsilon}\right)^2 \left(S^*_{ik} S^*_{kj} - \frac{1}{3} S^*_{kl} S^*_{kl} \delta_{ij}\right) + \alpha_3 \left(\frac{k}{\varepsilon}\right)^2
$$

$$
\left(S^*_{ik} \Omega^*_{kj} - \Omega^*_{ik} S^*_{kj}\right) + \alpha_4 \left(\frac{k}{\varepsilon}\right)^2 \left(\Omega^*_{ik} \Omega^*_{kj} - \frac{1}{3} \Omega^*_{kl} \Omega^*_{kl} \delta_{ij}\right)
$$

$$
+ \alpha_5 \left(\frac{k}{\varepsilon}\right)^3 \left(S^*_{ik} S^*_{kl} \Omega^*_{lj} - \Omega^*_{ik} S^*_{kl} S^*_{lj}\right) + \alpha_6 \left(\frac{k}{\varepsilon}\right)^3 \left(\Omega^*_{ik} \Omega^*_{kl} S^*_{lj}\right.
$$

$$
\left. + S^*_{ik} \Omega^*_{kl} \Omega^*_{lj} - \frac{2}{3} S^*_{lm} \Omega^*_{mn} \Omega^*_{nl} \delta_{ij}\right)
$$

$$(4.23)$$

In writing Eq. (4.23), we anticipate that the mean rate-of-rotation tensor Ω_{ij} may need to be modified to accommodate flows with streamline curvature [43] or coordinate system rotation [42]. Note that the angle-brackets denoting the ensemble mean have been dropped in writing Eq. (4.23). The multipliers α_1-α_6 are generally strain dependent and anisotropic. In some models, e.g., [28], an additional term may be included that is also cubic in the mean velocity gradients, i.e.: $\alpha_7 \left(\frac{k}{\varepsilon}\right)^3$ $\left(S^*_{kl} S^*_{kl} - \Omega^*_{kl} \Omega^*_{kl}\right) S^*_{ij}$. This term can be absorbed into the leading term on the RHS of Eq. (4.23), providing (additional) strain dependence to the multiplier α_1. Note the equivalence of the leading term in Eq. (4.23) with Eq. (4.18). It is important to appreciate that models of this form are not simply higher-order 'postulated' relations, but can be developed in a rigorous manner from algebraic Reynolds stress closures developed from the Reynolds-averaged momentum equations.

A key qualitative feature to recognize in Eq. (4.23) is that the quantity k/ε denotes the turbulent time scale, and hence $(k/\varepsilon)\left\langle S^*_{ij}\right\rangle$ represents a time scale ratio: the turbulent time scale by a time scale characteristic of the mean flow. When mean flow time scales are large compared to the turbulent time scale (the mean flow velocity gradients are small), then the higher order terms in Eq. (4.23) are of little importance. Accordingly, under these conditions, a stress model based on only the leading term can be expected to perform well. Various stress models of the form of Eq. (4.23) will be evaluated below to illustrate the potential benefits that they may offer in engine flows.

In addition to the stress modeling, the k-ε turbulence model and its variants rely on a modeled equation for the turbulent energy dissipation ε (e.g., [48]). No specific relationships in this model equation are amenable to quantitative assessment as will be performed with the stress model relationships described above. Hence, the appraisals made below will be based on a qualitative comparison of the trends in the computed length $\left(\frac{k^{3/2}}{\varepsilon}\right)$ and time scales (k/ε) with the measured quantities. It will be noted however, that when the measured length and time scales are scaled according to Eq. (4.21), very good quantitative agreement is observed over significant portions of the cycle.

4.3 Review of the Recent Literature

A general trend in experimental studies of in-cylinder fluid motion published in the last two decades is a tendency towards engine geometries and operating conditions which are more representative of production diesel engines. A large number of these studies have been devoted to the clarification of the effects of combustion bowl geometry, swirl level, and engine speed on the flow structures and turbulence. However, a necessary prelude to consideration of this recent work is a brief review of the formation of the mean flow structure in the late compression stroke and the subsequent 'squish/swirl' interaction. Our understanding of the structure of turbulent engine flows has also benefited from numerous experimental studies that have provided detailed information on the shear stresses and the anisotropy, as well as the length scales of the turbulence. The aforementioned topics form a general outline of this section, which begins with a brief overview of the induction stroke fluid dynamics that sets the stage for the later flow development.

4.3.1 Induction and Early Compression

A detailed understanding of the flow structure development and turbulence generation during the induction stroke is indispensable for engine technologies that rely on early in-cylinder mixing of fuel/air/residuals. For direct-injection diesel engines, however, the flow details during this period are less important, except to the extent that they influence the mean flow configuration and the in-cylinder turbulence level during the latter part of the compression stroke. Accordingly, only general observations are summarized below. Additional details can be found in [46] and in the individual references cited below.

1. Mean flow structures, such as the ring-vortex structures formed by the flow through the valve (e.g. [3, 52]) or double-vortex structures in horizontal, diametric planes (e.g. [8, 9]) generally decay by bottom center. The only mean flow structures surviving well into the compression stroke are tumble and swirl motions.
2. Discussion of tumble motion is entirely absent from the diesel engine literature, despite its important role in spark ignition engine applications. As will be seen below, to ensure high turbulence levels near TDC—even in large squish-area, reentrant, bowl-in-piston combustion chambers—it is important to maintain high levels of turbulence throughout the compression stroke. Tumble may play an important role here. However, flow tumble may be suppressed in high-swirl flows, as large centrifugal forces acting on high angular momentum fluid will resist inward fluid displacement by tumble motion. This is an area that requires additional exploration.
3. The position of the swirl center often rotates about the cylinder axis [9, 30, 96]. Bowl-in-piston engines promote swirl-centering and reduce the amplitude of this motion.

4. An engine speed dependency of the swirl flow development is often reported. Discussion of this speed dependency is deferred until a later section.
5. Some degree of axial stratification of the swirl velocity (or angular momentum) is frequently observed (e.g. [64, 70], refs. in [46]). Insufficient information is available to quantify the influence of port design on this phenomenon.
6. Modeling studies suggest that the bulk of the turbulent energy observed during induction is generated in-cylinder by the anisotropic stresses. Port-generated turbulence, convected into the cylinder, is a minor contribution.
7. Peak turbulence levels occur in the mid-induction stroke, and decay rapidly thereafter. Extrapolation of the decay rate suggests that induction generated turbulence will have substantially dissipated by intake valve closure. Additional numerical simulations [20, 33] and scaling analysis [62] support this observation.
8. The spatial distribution of turbulence energy is reasonably homogeneous, and is characterized by approximately equal normal stresses by the early part of the compression stroke.

Overall, as the last third of the compression stroke begins, the mean flow field in a swirl-supported, direct-injection engine is generally characterized by a single rotating vortex, with a center roughly aligned with the cylinder axis. The radial distribution of the swirl velocity in this vortex does not differ dramatically from a solid-body-like structure, although at the outer radii a flattened radial profile may be observed (e.g., Fig. 4.2). Some axial stratification of angular momentum may also be observed. Measurements of mean axial velocities are sparse, but the results of numerical simulations [72] suggest that they do not differ significantly from a linear variation from the piston surface to the head. The distribution of turbulent kinetic energy is approximately homogeneous, with the energy roughly equally distributed among the three component fluctuations. Velocity fluctuations are typically found to be 0.5–0.8 times the mean piston speed S_p. This flow description can be dramatically modified as the compression stroke proceeds, as will be described in detail below.

4.3.2 Near-TDC Mean Flow Structure: The Squish/Swirl Interaction

Gosman and co-workers ([46], and references therein) established through numerical modeling the existence of a strong interaction between the flow swirl and the squish flow that dominates the fluid mechanics of bowl-in-piston diesel engines near the end of compression. In contrast to cylindrical ('pancake') combustion chambers, wherein the solid-body-like form of the radial distribution of the swirl velocity is not appreciably modified during the compression and expansion processes, the squish flow significantly disrupts the radial distribution of the swirl velocity and causes large departures from a solid-body-like structure. This disruption creates r-z plane vertical structures that vary considerably with swirl level and bowl geometry, and which, in turn, change the convective transport of existing turbulence and generate additional turbulence.

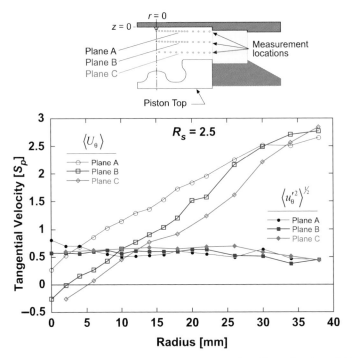

Fig. 4.2 Typical mean and rms swirl velocity profiles illustrating the solid-body-like mean flow structure and the homogeneity of the fluctuating velocity as the latter portion of the compression stroke begins (-55 CAD)

The process by which the squish flow interacts with the flow swirl is depicted schematically in Fig. 4.3. During compression, the vertical motion of the piston displaces fluid elements inward toward the cylinder centerline. Neglecting turbulent diffusion and assuming axisymmetric flow, the tangential momentum equation— Eq. (4.4a)— reduces to the statement that the angular momentum of the fluid element is conserved: $r\langle U_\theta \rangle$ is constant. Thus, as the element is displaced inward, its tangential velocity increases as the radius of gyration is reduced. This 'spin-up' process is analogous to the figure skater who increases/decreases his rotational speed through a corresponding decrease/increase of his rotational moment of inertia. As a consequence of the increased tangential velocity, the centrifugal

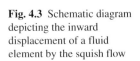

Fig. 4.3 Schematic diagram depicting the inward displacement of a fluid element by the squish flow

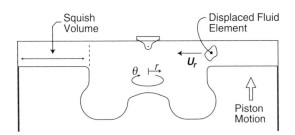

forces acting on the fluid element are increased. Eventually, as the fluid penetrates inward, the radial momentum imparted by the squish flow cannot overcome the increased centrifugal forces and further inward penetration is impeded. Because the squish flow is dominated by the changing combustion chamber geometry, the radial momentum imparted to fluid elements during the squish process is independent of the flow swirl level. In contrast, the centrifugal forces acting on a fluid element will increase approximately quadratically with the flow swirl ratio. The inward penetration of the squish flow, therefore, is strongly influenced by the swirl, with the greatest penetration occurring at the lowest swirl levels.

For low levels of flow swirl, the squish flow penetrates to nearly the cylinder centerline before it turns down into the bowl as required by symmetry constraints. As the flow swirl is increased, the inward penetration is reduced and the flow turns down into the bowl when the increasing centrifugal forces have overcome the initial radial momentum imparted by the squish process. For high swirl levels, the centrifugal forces are sufficiently great that the squish flow turns down into the bowl as soon as the combustion chamber geometry permits. After turning downward, the high-$\langle U_\theta \rangle$ fluid invariably attempts to return to the outer bowl radii due to the high centrifugal forces acting on it. The resulting near-TDC flow fields, as predicted by numerical simulation, are depicted in Fig. 4.4. Additional examples are found in references [1, 6, 13, 15, 33, 59, 65, 86, 91, 92, 98].

Note from Fig. 4.4 that the r-z plane flow structures formed by the squish/swirl interaction act to transport angular momentum to different locations within the bowl. Accordingly, the tangential velocity distribution, shown by the false color background, also varies dramatically with swirl level. As will be seen below, the spatial distribution of the swirl velocity $\langle U_\theta \rangle$ has an important influence on the generation of turbulence. Thus, it must be anticipated that differences in the spatial distribution and characteristics of the turbulence within the bowl will also be observed as the swirl level is varied.

Furthermore, note that at the highest swirl ratio, $R_s = 3.5$, a double vortex structure is developing within the bowl. The lower vortex develops as the high-momentum fluid in the bottom of the bowl is deflected upward by the bowl pip and attempts to return to the bowl periphery under the action of centrifugal force. A second, counter-rotating vortex subsequently forms above the pip. If the swirl ratio is increased further, the former vortex is confined to the lower, outer regions of the bowl, while the latter grows accordingly. In addition to turbulence generation effected by the spatial distribution of $\langle U_\theta \rangle$, we must also expect that the high level of mean flow deformation in the r-z plane at the interface between these two vortices will also elevate turbulence production.

An important aspect of the squish/swirl interaction process that is often overlooked is that as the high tangential velocity fluid is forced inward and downward into the bowl, the conserved quantity is angular momentum—not rotational kinetic energy. Arcoumanis et al. [6] remind us that, in an idealized situation, the rotational kinetic energy of the charge can increase by as much as the square of the cylinder-to-bowl diameter ratio. The source of this increased energy is ultimately associated with piston work performed during compression.

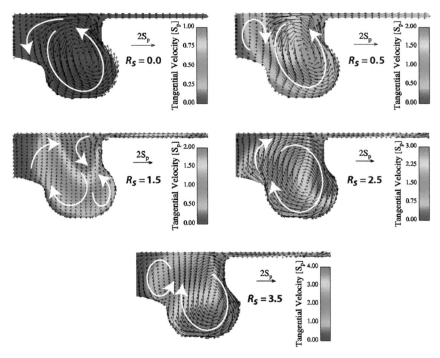

Fig. 4.4 Illustration of the changing r-z plane flow structures, and the differing spatial distribution of the swirl velocity, as the flow swirl ratio is increased. The velocity scales are expressed in units of mean piston speed S_p

Detailed experimental mappings of the r-z plane mean flow structures, such as might be obtained with particle image velocimetry, do not exist. However, the experimental data that do exist are collectively sufficient to validate the multi-dimensional model predictions and the physical ideas discussed above. Arcoumanis and co-workers [5] provided the first experimental validation of the effects of swirl–squish interaction on r-z plane flow structures. Their measurements of axial and radial velocity profiles clearly demonstrated the decreased penetration of the squish flow when flow swirl is present, and the subsequent formation of flow structures similar to those shown in Fig. 4.4. Additionally, the tangential (swirl) velocity measurements illustrate the role of these structures in transporting high angular momentum fluid to different locations within the bowl.

Further measurements supporting the predicted flow structures are reported by Rask and Saxena [80], who sketch streamlines based on radial velocity measurements that are reminiscent of the structures shown in Fig. 4.4 for the $R_s = 0.5$ swirl case. Their measurements also indicate that at higher swirl ratios, the squish flow turns down into the bowl at larger radii. Similarly, Fansler and French [35] report radial velocity measurements that are consistent with the above description of the r-z plane flow structures and the effect of increasing swirl on these structures. In addition, their tangential velocity measurements clearly illustrate the 'spin-up' of the fluid exiting the squish volume. Tangential velocity measurements reported by Sugiyama [91] plainly depict the increased inward penetration of high angular momentum fluid at the lower swirl ratios. Radial profiles of measured axial velocities reported by Arcoumanis et al. [9] are also consistent with the high swirl ratio flow structure depicted in Fig. 4.4. More recently, Miles et al. [67, 72] make a direct comparison between measured axial profiles of radial and tangential mean velocities and the predictions of a multi-dimensional simulation. The level of agreement observed suggests that the numerical predictions are capable of predicting the mean r-z plane structures with good quantitative accuracy—including such details as mean flow strain rates—provided the bulk swirl ratio is specified or predicted accurately.

4.3.3 Turbulent Flow Structure

4.3.3.1 Ensemble Averaging vs. Cycle-Resolved Analysis

A still unresolved issue in the study of turbulent engine flows is how one can separate and characterize quantitatively the 'turbulence' flow structures, when cycle-to-cycle variations in mean, bulk flow structures—with similar characteristic time and length scales—occur simultaneously. Discussion of this topic and selected applications can be found in references [36, 50, 60, 62, 106]. Methods to separate measured velocity fluctuations due to these two sources ('cycle-resolved' analysis) typically rely on temporal or spatial filtering techniques. Due to a lack of separation of length and time scales characterizing the turbulence and the cycle-to-cycle mean flow variations, these attempts invariably include some of the turbulence structures/energy in the mean flow results, and vice versa.

In the discussion that follows (and in the later sections of this manuscript), the statistical characterization of the turbulence is achieved through standard ensemble averaging techniques, unless otherwise specified. This choice has been made for the following reasons:

1. Cycle-to-cycle mean flow fluctuations are generally small when a strong, directed mean flow exists [60, 79]. In direct-injection diesel engines, such a mean flow structure is provided by flow swirl and by the pronounced squish flows present in engines with reentrant bowl geometries.
2. A significant part of the anisotropy of the turbulence is found in the large scales of the turbulence structures. If, due to the overlapping of scales, contributions from these large scales are removed from the turbulence, the anisotropic part of the turbulent stress tensor will be altered significantly. It is the anisotropic part of the turbulent stresses that is responsible for the transport of momentum that influences the mean flow development [78]. Additional perspective on this issue is provided by considering the Eulerian correlation functions that are obtained with filtered turbulent fluctuations [19, 39, 61]. Integral scales derived from these correlations would be near zero. Although the relationship between Eulerian and Lagrangian integral scales is not known, they are generally considered to be approximately proportional [93]. It is relevant to note here, then, that

a turbulence field with a Lagrangian integral length scale of zero is incapable of transporting momentum. Similarly, it is the anisotropic part of the stresses that are predominantly responsible for the production of additional turbulence. Altering (and, in some cases, nearly removing) the anisotropic stresses through application of cycle-resolved analysis techniques can thus fundamentally alter the very aspects of the turbulence that are most important to understand and accurately model.

3. As will be seen below, when the turbulent stresses are estimated from the measured mean velocity gradients using a stress modeling hypothesis and associated constants derived in canonical, ergodic flows, they agree very well with the experimentally measured stresses over significant portions of the cycle. A similar agreement is observed between measured turbulent time and length scales and the results of numerical simulations. This agreement would be unlikely to be observed if the measured stresses had significant contributions from cycle-to-cycle fluctuations.

Despite the above assertions, cycle-to-cycle fluctuations in the mean flow structure may be more important during some periods of the cycle or at some spatial locations than at others, and could result in significant error in the estimation of mean velocity gradients or turbulent stresses under some circumstances. Although these fluctuations will not be considered explicitly below, their possible existence must be borne in mind when the experimental data are employed for detailed validation of or selection of turbulent stress models, or for attempting to evaluate the dominant sources of turbulence production.

4.3.3.2 The Influence of Bowl Geometry and Flow Swirl

At first sight, there appears to be a broad range of views in the literature regarding the typical intensity of near TDC turbulence and its spatial distribution within the bowl. To organize and compare the various results, it is useful to segregate the various studies by bowl geometry, swirl ratio, and spatial locations investigated. Upon performing this segregation, a reasonably consistent picture of the effect of these variables on the turbulence structure emerges. This picture is summarized in Fig. 4.5.

An appropriate baseline against which to compare the turbulence structure obtained in bowl-in-piston engines is the flat-piston, or 'pancake' chamber engine. As discussed earlier, the intake-generated turbulence decays on a time scale significantly shorter than the duration of the engine cycle, and thus has little influence on the near-TDC fluctuations. In the absence of significant tumble motion, the near-TDC fluctuations are approximately 0.5 times the mean piston speed S_p [4, 5, 16, 46, 96], and are approximately homogenous. Swirl motion, in the form of solid-body-rotation, is not a source of turbulent kinetic energy, and there is evidence that increasing the flow swirl may slightly decrease the turbulent fluctuations near-TDC [16, 46, 96]. With swirl, however, increased fluctuations near the cylinder center are often observed (e.g. [102]), and are typically attributed to

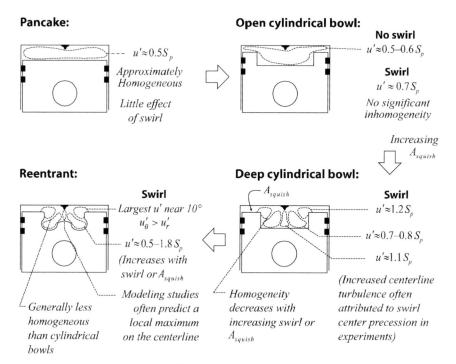

Fig. 4.5 Composite picture of near-TDC turbulence structure for various bowl types, summarizing results obtained from both measurements and numerical simulation as swirl ratio and squish area are varied

precession of the mean flow swirl center. In general, the component fluctuations are approximately equal, although there are several studies that provide (conflicting) evidence that some anisotropy develops. This will be examined more closely in a later section.

For non-reentrant, bowl-in-piston engines without flow swirl, both experimental [5] and numerical studies [46], indicate that the spatial distribution of turbulence within the bowl is generally fairly uniform, with an intensity comparable to that found in pancake chambers—$\frac{u'}{S_p} \approx 0.5 - 0.6$. With the addition of swirl, the numerical studies suggest that at low swirl ratio ($R_s \approx 1.0$) there is little change in the turbulence intensity or spatial distribution, while for higher swirl ratios ($R_s \approx 5.0$) a modest increase is seen ($\frac{u'}{S_p} \approx 0.7$). Experimental results [5, 6, 23, 64, 75] are consistent, with swirl ratios at intake valve closure (IVC) of 2–3 providing perhaps a marginal increase in the turbulent kinetic energy but not significantly affecting the spatial homogeneity. The increased turbulence observed with swirl—not seen in pancake chambers—is indicative of the disruption of a solid-body-like swirl velocity distribution by the squish flow.

In the above referenced work, a small squish area ratio A_{sq}, defined as the ratio of the squish area to the cylinder bore area, was employed—hence the designation 'open cylindrical bowl' in Fig. 4.5. The squish flows are therefore of modest strength. Consequently, no significant increase in turbulent fluctuations near the bowl lip is seen as TDC is approached [5, 46, 75, 100], even with modest levels of flow swirl.

With the addition of both swirl and the larger squish area ratios associated with deeper cylindrical bowls, an increase in turbulence fluctuations near the bowl lip to levels of roughly $1.2\,S_p$ is typically observed. This peak is generally attributed to large amounts of r-z plane flow deformation ('shear') associated with the squish flow and dominated by $\frac{\partial \langle U_r \rangle}{\partial z}$. Stronger squish flows also are expected to result in larger departures of the swirl velocity from a solid-body-like form, however, due to the 'spinning-up' of high angular momentum fluid as it is displaced inward. Thus, large r-θ plane and z-θ 'plane' flow deformation can also exist, as measured by $r\frac{\partial}{\partial r}\left(\frac{U_\theta}{r}\right)$ and $\frac{\partial \langle U_\theta \rangle}{\partial z}$ (see the mean flow fields shown in Fig. 4.4). Both experimental [26, 34, 35, 74, 84] and numerical studies [6], exhibit this squish-flow driven increase in turbulent fluctuations, although exceptions can be found [59]. An increase in R_s generally increases the intensity of the squish-generated fluctuations [35, 84]. Because the magnitude of the squish flow velocity near the bowl lip is not sensitive to R_s, this observation implies that radial and axial gradients in $\langle U_\theta \rangle$ identified above play a role in generating the increased turbulence. In addition to turbulence generation by mean flow deformation, numerical studies indicate that flow separation at the bowl lip may also contribute to the enhanced fluctuations observed in that region [20]. Finally, studies conducted with higher swirl and squish area ratios [34, 74] show less homogeneity within the bowl.

Besides the localized peaks in the turbulent fluctuations seen near the bowl lip, peaks are often seen near the bowl centerline as well. Experimentally, these are often attributed to swirl center precession [5, 96]. In one instance [36], cycle-resolved analysis has demonstrated that the measured increase in fluctuations seen on the bowl centerline is associated with low-frequency motions that are self-correlated over appreciable periods (30–100 CAD), lending credence to this suggestion. Nevertheless, as discussed in [36], multi-dimensional models often predict increased fluctuations on or near the bowl centerline [6, 59] that can exceed $1.1\,S_p$. These numerical studies do not identify the mechanism by which the increased fluctuations are produced.

Reentrant bowls have still higher squish area ratios (resulting in increased tangential velocities for inwardly displaced fluid), and the trends observed above for non-reentrant bowls as A_{sq} and R_s are increased continue to hold. Overall, the average level of measured turbulent fluctuations within the bowl is increased over non-reentrant bowls, although the specific measured levels vary between roughly 0.5 and $1.5\,S_p$ [9, 13, 67, 72, 75, 80, 96]. On the whole, higher fluctuations reported in these studies correlate well with higher swirl ratios and squish area ratios. Rask and Saxena [80], in a study specifically designed to investigate the effects of A_{sq}, corroborate this finding, although they observe only a minor influence of swirl ratio. In general, the homogeneity of the turbulent fluctuations within the bowl suffers with reentrant bowls, although the available measurements suggest that away from the walls the spatial variation in the turbulent fluctuations is within a factor of 2

[9, 13, 67, 80, 96]. Numerical simulations further support this estimate [86], with the exception of local peaks computed on the bowl centerline [15, 33, 92].

Due to their characteristically large A_{sq}, regions of high turbulent fluctuations are observed numerically in the lip region of reentrant bowls, even with no swirl [33, 92]. With swirl, both experimental [11, 26, 27, 35, 80, 84] and numerical [33, 46, 86, 91, 92] studies indicate high turbulence regions near the lip. Studies which provide a crank-angle evolution of these fluctuations indicate that they are maximized near the time of the peak squish flow, roughly 10 CAD prior to TDC.

An interesting attribute of reentrant bowl geometries is the presence of a pronounced peak in turbulence intensity near the bowl lip region during the reverse-squish period, 10–20 CAD ATDC. This reverse-squish peak is observed in both experiments [4, 13, 35, 80, 84] and simulations [46, 59, 91]. Fansler and French [35] have provided an explanation for the occurrence of this peak. With a re-entrant bowl, fluid beneath the bowl rim must flow inward to pass through the bowl throat as the piston descends. However, the expanding squish volume dictates that the flow be outward at and above the bowl throat. The radial flow must therefore undergo a reversal near the bowl lip, leading to regions of large $\frac{\partial \langle U_r \rangle}{\partial z}$. This effect is absent— or greatly diminished—in non-reentrant bowls, although numerical studies suggest there may be modest reverse-squish turbulence production near the lip for these geometries also [6].

Two additional phenomena that occur during the reverse squish period merit discussion. Firstly, large amounts of reverse squish turbulence have been observed near the bowl centerline [19] in an engine with a non-axisymmetric bowl. Concurrently, large mean velocities are observed. Similar levels of turbulent and mean velocities are not seen during the squish period. A similar increase has been observed near the bowl centerline in the LDV experiments of Beard et al. [13]. This phenomenon bears further examination.

Secondly, turbulence is expected to be generated in the reverse-squish period by large velocity gradients above the piston face. This is distinct from the reverse-squish peak discussed above, which occurs within the bowl throat. In addition to shear generated turbulence, numerical studies suggest that unsteady flow separation from the bowl edge is a significant source of flow turbulence [6, 20]. Measurements within the squish region near TDC are scarce, but radial profiles of tangential and axial fluctuations obtained 40 CAD ATDC [67] show no evidence of intense turbulence in the squish region at this time. Measured fluctuations are radially homogeneous at a level of approximately $0.5\,S_p$.

A final observation related to the effects of bowl geometry can be found in the measurements of Beard et al. [13], who find an enhancement of fluctuations when the bowl floor has a pronounced central pip. Measured radial velocity fluctuations reported by Cipolla et al. [23] similarly exhibit a modest increase in the lower-central region of the bowl—near the pip—as do simulation results reported in [6] and [59]. The mechanism for the production of these enhanced fluctuations has not been clarified.

4.3.3.3 Anisotropy and Shear Stresses

As described in the theoretical background section, there is no inherent distinction between turbulent shear stresses and normal stresses, since the turbulent stress tensor can always be diagonalized by choice of an appropriate coordinate system (the principal axes)—the fundamental distinction is between the isotropic and the anisotropic part of the stress tensor. In an incompressible flow, the anisotropic stresses are responsible for generating turbulence—the isotropic stresses play no role. Furthermore, the anisotropic stresses are responsible for the transport of momentum, thus influencing the mean flow development. The anisotropy of the turbulence is thus an essential feature, and accurate prediction of the anisotropic stresses is central to turbulence modeling.

Shear stresses are always related to the anisotropic part of the stress tensor, and knowledge of their magnitude is clearly desirable. The anisotropy in the normal stresses is of equal importance, however. The dominant source of normal stress anisotropy is associated with anisotropy in the production of the individual stresses[1]. Examination of the stress anisotropy, therefore, can provide valuable information on the sources of turbulence even in the absence of knowledge of the shear stresses.

Measurement of the shear stresses, along with the corresponding normal stresses, provides a complete characterization of the anisotropic stresses. Typically, this requires simultaneous measurement of two-components of velocity. The shear stresses in a given plane can be extracted algebraically, however, from independent measurements of the normal stresses along three-different directions in the plane of interest, e.g. [40]. This technique requires highly repeatable flow conditions, and attempts to apply it to engine flows by the current author have failed to yield reliable results. Shear stresses reported in the engine literature have all been obtained via simultaneous measurement of multiple components of velocity.

Foster and Witze [37, 103] report the first measurements of shear stress in an engine flow, having measured $\langle u'_r u'_z \rangle$ (or, equivalently, $\langle u'_\theta u'_z \rangle$) on the centerline of a pancake-shaped chamber. Although this engine geometry is not representative of typical direct-injection diesel engines, these measurements are examined in some detail in the following paragraphs—due to the clear illustration they provide of the enhanced understanding that may be gained through a more complete characterization of the turbulent stress tensor.

To examine the evolution of the underlying structure of the stress tensor reported in [37] in greater detail, it is useful to normalize it by $2k$, such that the sum of the normal stresses equals 1 [v. Eq. (4.23)]. In this manner, changes in the stress tensor caused by the changing turbulence energy—which peaks approximately 15 CAD before TDC—are removed. The components of the normalized stress tensor are depicted in Fig. 4.6, from which the following observations are made:

1. The normalized stress tensor is essentially identical for both engine speeds.
2. The normal stresses follow a well-defined trend. Prior to –35 CAD the majority of the turbulence energy is in the z-component fluctuations (perpendicular to

---

[1] Close to walls, however, other factors may affect the isotropy of the normal stresses.

Fig. 4.6 Crank angle
evolution of the components
of the normalized stress
tensor reported by Foster and
Witze [37]

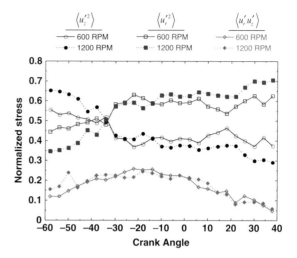

the piston surface). The fraction of energy in the z-component fluctuations then decays approximately monotonically, while the r- (or θ -) component fluctuations increase.

3. The shear stresses exhibit a smooth variation, showing a modest peak at approximately −15 CAD.

The gradual transfer of turbulence energy from one normal stress component to the other, and the maximum in the normalized shear stresses in the midst of this transfer, suggest a gradual rotation of the principal axes that characterize the r-z plane stress. To examine this more closely, the magnitude of the normal stresses, with the normalized stress tensor expressed in principal axes, and the angle made by the major axis of the stress tensor with the horizontal plane are plotted in Fig. 4.7.

From Fig. 4.7, the evolution of the structure of the r-z plane turbulent stresses is much clearer. The energy in the turbulent normal stresses, expressed in principal axes, remains approximately equally distributed at all crank angles, with approximately 70% of the energy in one component, and 30% in the other. The principal axes slowly rotate (in a manner that is nearly linear with crank angle), such that the major axis goes from being nearly aligned with the vertical direction to being nearly aligned with the horizontal direction. This process is depicted pictorially in Fig. 4.8.

Foster and Witze do not report mean velocity gradients, and a detailed assessment of the stress modeling is not possible. Nevertheless, some general observations can be made. First, if one idealizes the flow in the cylinder as undergoing a simple one-dimensional compression, then the only significant mean normal strain rate is $\frac{\partial \langle U_z \rangle}{\partial z}$, and all turbulence production by the isotropic stresses feeds $\langle u'^2_z \rangle$. Consequently, the turbulence is expected to be anisotropic, with the major axis of the stress tensor roughly aligned with the z-direction. During expansion, the opposite effect is anticipated. A one-dimensional expansion process will extract energy from $\langle u'^2_z \rangle$,

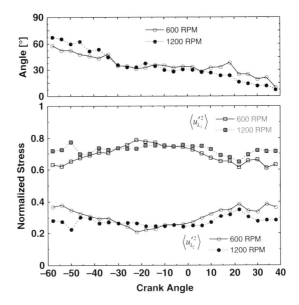

Fig. 4.7 Evolution of the turbulent normal stresses expressed in principal axes, and the angle of the major axis with the horizontal plane

and the major axis of the turbulent stresses can be expected to be approximately perpendicular to the z-direction. This simple picture is consistent with the behavior of the measured stresses, with the exception that the measured rotation of the principal axes is not symmetric about TDC—the principal axes are oriented at $45°$ to the r- and z-coordinate axis at approximately -35 CAD.

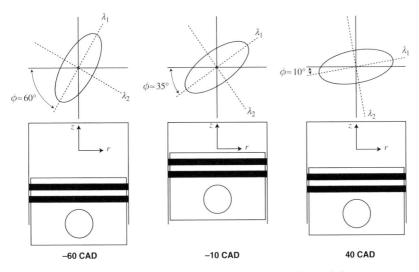

Fig. 4.8 Pictorial description of the evolution of the major axes of the turbulent stress tensor as the piston passes through TDC

   This asymmetry with respect to TDC is consistent with the effects of tumble motion in the r-z plane, which is identified by Foster and Witze as being the dominant mean flow structure present in their engine. Measurements of the evolution of a tumble vortex as it is compressed [7] indicate that by about -30 CAD $\frac{\partial \langle U_r \rangle}{\partial z} >> \frac{\partial \langle U_z \rangle}{\partial r}$. That is, the mean flow in the r-z plane is dominated by $\frac{\partial \langle U_r \rangle}{\partial z}$. Consequently, turbulence generation associated with the mean r-z plane velocity gradients primarily enhances the radial component fluctuations, leading to a rotation of the principal axes of the turbulent stress tensor. Thus, while a quantitative analysis of the coupling of the mean velocity gradients with the turbulent stress tensor is not possible, qualitatively the behavior of the stress tensor is fully consistent with expectations based on the flow physics.

   Dimopoulos and Boulouchos [31, 32] measure the full turbulent stress tensor in a pancake chamber for various swirl levels. Significant differences in the evolution of both the anisotropic normal stresses and the shear stresses are observed as the flow swirl is varied. For the high-swirl flow measurements [32], mean velocity gradients are also reported and employed to qualitatively assess the validity of stress modeling via an eddy-viscosity model. The results of this assessment show that while the eddy viscosity is generally positive, it is anisotropic and varies significantly with crank-angle.

   Assuming axisymmetry, it is also possible to estimate the magnitude of the various sources of turbulence production from the data of Dimopoulos and Boulouchos. These estimates indicate that the radial gradients in the swirl velocity $\langle U_\theta \rangle$, coupled with the r-θ plane stress, do not provide a significant enhancement to the turbulent kinetic energy k—energy in the radial fluctuations is removed at approximately the same rate that energy is added to the tangential fluctuations. Flow deformation in the r-z plane, dominated by $\frac{\partial \langle U_r \rangle}{\partial z}$, is the largest net source of turbulence production by the anisotropic stresses. Like the Foster and Witze study discussed above, this source is associated with tumble motion within the cylinder and enhances the radial component energy. Interestingly, for this pancake-shaped chamber, production via the isotropic stresses is the dominant source of k, and appears to be approximately two to three times the size of the turbulence production associated with the mean tumble motion at the time of peak k (-20 CAD). Recent results in the author's laboratory, discussed below, show a similar dominance of the isotropic production terms in low-swirl, bowl-in-piston engines. It must be noted, however, that rotation of the tumble vortex around the cylinder axis by the flow swirl could be subject to cycle-to-cycle fluctuations that lead to underestimation of $\frac{\partial \langle U_r \rangle}{\partial z}$ and, hence, underestimation of the tumble induced turbulence production.

   Auriemma and co-workers [11] present measurements of $\langle u'_r u'_\theta \rangle$ and the corresponding radial and tangential fluctuations near the lip of a re-entrant, bowl-in-piston combustion chamber with swirl. During the latter part of the compression stroke the component fluctuations are approximately equal, but they begin to diverge and rapidly increase beginning at about -15 CAD. Shortly before TDC, the average fluctuation level has increased by over 50%, and the tangential fluctuations exceed the radial fluctuations by approximately 20%. Fansler and French [35] observed very similar behavior, though the increase in fluctuations and the degree of anisotropy was larger.

Based on the available measurements it is possible to generate a rational—albeit incomplete—picture of the dominant turbulence generation processes near the lip. The shear stress $\langle u'_r u'_\theta \rangle$ measured in [11], which was small prior to –15 CAD, shows an attendant increase to levels corresponding to a correlation coefficient of 0.3 by TDC. Based on complementary measurements reported in [27], the radial gradient in $\langle U_\theta \rangle$ is positive at this time. According to Eq. (4.11), the action of the increased $\langle u'_r u'_\theta \rangle$ (changing signs to account for the different coordinate system used in [11]) will be to produce tangential fluctuations while decreasing the radial fluctuations. Thus, it is probable that the increased $\langle u'_r u'_\theta \rangle$, in conjunction $\langle U_\theta \rangle$ and its radial gradient, is a major contributor to the observed anisotropy in the tangential and radial fluctuations. However, like the study of Dimopoulos and Boulouchos [32] described above, estimates of $\langle U_\theta \rangle$ and $\frac{\partial \langle r U_\theta \rangle}{\partial r}$ suggest that the net influence on k of the turbulent fluctuations produced by $\langle u'_r u'_\theta \rangle$ is considerably smaller than the influence on the individual component energies. Based on the anticipated large values of $\frac{\partial \langle U_\theta \rangle}{\partial z}$ associated with the 'spinning-up' of fluid displaced by the squish process, $\langle u'_\theta u'_z \rangle \frac{\partial \langle U_\theta \rangle}{\partial z}$ can also be expected to be an important source of tangential fluctuations and increased k. The observed sensitivity of near-lip turbulence to R_s, cited above, further indicates the probable importance of turbulence production by these radial and axial gradients in $\langle U_\theta \rangle$.

Fansler and French [35] have identified $\langle u'_r u'_z \rangle \frac{\partial \langle U_r \rangle}{\partial z}$ as another potentially important production term near the lip. This term will enhance the radial fluctuations. Note, however, that this enhancement may be masked by the reduction in the radial fluctuations associated with $\langle u'_r u'_\theta \rangle$ discussed above. Another potential source of k is production during the compression process by the isotropic stresses. This source, however, generally peaks near –20 CAD and is thus an unlikely contributor to the enhanced fluctuations in the vicinity of the lip near TDC. Overall, the picture that emerges is one in which squish-generated axial gradients in $\langle U_\theta \rangle$ and $\langle U_r \rangle$ enhance both the tangential and radial fluctuations. Production terms associated with the radial distribution of $\langle U_\theta \rangle$ also likely generate turbulence energy, but mainly redistribute energy from the tangential fluctuations into the radial fluctuations, thereby enhancing the anisotropy.

Although some degree of anisotropy in the turbulent normal stresses is reported in several additional studies employing bowl-in-piston geometries (e.g. [9, 13, 67, 80]), the measurements are generally insufficient and the degree of anisotropy too slight to draw well-founded conclusions. Even in pancake-chambers, the available experimental results are often conflicting. Late in the compression stroke, some studies exhibit lower axial fluctuations [32, 37], others report higher fluctuations [36], and still others indicate almost no anisotropy [5, 7, 16] or show location dependent behavior [31]. Evidently, additional detailed measurements that characterize potential sources of anisotropy associated with the turbulence production terms will be required if these complex flows are to be better understood. A first step in this direction, in which the anisotropy of the normal stresses is correlated with the dominant production terms [72], is described briefly below.

4.3.3.4 Length Scales

There are two broad areas in which improved knowledge of the turbulent length scale can improve our understanding and ability to model turbulent engine flows. Firstly, turbulent viscosity based stress models can be viewed as mixing length models. That is, the viscosity can be expressed as a product of a characteristic turbulent velocity scale and a characteristic length scale over which the turbulence can effectively transport momentum. This length scale can be expressed in terms of a Lagrangian integral scale[2], which, in turn, can be related to Eulerian integral scales [93]. Evaluation of Eulerian length scales is thus central to the assessment of various turbulent stress models and the benefits that might be achieved by adopting more complex formulations. Examples of more complex viscosity based stress models include those with non-isotropic viscosities or with strain-dependent coefficients.

Secondly, some specification of a turbulent length or time scale is required to connect the large-scale, energetic turbulent motions with the small-scale motions that dissipate the turbulent energy. Typically, this specification takes the form of Eq. (4.21), which assumes that the dissipation of turbulence energy is determined by the rate $\left(u'/\ell \right)$ at which turbulence energy $\left(u'^2 \right)$ is provided from the larger scales. Thus, evaluation of length scales and their evolution allows the examination, albeit qualitatively, of the turbulent dissipation rate. Furthermore, analysis based on spectral similarity considerations [93] indicates that when ℓ is defined by Eq. (4.21), it is directly proportional to the Lagrangian integral scale which arises in the mixing length considerations discussed in the preceding paragraph.

The foregoing discussion is intended not only to motivate the study of turbulent length scales, but also to provide some perspective on their interpretation. Turbulent length scales are often judged to be a measure of a distance over which a flow remains self-correlated. While this viewpoint is certainly valid, the additional perspective offered by noting the relationship of the integral scales to the momentum transfer (stress modeling) and to the dissipation rate adds physical understanding. For example, increased turbulent dissipation will act not only to dissipate turbulence directly, but, all other factors being equal, will also reduce production through a diminished mixing length. Further, this additional perspective indicates that some common data reduction practices are likely to produce results which no longer reflect the physical significance of the integral scale in the momentum transport process. Determination of integral scales through integration of a positive envelope delimiting the magnitude of the measured correlation functions is one example of such a practice. Although an improved estimate of the length over which a turbulent flow is self-correlated may be obtained, the result is unlikely to correctly reflect the

---

[2] Integral time or length scales are determined by integration of temporal or spatial correlation functions, e.g. [51]

distance over which the fluctuating velocities can transport momentum. Similarly, as noted in Sect. 4.3.3, high-pass filtering of velocity data can significantly impede the ability of the flow to transport momentum, and integral scales computed from such data are typically small.

In a general turbulent flow many different spatial correlations can be defined, depending on the orientation of both the spatial separation vector as well as the velocity components considered. A corresponding number of different integral length scales exist. For isotropic turbulence, however, there is a significant simplification: only two distinct length scales ℓ_f and ℓ_g exist—based on the longitudinal and lateral velocity autocorrelation functions $f(r)$ and $g(r)$, respectively. The longitudinal correlation corresponds to fluctuating velocities parallel to the separation vector, while the lateral correlation corresponds to perpendicular fluctuations. Note that in an isotropic flow there can be no dependency on the orientation of the separation vector and the correlation between orthogonal fluctuating velocities must, like the shear stress, be zero. Furthermore, ℓ_f and ℓ_g are not independent, but $\ell_f = 2\ell_g$.

Preferably, integral length scales are obtained directly using spatial correlation functions calculated from multi-point experimental data, such as is provided by particle image velocimetry (PIV) or multiple probe volume laser doppler anemometry (LDA) techniques. Spatial correlations obtained by rapidly traversing a measuring instrument through the flow can also provide accurate scale estimates—provided the traverse time is small as compared to the time scales of the flow. 'Flying' hot-wire anemometry (HWA) or scanning LDA are examples of this latter method. Most commonly, however, integral length scales are estimated from single point experimental data using Taylor's hypothesis, which relates ℓ to the integral time scale τ via

$$\ell = \langle U \rangle \, \tau \qquad (4.24)$$

The validity of Taylor's hypothesis rests on the assumption that the characteristic time scale of a turbulent eddy is much larger than the time required for that eddy to be convected through the measurement location by the mean flow, or equivalently, $u' << \langle U \rangle$. In many engine flows the fluctuating velocity can often exceed the mean flow velocity and Eq. (4.24) cannot be employed. However, for flows with a strong swirl or squish component Eq. (4.24) is often approximately valid—and scales estimated via Taylor's hypothesis prove to be reasonably accurate estimates of the integral length scale. Support for this statement is provided by the following observations:

1. Little variation of estimated length scales is observed even when large changes in the mean velocity occur [72].
2. Comparison of directly measured integral length scales from multi-point data with estimates obtained via Taylor's hypothesis exhibit reasonable correspondence in both magnitude and evolutionary trends [97][3].

---

[3] The cited comparison relies on the assumption of isotropy and thus does not solely assess Taylor's hypothesis.

3. Similar values of integral length scales are obtained in numerous experimental
   studies—some employing direct methods and others employing Taylor's
   hypothesis—as demonstrated in Table 4.1.

Table 4.1 presents an overview of published length scale measurements. Additional
results can be found in the literature, but have not been included if non-standard
data reduction practices were employed or if the application of Taylor's hypothesis
was deemed inappropriate. The length scales reported are reasonably consistent,
and show little variation with the specific combustion chamber geometry employed.
Generally, lateral length scales ℓ_g measured near TDC are in the range of 1–3 mm,
while longitudinal scales ℓ_f range from roughly 2–8 mm. In studies that report
both scales, the ratio ℓ_f/ℓ_g is often approximately 2, indicating that approximate
isotropy of the length scales may frequently exist. Near equal longitudinal scales
measured for two different separation vector orientations further support this view
[63]. However, the lateral scales reported in [39] demonstrate that near-TDC length
scales measured for different velocity components (but the same separation vector
orientation) can vary by as much as a factor of 2. To date the evidence suggests that,
for pancake chambers, isotropy holds for separation vectors and velocity components lying in the plane perpendicular to the cylinder axis, while scales involving
velocity components aligned with the cylinder axis may differ considerably.

Most of the studies reporting the temporal evolution of the length scale [38, 39,
47, 53, 72, 97] indicate that it exhibits a broad minimum near TDC, although this
minimum may not be apparent if measurements are not conducted sufficiently far
into the expansion stroke [39, 55, 57]. The temporal evolution, as well as the magnitude of the length scale, has been shown to be reasonably insensitive to engine
speed [12, 39, 53, 57, 97], a conclusion which is also collaborated by unpublished
measurements in the author's laboratory.

There are few reported results regarding spatial variations in the length scale.
At low swirl, the length scale is found to be approximately uniform radially [53].
This behavior is expected given the dominance of turbulence production by bulk
compression (which is uniform throughout the combustion chamber) identified
below in Sect. 4.4.3. However, with a modest swirl ratio of 2.1, large radial variations in ℓ are observed, such that ℓ is largest near the cylinder centerline for the
pancake chamber employed. Similar trends can be deduced from the correlation
functions presented in [44], though the variation appears to be less pronounced.
Axial variations in ℓ, measured in both pancake and bowl-in-piston chambers are
modest [53, 72].

The aforementioned change in the spatial behavior of ℓ as the swirl level is
increased is surprising, given the relative insensitivity to swirl of the near-TDC
velocity fluctuations in pancake chambers. However, changes in the magnitude
and/or evolution of ℓ with swirl have also been observed by others in pancake
chamber geometries [41, 55, 57]. Similarly, marked variations in ℓ with swirl level
have been observed in bowl-in-piston engines [72].

Finally, it should be noted that while integral scales are often comparable to the
largest scales found in a flow, and while physical flow boundaries clearly limit the

Table 4.1. Length scale measurements in the literature

Chamber Geometry	Method	Crank Angle	ℓ_f [mm]	ℓ_g [mm]	Swirl Ratio	Reference	Comments
Pancake	Taylor's Hyp. (HWA)	−22.5	≈6[a]	–	≈3	Lancaster [57]	ℓ_g initially increases during compression, then decreases as TDC is approached
Pancake	Direct (Multipoint HWA)	TDC	2.7 3.0	–	≈0 ≈1 (est.)	Kido, et al. [55]	Evolution of ℓ_f with swirl similar to the evolution of ℓ_g measured in [57]
Pancake	Taylor's Hyp. (LDA)	TDC	≈4	–	3.4 (near TDC)	Dimopoulos & Boulouchos [32]	–
Pancake	Direct (Laser Homodyne)	TDC	–	1.1–3.2[b]	0 & 2.1	Ikegami, et al. [53]	Minimum near TDC for $R_s = 0$. Little speed dependency, modest clearance height dependency. Large radial inhomogeneity for $R_s = 2.1$
Pancake	Direct (Multipoint LDA)	TDC	–	0.9–2.4[c]	≈4	Fraser & Bracco [38], [39]	Multiple lateral scales measured, exhibiting different clearance height dependency and evolutionary trends. Modest anisotropy/inhomogeneity
Pancake	Direct (Scanning LDA)	−30° 30°	6.9 8.3[d]	3.4 3.8[d]	3.6	Glover, et al. [44]	–
Pancake	Direct (Flying HWA)	≈−30°	6.0	–	–	Collins, et al. [24]	–
Pancake	Direct(PIV)	TDC	≈8 ≈6	–	0.7 5.2	Funk, et al. [41]	–
Pancake(Square)	Direct(PIV)	TDC	8.0–8.4	–	≈0	Marc, et al. [63]	ℓ_f measured for two separation vector locations. Results indicate approximate isotropy

Table 4.1. (continued)

Chamber Geometry	Method	Crank Angle	ℓ_f [mm]	ℓ_g [mm]	Swirl Ratio	Reference	Comments
Wedge	Taylor's Hyp. (HWA)	TDC	–	2.0	Unknown $\langle U \rangle \approx 10u'$	Dent & Salama [29]	–
Pent-roof	Direct(Scanning LDA)	TDC	≈6	≈3	≈0	Hadded & Denbratt [47]	Modest scale dependency on tumble ratio. ℓ_f and ℓ_g exhibit minimum near TDC
Pent-roof	Direct(PIV)	−60°	≈9	≈7	1.8	Li, et al. [58]	ℓ_f and ℓ_g measured at multiple locations and two separation vector orientations. Moderate anisotropy and inhomogeneity
Bowl-in-Piston	Direct(Multi-point LDA)	TDC	3.2	1.1	≈4 (near TDC)	Valentino et al. [10],[97]	Broad minimum near TDC
Bowl-in-Piston	Taylor's Hyp. (LDA)	TDC	6.2−9.9	–	4.7 (est. TDC)	Ball & Pettifer [12]	Range in scales due to modest speed dependence
Bowl-in-Piston	Taylor's Hyp. (LDA)	TDC	2.2−4.0	0.8−2.5	1.5−3.5	Miles et al. [70],[72]	ℓ_g decreases with increasing R_s. Modest spatial variations. ℓ_f and ℓ_g increase during expansion

[a] Average over all tests. Differs from summary in [38] due to choice of mean velocity when applying Taylor's hypothesis.
[b] ℓ_g is measured. Scales reported as ℓ_f are computed as $\ell_f = 2\ell_g$.
[c] Length scales listed are "fluctuation" scales in Fraser & Bracco's terminology.
[d] Corrected for cycle-to-cycle fluctuations.

integral scale, there is no clear reason that the integral scale must be proportional to instantaneous engine geometry (such as clearance height), provided that the physical dimensions are sufficiently large. Generally, studies reporting the evolution of the integral scale suggest that it is not simply proportional to the instantaneous clearance height. Fraser and Bracco [39] have measured several integral scales near TDC. They find that while some scales are independent of clearance height, the integral scale computed from axial velocity fluctuations is not.

4.3.4 Engine Speed Scaling

The near-linear variation of the duration of the combustion process with engine speed in internal combustion engines is the single most important factor enabling their operation over a wide speed range. This engine speed scaling of the combustion duration is due to turbulent transport processes; consequently, the variation of the turbulent flow structure with engine speed has been the subject of a large number of studies. Bopp et al. [16] provide a summary of work performed in this area through 1986. Here, attention is focused more specifically on studies conducted in bowl-in-piston engine geometries representative of direct-injection diesel engines.

4.3.4.1 Mean Flow Structures

Interest in the variation in the mean flow structures with engine speed is warranted due to their role in bulk transport of momentum and species, as well as in production of turbulent fluctuations. Non-dimensionalizing the Reynolds-averaged Navier-Stokes (RANS) equations by defining appropriate length, time, and velocity scales, one sees that a change in engine speed impacts the relative importance of only the viscous diffusion terms. In a turbulent flow, these terms are negligible in comparison to the turbulent diffusion, and no significant effect of engine speed on the mean in-cylinder flow structures is expected, provided the flow boundary and initial conditions are the same. With a single exception [64], consideration of the impact of engine speed-related changes in the flow boundary conditions—and, consequently, on the in-cylinder flow structure—is absent from the experimental literature. Despite this neglect, these speed-related changes, which are caused by the influence of pressure waves and the ram effect in the intake system, appear to be the key to reconciliation of the disparate results reported in the literature. In considering the effects of intake manifold dynamics the simplifying assumption of a spatially uniform density made in Sect. 4.2 clearly does not apply.

Although the majority of studies report a near-linear scaling of the mean flow velocities with engine speed [13, 16, 19, 27, 65, 106], non-linear behavior is often observed—chiefly in those studies that report spatial profiles of the mean velocities. Monaghan and Pettifer [74] performed a detailed, early study of speed effects on the in-cylinder flow structure. Significant differences in both the radial and axial profiles of the swirl velocity (normalized by S_p) were observed at the end of induction as engine speed was varied. Similar results are reported by Arcoumanis et al. [8]. In contrast, the IVC flow structure measured in an engine in which the influence of the intake system was minimized [64] shows almost no variation with engine speed.

Induction generated differences in mean flow structure are found to survive compression, resulting in a significant speed dependence of the near-TDC flow structure [64, 74, 104, 106]. Perhaps most surprising are the large differences observed in the squish-driven radial flows reported by Yianneskis et al. [104] at −30 CAD. Apparently, flow patterns dictated by piston motion and geometry (i.e., squish) are not always sufficiently dominant to overcome mean flow differences induced by other factors. Like the behavior seen in the IVC flow structure, however, near-TDC flow structures measured when the influence of the ram effect and pressure waves in the intake system have been minimized show very little sensitivity to engine speed [16, 64]. Unpublished measurements obtained in the author's laboratory further support this finding, as do multi-dimensional, compressible flow simulations of this engine. Examples of these simulations, based on full-induction stroke calculations, are shown in Fig. 4.9—and exhibit nearly indistinguishable mean flow structures (and turbulent velocity distributions) when velocity is normalized by S_p. These simulations were conducted using a k-ε turbulence model.

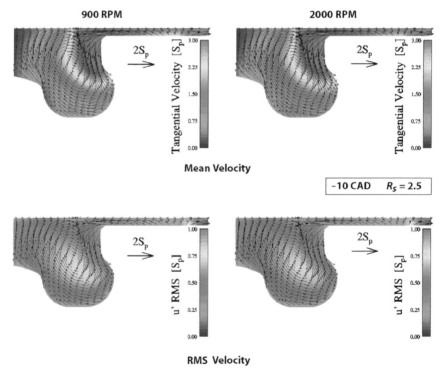

Fig. 4.9 Mean and rms simulated velocity fields, exhibiting the insensitivity of the numerical results to engine speed

Finally, Margary et al. [64] demonstrate that the magnitude of the fluctuations in mass flow rate and pressure in the inlet port increases with engine speed. Consequently, departures from a linear speed scaling can be expected to increase with increased engine speed. Such increased departures from linear speed-scaling at high engine speeds has been observed by Corcione and Valentino [27], providing additional support for the suggestion that intake effects are primarily responsible for speed-related differences in flow structure.

In production multi-cylinder engines the ram effect and pressure wave dynamics in the intake system are intentionally exploited to tailor the volumetric efficiency over the engine speed range. It is unlikely that these effects can be perfectly matched from cylinder-to-cylinder. Thus, there is likely to be both a speed dependency and cylinder-to-cylinder variations in the mean flow structure. A rough estimate of the magnitude of these flow structure differences can likely be obtained from the differences observed in the above referenced studies as engine speed is changed.

4.3.4.2 Turbulent Fluctuations

Unlike the mean flow, the effects of viscosity on the turbulent stresses cannot be neglected on a theoretical basis, due to the importance of the viscous dissipation in balancing the RANS equations governing the evolution of these stresses. Consequently, one might expect changes in engine speed (with an associated change in flow Reynolds number) to exert an influence on the magnitude of the normalized turbulent fluctuations. However, because the large-scale turbulent eddies must have a time-scale comparable to the mean flow deformation, and because—like the mean flow structures—the large-scale motions are constrained near TDC by the combustion chamber geometry, a near-linear engine speed scaling of the large-scale turbulent velocities is unavoidable. Like canonical laboratory flows, the effect of increased Reynolds number is principally to change the smaller scales of the flow, leaving the large-scale structure unchanged. Since the large-scale motions are the dominant contributors to measured turbulent fluctuations, a near-linear scaling of these fluctuations must be expected.

Measurements of fluctuating velocities in bowl-in-piston diesel engine generally support this linear speed scaling [13, 19, 27, 106], which is observed even at high engine speeds when relatively poor scaling of the mean velocities is observed [27]. However, deviation from perfect linearity can be found. Bopp et al. [16] provide an excellent pictorial summary of early attempts to demonstrate the linearity of the speed scaling of the turbulent fluctuations. Furthermore, their measurements suggest a slightly stronger increase in the turbulent fluctuations than a linear scaling would admit. Additional evidence of a stronger increase is reported by Yianneskis [104]. Rask and Saxena [80], however, report fluctuating velocities that increase more slowly than engine speed. Overall, the departures from linearity are relatively small, and the experimental evidence does not support the adoption of a scaling law that is other than linear.

In closing, it should be noted that because there is no inherent distinction between normal stresses and shear stresses, a linear scaling of the RMS fluctuations with engine speed (or S_p) implies that the shear stresses will scale with S_p^2. Approximate scaling of the shear stress with S_p^2 has been demonstrated in pancake chambers by Foster and Witze [37] and by Dimopoulous and Boulouchos [31]. For bowl-in-piston geometries, this approximate scaling can also be inferred from the data reported by Auriemma et al. [11].

4.3.5 Studies with Fuel Injection and Combustion

Prior to the work described below, measurement of the flow velocity with fuel injection and combustion was confined to only two earlier studies. In the first, Wigley and co-workers [100] obtained measurements in a modified production engine with a non-reentrant bowl. This work showed that the occurrence of fuel injection and combustion could result in a significant increase in the tangential velocity fluctuations, to levels approaching $\frac{u'}{S_p} \approx 1.8$. More importantly, however, the mean tangential velocity distribution was modified by the injection/combustion event. This was the first measurement of the ability of the fuel injection event to re-distribute the angular momentum within the cylinder—a process that is central to the development of post-injection flow structures. These observations were reinforced in the second study [90], wherein fuel injection and combustion were again observed to a) increase the post-injection turbulence fluctuations to levels of $\frac{u'}{S_p} \approx 2 - 3$, and b) modify the mean tangential velocity.

4.4 Recent Progress

The latter half of this exposition is focused on recent progress made toward clarifying the turbulent flow structure and its evolution through measurements made at Sandia National Laboratories and through numerical simulations performed at the Engine Research Center of the University of Wisconsin at Madison. Portions of this work have been discussed above in conjunction with the review of recent literature—e.g., the comparison of the mean flow development with model predictions. However, there are three aspects of this work that represent a significant departure from past work, and which we focus on in this section:

First, the components of the stress tensor in the horizontal plane, both normal and shear stresses, are measured along with the gradients of the mean velocities. Accordingly, the production of turbulent stresses associated with horizontal plane flow deformation is evaluated directly. Measurement of mean velocity gradients in the axial direction further enables estimation of additional turbulence production terms in Eq. (4.11). The measurement results largely substantiate the predictions of the numerical simulations, which are then employed to evaluate the likely significance of any unmeasured terms. Thus, the bulk flow structures and mechanisms primarily responsible for the generation of turbulent kinetic energy are clearly identified.

Second, the experimental work is extended to include engine operation with fuel injection and combustion, and the resulting flow fields are probed with sufficient detail to allow the effects of the injection and combustion processes on the flow structure to be clarified.

Third, additional characterization of the turbulence structure by estimation of Eulerian length scales permits a semi-quantitative evaluation of the turbulent stress modeling embodied in Eqs. (4.18) and (4.22). Furthermore, the evolution of the turbulent length scale can be employed to assess qualitatively the modeling of the effect of fuel injection and combustion on the dissipation of turbulent kinetic energy.

4.4.1 Engine Geometry

The engine employed in this work has characteristics typical of engines intended for light-duty automotive applications. These include: an axisymmetric, reentrant bowl; a central, vertical common-rail fuel injector; 4-valves; and a displacement of $422\,cm^3$ in a single-cylinder configuration. The bowl geometry is shown in the inset to Fig. 4.10. Note that the realistic bowl geometry was not compromised by the optical access requirements. The coordinate system employed throughout this exposition is also defined in Fig. 4.10.

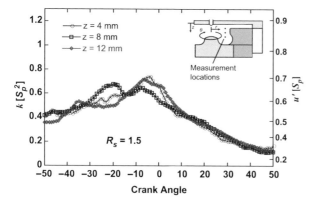

Fig. 4.10 Turbulent kinetic energy measured at three axial locations for a swirl ratio $R_s = 1.5$

All experimental velocity data were obtained with a two-component laser Doppler velocimetry system. Details of the system and of the data acquisition and reduction procedures can be found elsewhere [68, 72].

Variable levels of flow swirl are generated through a pair of helical ports which can be independently throttled. The swirl ratios quoted herein are based on steady-state flow bench testing and are derived using the Ricardo method [74]. Measurement of the swirl velocity radial profiles in the operating engine yield swirl ratios of 2.1 [67] and 3.5 for port throttle settings yielding Ricardo swirl ratios of 2.5 and 3.5, corroborating the flow bench results.

4.4.2 Numerical Methodology

Modeling of the in-cylinder flow processes was performed in two discrete phases due to the complex geometry of the intake ports. First, the induction stroke was calculated using the STAR-CD code, employing a detailed mesh of the cylinder, valve, ports, and runner geometries. At IVC, the solution was mapped to a block-structured sector grid, and the calculations proceeded using the KIVA-3 V code [2]. The KIVA computations were performed using a $60°$ sector mesh which contained approximately 42,000 cells. Flow turbulence was modeled using a variant of the RNG k-ε model adapted for engine flows [48]. Additional details are available in references [14] and [72].

4.4.3 Pre-TDC Turbulent Flow Structure Evolution

As is apparent from the discussion of Sect. 4.3, there are numerous studies in the literature characterizing the magnitude of the turbulent velocity fluctuations in bowl-in-piston engines, far fewer studies which characterize the shear or anisotropic stresses, and only a single previous study [35] in which the probable source of increased turbulence energy was identified. Eq. (4.8) indicates that the change of turbulent kinetic energy at any particular location is governed by production, viscous dissipation, and transport by the turbulent fluctuations and the mean flow. Of these factors, production and mean flow transport are anticipated to be the dominant contributors to rapid changes in the turbulence energy. Further, for the axisymmetric flows existing prior to injection, mean flow transport is limited to convection by the r-z plane flow structures—the importance of which can often be estimated from knowledge of the mean flow streamlines and the spatial distribution of turbulence energy within the bowl. Accordingly, the magnitude of the various turbulence production terms identified in Eq. (4.11), when coupled with the measured anisotropy in the normal stresses, can frequently identify the source of changes observed in the turbulence energy.

To identify the dominant sources of turbulence energy, and to clarify the flow physics responsible for the trends summarized above as bowl geometry or swirl ratio are varied, the local production terms and the normal stress anisotropy have been measured and predicted numerically for various swirl ratios and locations within the bowl, and are described in this section. All of the results reported correspond to motored engine operation at a speed of 1500 rpm.

4.4.3.1 Low Swirl Ratio

A surprising result, given the reentrant bowl geometry and the anticipated significant production of turbulence energy by the squish flow, is the spatial homogeneity of the turbulence observed at the lowest swirl ratio investigated, $R_s = 1.5$. Figure 4.10 presents the turbulent kinetic energy, estimated from $\left\langle u'^2_r \right\rangle$ and $\left\langle u'^2_\theta \right\rangle$, measured at three locations within the bowl as a function of crank angle. The right-hand axis provides the corresponding characteristic turbulence velocity, $u' = \sqrt{2k/3}$. The cause of this spatial homogeneity is most easily ascertained by examining the major turbulence production terms, which are shown for the $z = 8$ mm location in Fig. 4.11.

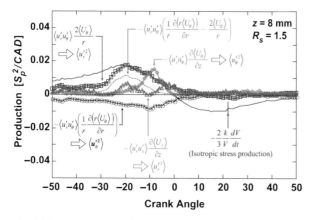

Fig. 4.11 Relevant production terms measured at $z=8$ mm. The shear stresses multiplying the axial direction mean velocity gradients were not measured directly, but estimated via Eq.(4.2)

Although the production of $\langle u_r'^2 \rangle$ and $\langle u_\theta'^2 \rangle$ [see Eq. (4.11)] by the shear stress in the r-θ plane can be significant, the two sources largely balance one another, leading to little net production of k. That is,

$$\frac{1}{r}\frac{\partial \left(r \langle U_\theta \rangle \right)}{\partial r} \approx \frac{2 \langle U_\theta \rangle}{r} \tag{4.25}$$

over much of the cycle, and the net production is only significant for a brief period. This result is consistent with estimates which can be derived from the measurements reported in [11]. The implication of Eq. (4.25) is that locally the flow has not departed appreciably from solid-body rotation.

A second surprising result is the relatively minor turbulence production at this location associated with the axial gradient of the mean squish velocity, $\frac{\partial \langle U_r \rangle}{\partial z}$. Production from this source exhibits only small peaks near -20 CAD and near TDC. The former peak is due to rapid deceleration of the inward squish velocity as the squish flow enters the bowl, and the latter is due to a large rotational mean flow structure in the r-z plane, and is thus likely counter-balanced by negative production associated with $\frac{\partial \langle U_z \rangle}{\partial r}$.

The most significant production associated with the anisotropic stresses is due to $\frac{\partial \langle U_\theta \rangle}{\partial z}$, as might be expected from the near-TDC distribution of the tangential velocity seen in Fig. 4.4. However, this is the dominant source of turbulence production for only a very brief period near -10 CAD.

Overall, the production due to the isotropic part of the normal stresses, or bulk compression, is the dominant source of turbulence energy. A similar observation was made earlier from analysis of the pancake-chamber measurements of Dimopoulos and Boulouchos [32]. Production from this source is also dominant at the $z=4$ and $z=12$ mm measurement locations. Accordingly, it is not surprising that the spatial distribution of k is approximately homogeneous: the production by the isotropic stresses does not depend on the details of the mean flow structure. If the initial distribution of k is near homogeneous—as was observed in Fig. 4.2—we expect that homogeneity will be approximately maintained, provided that production by the anisotropic stresses is small.

Although the net production of k by the isotropic stresses is not dependent on the details of the flow, the amount of energy directed into the individual fluctuating velocity components will be proportional to the corresponding normal component of the mean rate-of-strain tensor, as dictated by Eq. (4.12). Because the isotropic production is the dominant source of turbulence energy, we expect that it may also exert a dominant influence on the anisotropy on the individual normal stresses. Figure 4.12, which shows the normal stresses in the upper portion, and the components of the net isotropic production in the lower, largely confirms this expectation. Although the correlation is imperfect, the inequity in the individual normal stresses generally parallels the relative magnitude of the components of the isotropic production. In particular, the sharp decrease in the production of $\left\langle u_r'^2 \right\rangle$ near -25 CAD is mirrored by a decrease in the measured $\left\langle u_r'^2 \right\rangle$ at this time, despite the concurrent positive production of $\left\langle u_r'^2 \right\rangle$ by $2\left\langle u_r' u_\theta' \right\rangle \left\langle \frac{U_\theta}{r} \right\rangle$ seen in Fig. 4.11. Similar correlations are also seen at the remaining measurement locations [72].

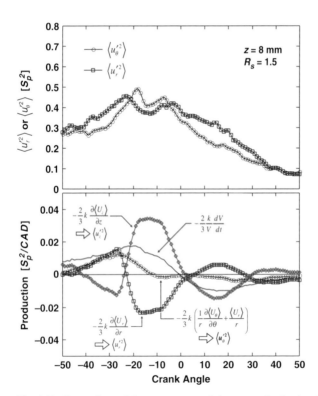

Fig. 4.12 Comparison of the components of the net production by the isotropic stresses with the anisotropy in the measured normal stresses

Fig. 4.13 Simulated mean flow field illustrating the mean flow structures giving rise to the turbulence production terms shown in Figs. 4.11 and 4.12. $R_s = 1.5$

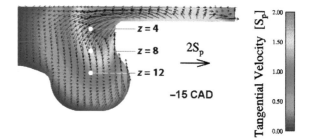

Axial fluctuations $\langle u'^2_z \rangle$ were not measured. However, Fig. 4.12 indicates that for a brief period prior to TDC large production of $\langle u'^2_z \rangle$ by the isotropic normal stresses exists, coincident with the decreased production of $\langle u'^2_r \rangle$ discussed above. The simulated mean flow field at this time, shown in Fig. 4.13 for -15 CAD, illustrates the physical cause of this behavior. At the $z = 8$ mm measurement location, the squish flow trajectory that produced the modest $\frac{\partial \langle U_r \rangle}{\partial z}$ also produces a positive $\frac{\partial \langle U_r \rangle}{\partial r}$. This has the effect of stretching vortices aligned with the radial direction, leading to increased $\langle u'^2_z \rangle$ and $\langle u'^2_\theta \rangle$ relative to $\langle u'^2_r \rangle$. Simultaneously, the rising piston motion opposes the descent of the squish flow, and the fluid is compressed axially. Due to this axial compression, $\langle u'^2_r \rangle$ and $\langle u'^2_\theta \rangle$ are decreased relative to $\langle u'^2_z \rangle$. The net result is a relative decrease in $\langle u'^2_r \rangle$, an increase in $\langle u'^2_z \rangle$, and little change in $\langle u'^2_\theta \rangle$.

The opposite situation occurs at $z = 4$ mm. At this location the decelerating squish flow is compressing the fluid radially, while the downward bending streamlines result in a net expansion in the axial direction. Here we expect that radial fluctuations will be enhanced and the axial fluctuations will be reduced.

4.4.3.2 Higher Swirl Ratios

At the higher swirl ratios, significantly less spatial homogeneity is seen as production by the anisotropic stresses becomes dominant at some spatial locations. Figure 4.14 illustrates that, in the upper region of the bowl near the lip, a significant increase in k is seen near -10 CAD for $R_s = 2.5$, in agreement with numerous previous studies cited above. A similar increase is not seen at locations lower in the bowl. The individual radial and tangential component energies, and the dominant sources of turbulence production measured at the upper location are shown in Fig. 4.15. As was seen at the lower swirl ratio, production by the isotropic stresses is again the dominant source of k until nearly -20 CAD. In this early period, the only other term of significance is $-\langle u'_z u'_\theta \rangle \, \partial \langle U_\theta \rangle / \partial z$, which is the probable source of the higher level the tangential component energy seen in the upper portion of Fig. 4.15 prior to -25 CAD.

Beyond -20 CAD, turbulence production associated with the radial distribution of the swirl velocity (or angular momentum) is clearly dominant. Although

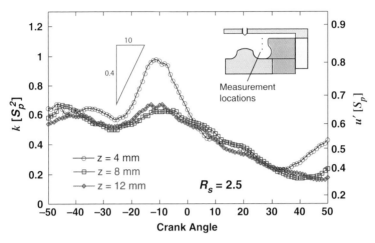

Fig. 4.14 Turbulent kinetic energy measured at three axial locations for $R_s = 2.5$

production of $\left\langle u'^2_\theta \right\rangle$ continues to be countered by negative production of $\left\langle u'^2_r \right\rangle$, the local flow structure has clearly departed from a solid-body-like form and the net production associated with the r-θ plane shear stress is large. Note that the peak magnitude of this net r-θ plane production, approximately 0.04 S_p^2/CAD, corresponds closely to the measured dk/dt seen in Fig. 4.14. Furthermore, it is clear that production by these sources is the cause of the normal stress anisotropy seen in the upper portion of Fig. 4.15 at this time.

The evolution of the mean flow structure at this time is shown in Fig. 4.16 with a sequence of numerically predicted flow fields. Although some minor differences exist between the numerical predictions and the measurements, the simulation results serve to clearly identify the relevant physical processes. Complementary, experimentally determined components of the mean horizontal plane strain rate $\langle S_{r\theta} \rangle$ are shown in Fig. 4.17. Jointly, these figures illustrate the mechanism responsible for the increased r-θ plane production.

Prior to -20 CAD, as the squish flow transports high angular momentum fluid from the squish volume inward, the mean tangential velocity at the $z = 4$ mm measurement location increases, as does the radial gradient in angular momentum. Both quantities increase proportionally, however, such that there is little local departure from a solid-body-like flow structure and $\langle S_{r\theta} \rangle$ remains near zero. The peak tangential velocity occurs at approximately -20 CAD, after which the high momentum fluid begins to turn more sharply down into the bowl and the clockwise-rotating r-z plane flow structure begins to form. This structure subsequently transports low momentum from the central regions of the bowl outward, resulting in a decreased swirl velocity at the measurement location—as seen in Fig. 4.17. The outward transport of low momentum fluid is opposed, however, by continued transport of high momentum fluid inward by the squish flow. These opposing transport processes result in increased radial gradients in angular momentum, even while the local swirl velocity is decreasing. Accordingly, the local flow structure exhibits a significant departure from a solid-body-like form, and the mean flow deformation in

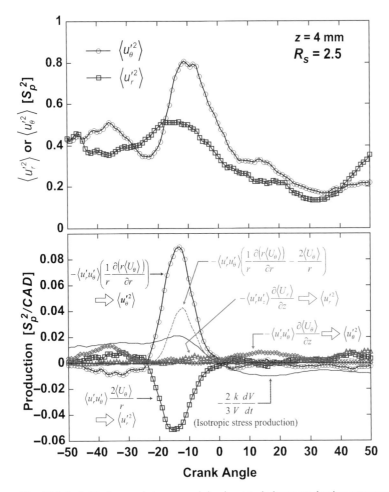

Fig. 4.15 Individual normal stresses and dominant turbulence production terms measured in the upper region of the combustion chamber

the r-θ plane increases significantly—as shown by the behavior of $\langle S_{r\theta}\rangle^4$. Concurrently, for approximately equal normal stresses, Eq. (4.15) or (4.16) suggests that $\langle u_r' u_\theta'\rangle \propto -\langle S_{r\theta}\rangle$. The shear stress shown in the lower portion of Fig. 4.17 supports this expectation, although some deviations from this form can be expected due to the finite axial gradients in $\langle U_r\rangle$ and $\langle U_\theta\rangle$, as well as the net radial compression $\langle S_{rr}\rangle$. Together, the increased $\langle S_{r\theta}\rangle$ and $|\langle u_r' u_\theta'\rangle|$ produce the increased r-θ plane production.

In contrast, the production due to mean shear associated with the squish flow $-\langle u_r' u_z'\rangle \frac{\partial \langle U_r\rangle}{\partial z}$ is estimated to be negligible, as shown in Fig. 4.15. The results of the

---

[4] These opposing flows also lead to a net radial compression of fluid elements at the interface, with resulting generation of $\langle u_r'^2\rangle$ by the isotropic normal stresses [72].

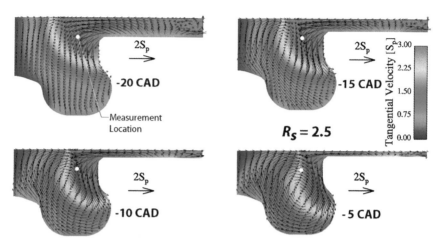

Fig. 4.16 Numerically simulated mean flow fields illustrating the evolution of the flow structures and swirl velocity distribution leading to enhanced r-θ plane turbulence production

numerical simulation at this location further substantiate this assertion. Accordingly, the 'squish-generated' turbulence seen here is dominated by squish-induced changes in the tangential velocity distribution, not the squish velocity $\langle U_r \rangle$ or its gradients directly. This latter point bears further examination, particularly as production associated with $\frac{\partial \langle U_r \rangle}{\partial z}$ is commonly thought to be the dominant source of squish generated turbulence. At the time of the peak squish flow near -10 CAD, the average

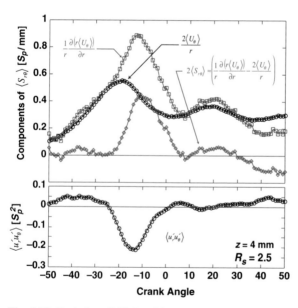

Fig. 4.17 Evolution of $\langle S_{r\theta} \rangle$ and its component terms, with the corresponding r-θ plane shear stress

squish velocity at the bowl rim exceeds $6\,S_p$. Surely large axial gradients in the squish velocity—and attendant production of turbulence—must exist where this flow passes over the bowl lip.

To explore this further, $\frac{\partial \langle U_r \rangle}{\partial z}$ and the turbulence production $-\langle u_r' u_z' \rangle \frac{\partial \langle U_r \rangle}{\partial z}$ predicted in the numerical simulations were examined at a radial location 1 mm from the bowl lip at axial locations of $z = 1$, 2, 3, and 4 mm [81]. In all cases, $\frac{\partial \langle U_r \rangle}{\partial z}$ reached a maximum level of about $0.6\,S_p$/mm, resulting in brief periods (≈ 10 CAD) during which turbulence production rates achieved levels of roughly $0.02\,S_p^2$/CAD. These simulation results thus suggest that, in general, squish generated turbulence is dominated by production associated with the radial gradients of $\langle U_\theta \rangle$. However, axial gradients in the squish velocity do appear to be responsible for the increased turbulence energy seen late in the expansion stroke in Fig. 4.14. Similar late-cycle increases in turbulence have been observed by Christensen and Johansson [22].

At still higher swirl ratios, the fluctuations near the bowl lip prior to TDC are increased further, and the anisotropy in the normal stresses also increases, such that the energy in the tangential fluctuations can exceed twice the energy in the radial fluctuations. The physics of the flow structure development is the same as was seen at $R_s = 2.5$, however. Of greater interest is the sharp, pre-TDC increase in k that is now also seen deep within the bowl, as shown in Fig. 4.18.

The dominant production terms and the individual component energies measured at the lowest location within the bowl, $z = 12$ mm, are shown in Fig. 4.19. At first glance, the production terms of Fig. 4.19 look similar to those seen in Fig. 4.15. Production of $\langle u_r'^2 \rangle$ and $\langle u_\theta'^2 \rangle$ by r-θ plane flow deformation is dominant in the period approaching TDC, with production of one component largely offsetting production of the other. However, in this case, the production of $\langle u_r'^2 \rangle$ is positive, while the production of $\langle u_\theta'^2 \rangle$ is negative. Accordingly, Eq. (4.11) implies that the shear stress $\langle u_r' u_\theta' \rangle$ must be positive, and Eq. (4.16) suggests that $\langle S_{r\theta} \rangle$ is likely negative. These

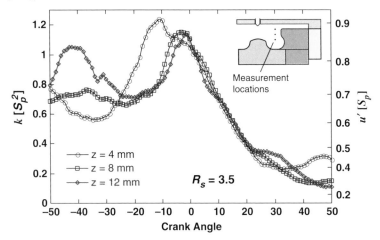

Fig. 4.18 Turbulent kinetic energy measured at three axial locations for $R_s = 3.5$

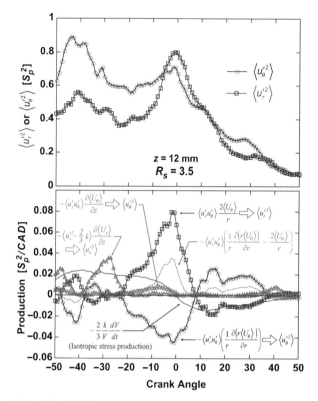

Fig. 4.19 Individual normal stresses and dominant turbulence production terms measured at $R_s = 3.5$

expectations are supported by Fig. 4.20, which shows the evolution of $\langle u_r' u_\theta' \rangle$ as well as $\langle S_{r\theta} \rangle$ and its components.

In the period during which significant r-θ plane production is seen (between about -10 CAD and TDC), Fig. 4.20 indicates that the angular momentum gradient is near a local minimum, while the tangential velocity is increasing rapidly. The flow structure responsible for this behavior is shown in Fig. 4.21 for a crank angle of −5 CAD. At this higher swirl ratio, high angular momentum fluid has been transported along the bowl periphery to the bowl floor by the clockwise-rotating r-z plane flow structure. The increasing $\langle U_\theta \rangle$ beyond −25 CAD is evidence of this high momentum fluid passing through the measurement location. As this fluid continues to be transported inward across the bowl floor, its angular momentum is approximately conserved ($\langle U_\theta \rangle$ increases), resulting in a minimal radial gradient in angular momentum $r\langle U_\theta \rangle$. The minimum in the angular momentum gradient, coupled with the increasing $\langle U_\theta \rangle$, results in the negative $\langle S_{r\theta} \rangle$ seen in Fig. 4.20, with a corresponding positive peak in the shear stress $\langle u_r' u_\theta' \rangle$. Unlike the evolution of $\langle S_{r\theta} \rangle$ seen in Fig. 4.17, at this location and swirl ratio the flow exhibits departures from solid-body-like behavior (non-zero $\langle S_{r\theta} \rangle$) at almost all crank angles. However, the shear stress does not always respond in kind, and the net r-θ plane turbulence production

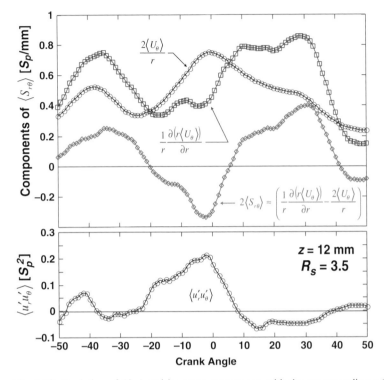

Fig. 4.20 Evolution of $\langle S_{r\theta} \rangle$ and its component terms, with the corresponding r-θ plane shear stress at $R_s = 3.5$

is not always large. A more detailed examination of the behavior and modeling of the shear stress will be provided below.

Also shown in Fig. 4.19 is the production of k associated with $\frac{\partial \langle U_\theta \rangle}{\partial z}$, which provides modest production of $\langle u_\theta'^2 \rangle$ just prior to TDC, and production of $\langle u_z'^2 \rangle$ estimated from $\frac{\partial \langle U_z \rangle}{\partial z}$. The latter source, as well as concurrent large production of

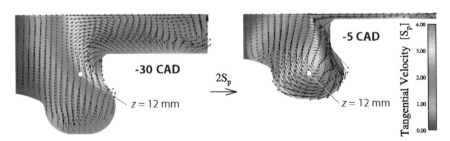

Fig. 4.21 Simulated flow fields illustrating the large axial flow compression $(\partial \langle U_z \rangle / \partial z)$ at the $z = 12$ mm measurement location occurring near -30 CAD, and the distribution of the swirl velocity leading to turbulence production deep in the bowl just prior to TDC. $R_s = 3.5$

$\langle u'^2_z \rangle$ by the isotropic portion of the normal stresses, can be expected to generate large axial component fluctuations near –30 CAD. As discussed in the context of Fig. 4.13, this is due to the opposing effects of the descending squish flow and the rising piston motion, giving rise to compressive strain in the axial direction. At $R_s = 3.5$ this effect is more pronounced, however, as can be seen in the predicted flow field at –30 CAD shown in Fig. 4.21. The axial compression results in large positive $\langle S_{rr} \rangle$ as well as $\langle S_{\theta\theta} \rangle$ (through the term $\frac{\langle U_r \rangle}{r}$). As a result, production of $\langle u'^2_r \rangle$ and $\langle u'^2_\theta \rangle$ by the isotropic normal stresses is negative, and is the probable cause of the decreasing component energies seen at this time in Fig. 4.19.

As evidenced from the above discussion, many of the features of the turbulence energy evolution can be explained through evaluation of the individual production terms in Eq. (4.11). This approach is not entirely adequate, however. For example, there is no obvious source among the various production terms of the increased $\langle u'^2_\theta \rangle$ seen in the upper portion of Fig. 4.19 near –40 CAD. As the flow inhomogeneity increases with increasing swirl, however, mean flow transport of turbulence generated elsewhere in the combustion chamber increasingly influences the evolution of the turbulent stresses—and is the probable source of this behavior.

A final observation regarding the production of turbulent energy at the higher swirl ratios is relevant: examination of Figs. 4.15 and 4.19 indicates that, despite the increased importance of turbulence production by the anisotropic stresses, the dominant source of k integrated over the latter 50 CAD of the compression stroke is production by the isotropic normal stresses. Similarly, while production by the anisotropic stresses is clearly influencing the homogeneity and individual component energies, many aspects of the component energy evolution still appear to be dominated by production associated with the isotropic stresses. Furthermore, during the expansion stroke, the attenuation of the turbulence energy by the isotropic stresses generally dominates over the production by the anisotropic stresses.

Other than the isotropic normal stresses, production associated with the radial distribution of the swirl velocity has been shown to be the most important source of turbulence energy and normal stress anisotropy, followed by production associated with $\frac{\partial \langle U_\theta \rangle}{\partial z}$. Overall, the detailed measurements of production associated with the r-θ plane stresses has substantiated the picture of turbulence production near the bowl lip that can be inferred from previous measurements in the literature, and highlighted additional production mechanisms that occur deeper in the bowl. The swirl velocity and its spatial distribution are perhaps the most easily influenced flow variables that significantly impact the turbulence levels within the bowl.

4.4.4 Operation with Fuel Injection and Combustion

Fuel injection exerts a profound influence on the in-cylinder mean flow structures. These structures not only influence the spatial distribution of fuel within the cylinder, but also impact the post-injection turbulence energy and its spatial distribution dramatically. Subsequently, combustion is also expected to generate additional flow structure changes. In the material presented below, experimental and numerical studies have been performed with fuel injection and combustion, as well as with fuel injection alone. Unless otherwise noted, these results have been obtained at a simulated idle condition of 900 rpm with a fuel load of approximately 6 mg per injection.

4.4.4.1 Mean Flow Structure: The Spray/Swirl Interaction

The interaction of the swirling flow with the injected fuel sprays is closely analogous to the squish-swirl interaction described above [71], as it involves a competition between radial momentum induced by the fuel spray and centrifugal forces that increase as high-angular momentum fluid elements are transported inward. The inward transport of this fluid is accomplished through two principal mechanisms: First, the fuel jets entrain nearly quiescent, low angular momentum fluid from the central portion of the cylinder—which is subsequently carried to the bowl periphery by the jets. Second, the fuel jets are a large source of radial flow momentum, which is transferred to the in-cylinder fluid via entrainment and turbulent diffusion. Deflection of the fuel jets and accompanying fluid by the walls of the combustion chamber results in energetic rotating flow structures in the r-z plane. Both displacement by the low-momentum fluid transported from the center of the cylinder to the bowl periphery and convection by the rotating r-z plane structures act to re-distribute the angular momentum within the bowl, generally displacing high-momentum fluid inward in a manner very similar to the inward transport of high-$\langle U_\theta \rangle$ fluid by the squish flow.

To illustrate this process, post-injection flow fields determined via numerical simulation are shown in Fig. 4.22 at 12° ATDC. For this operating condition, fuel injection begins just prior to TDC and ends at approximately 2.5° ATDC. No combustion was permitted to take place in these simulations. Flow structures predicted with combustion are similar, however. The figure depicts the flow structures in an r-z plane that approximately bisects the angle between two fuel jets. The numerical results indicate, though, that fuel-jet-induced azimuthal asymmetry has substantially disappeared by this crank angle. At the lowest swirl ratio shown, $R_s = 1.5$, the flow structure in the bowl consists of a single r-z plane vortex, similar to that seen in Fig. 4.4 for $R_s = 2.5$, which has been formed by the deflection of the fuel jet at the bowl wall. High-$\langle U_\theta \rangle$ fluid, displaced from the bowl periphery, has been transported inward along the bowl floor as far as permitted by the bowl pip. The pip deflects the flow upward, but the fluid continues to follow an r-z plane trajectory that is directed radially inward. This is consistent with the relatively low centrifugal forces acting on the fluid, which do little to impede its inward penetration. As the swirl ratio is increased, the high tangential velocity fluid at the bottom of the bowl still penetrates inward to the base of the pip. Upon being deflected upward, however, it attempts to return to the bowl periphery more rapidly due to the higher centrifugal forces. Consequently, at $R_s = 2.5$, a double-vortex structure reminiscent of that seen in Fig. 4.4 for $R_s = 3.5$ has been generated in the r-z plane. At $R_s = 3.5$, the centrifugal forces deflect the high-$\langle U_\theta \rangle$ fluid outward so rapidly that the once

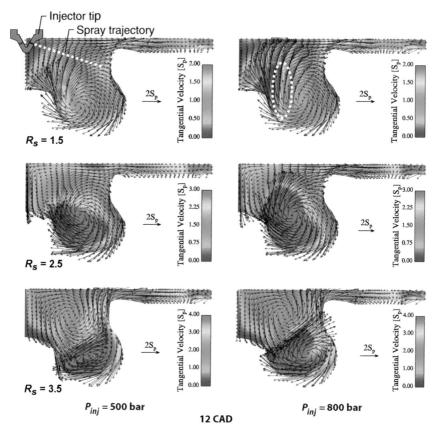

Fig. 4.22 Mean velocity fields illustrating the interaction between the fuel spray and the flow swirl. The injector position and the fuel spray trajectory are shown for reference in the upper left image. At the time of injection (near TDC) the piston is positioned approximately 1.3 mm higher in the cylinder than at the crank angle shown above

dominant clockwise-rotating vortex is now confined to the lower portion of the bowl, and the counter-clockwise vortex in the upper central bowl now dominates.

With increasing injection pressure, the momentum of the fuel and entrained air is increased, as is the momentum transfer to the air at the edges of the spray. Consequently, the flow structures developed in the r-z plane are more energetic, and the balance between radial momentum and centrifugal forces acting on high-$\langle U_\theta \rangle$ fluid changes. As discussed above, at a fixed injection pressure the high-$\langle U_\theta \rangle$ fluid leaving the bottom of the bowl bends outward more rapidly at the higher swirl ratios. With higher injection pressures, however, the more energetic flow structures transport the high-$\langle U_\theta \rangle$ fluid further inward and upward before centrifugal forces turn it toward the bowl periphery. At the higher swirl ratios, when a double-vortex structure is formed, the injection pressure thus affects the size and relative strength of the two vortices. Higher injection pressures enhance the outer, lower vortex due to the increased inward/upward flow penetration. With lower injection pressures, however, this vortex is confined deeper within the bowl.

It is important to recognize that the r-z plane flow structures and spatial distribution of high-$\langle U_\theta \rangle$ fluid seen in Fig. 4.22 are long-lived. At $R_s = 1.5$ the flow continues to be dominated by a large clock-wise rotating vortex and the double-vortex structure formed at $R_s = 2.5$ and $R_s = 3.5$ is still evident well into the expansion stroke [71]. Overall, it is evident that the balance between injection pressure and swirl—or injection imparted momentum and centrifugal forces—is the dominant factor influencing the post-injection bulk flow structures.

Although additional simulations and experiments have not yet been carried out to clarify the effects of varying speed and load on the spray/swirl interaction, the general trends to be expected are clear from our understanding of the physics of the interaction. If engine speed is increased, the r-z plane flow structure and the relative swirl velocity distribution in the absence of fuel injection remain substantially unchanged, due to the invariance of the flow with engine speed discussed in Sect. 4.3.4. However, the centrifugal forces on fluid elements displaced inward by the spray-induced flow structures will be larger, with the net effect being similar to an increase in swirl ratio or a decrease in injection pressure. That is, the outer, lower r-z plane vortex will be reduced in size. Conversely, for higher loads—at fixed engine speed—we expect that the spray induced structures will be more energetic and will result in enhanced outer, lower vortical structures, similar to those seen with increased injection pressure in Fig. 4.22.

The analogy between the squish/swirl interaction and the spray/swirl interaction continues to hold when the kinetic energy of the mean flow is considered. Just as the mean flow rotational kinetic energy increases as the rising piston forces the swirling flow into the bowl, so the fuel injection process must also increase the rotational kinetic energy of the mean flow. This energy increase is due to work performed on the high angular momentum fluid elements in the process of displacing them inward against the centrifugal forces. In this instance, the source of the increased rotational energy is the kinetic energy of the fuel spray. Thus, a mechanism exists wherein some fraction of the energy of the injection event can be stored in the mean flow, and potentially released later as turbulent kinetic energy.

Simple arguments [71] suggest that the kinetic energy increase could be substantial if a significant fraction of the low momentum fluid from the central region of the cylinder is displaced to the bowl periphery. The results of numerical simulations, however, indicate that, at the idle condition considered, the increase in the mean rotational kinetic energy within the bowl is approximately 10%. Further work needs to be performed to quantify the magnitude of this effect at higher loads.

An assessment of the effects of combustion on the mean flow structures described above is in order. Accordingly, Fig. 4.23 depicts the mean flow fields predicted both with and without combustion for $R_s = 2.5$ and an injection pressure of 600 bar. Although some differences in the details of the flow structures clearly exist, the main features remain unchanged. In particular, the double vortex structure in the r-z plane is formed in much the same manner, and high-angular momentum fluid is transported inward and upward in a similar fashion for both cases. As will be seen below, the spatial distribution of the high angular momentum fluid and the double vortex r-z plane structure are important flow features that markedly influence the generation of turbulence in the expansion stroke.

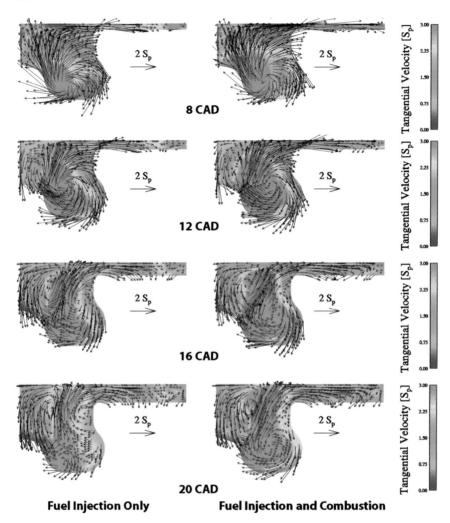

Fig. 4.23 Simulated flow fields illustrating the parallel development of the mean flow fields both with and without combustion. In the combusting case, the peak heat release occurs at 5 CAD

To illustrate the development of the post-injection mean flow structures in this section, the results of numerical simulations have been employed exclusively. The accuracy of these predicted mean velocity fields has been demonstrated through comparison of the axial profiles of mean radial and tangential velocities for fired engine operation [67]. The influence of the fuel injection event on the r-z plane flow structures and on the axial distribution of $\langle U_\theta \rangle$ is faithfully captured well into the expansion stroke. Furthermore, radial profiles of $\langle U_\theta \rangle$ confirm the transport of high angular momentum fluid to the inner portions of the bowl, resulting in negative radial gradients in $\langle U_\theta \rangle$ (and in angular momentum $r \langle U_\theta \rangle$).

Additionally, detailed comparisons of measured and simulated turbulent flow quantities have been made for operation with fuel injection, but without combustion

[70]. The simulation results replicate trends in the mean flow r-θ plane strain rates, turbulent kinetic energy, and turbulence production by r-θ plane flow deformation as both swirl ratio and injection pressure are varied. Overall, experimental validation of the simulation results provides strong support for the accuracy of the gross behavior of the predicted flow structures and their evolution with swirl ratio and injection pressure.

4.4.4.2 Turbulent Kinetic Energy

In agreement with the two earlier studies in fired engines cited in Sect. 4.3.5, fuel injection and combustion are found to exert a large influence on the turbulent kinetic energy within the combustion chamber. Moreover, the changes in turbulent kinetic energy are found to be strongly dependent on swirl ratio, injection pressure, and measurement location. The evolution of the turbulent kinetic energy, measured with

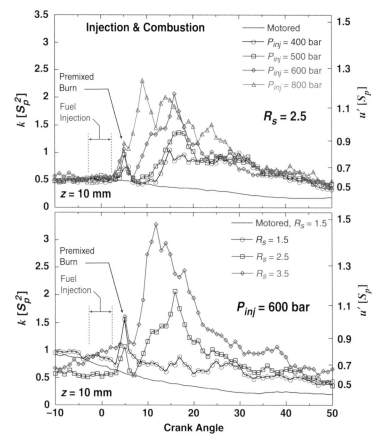

Fig. 4.24 Measured evolution of turbulent kinetic energy for fired engine operation, estimated from $\langle u_r'^2 \rangle$ and $\langle u_\theta'^2 \rangle$

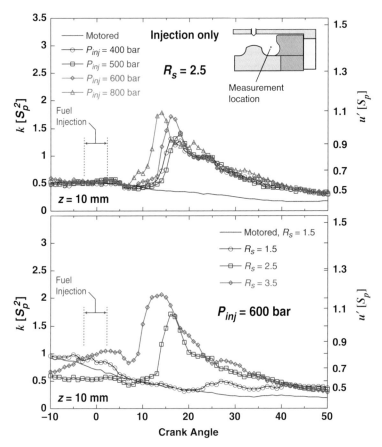

Fig. 4.25 Evolution of turbulent kinetic energy measured with combustion inhibited (fuel injection only), for the same operating condition as the fired measurements of Fig. 4.24

the engine firing, is shown in Fig. 4.24. These measurements were made in a vertical plane approximately midway between two adjacent sprays, at a radius of 13.6 mm and an axial location 10 mm below the head (see Fig. 4.25).

The upper portion of the figure examines the effect of injection pressure at fixed swirl ratio. At the measurement location considered, there is no immediate influence of the fuel injection event on the turbulent fluctuations. Later, near 5 CAD, a brief increase in turbulence is observed, after which the turbulence energy drops back to a level characteristic of the motored fluctuations. It is noteworthy that the timing and duration of this brief increase corresponds closely to the timing and duration of the premixed burn portion of the combustion event. After a short, injection-pressure-dependent delay, the measured turbulence energy again increases and is maintained at an elevated level until well into the expansion stroke. At the lowest injection pressure, this elevated 'late-cycle' turbulence achieves an energy level more than double that of the motored flow. A five-fold increase in energy is observed at the highest injection pressure.

The lower portion of Fig. 4.24 presents the variation in turbulence energy observed as the swirl ratio is varied at fixed injection pressure. In this instance, the most notable feature is that at the lowest swirl ratio, $R_s = 1.5$, no significant increase in late-cycle turbulence energy is observed, although the decay is slower than is seen for motored engine operation. At the higher swirl ratios, the late-cycle increase in k is larger, but occurs at the same time.

In this chapter the position that the turbulence is well represented by the ensemble, phase-averaged statistics of the measurements has been adopted. For fired engine operation, however, it is important to establish that cycle-to-cycle fluctuations in the fuel injection event or in the subsequent combustion do not result in a false indication of increased turbulence. To examine this issue, a cycle-resolved analysis of the fired engine data has been performed [69]. In this analysis, the turbulent fluctuations were evaluated using only the high-frequency portion of the measured velocity records. Two limiting frequencies were considered: the first reduced the turbulent kinetic energy to approximately 25% of the full-bandwidth energy, and the second resulted in a further 50% drop in energy. In both cases, a relative increase in the late-cycle turbulence comparable to that seen in the full bandwidth results was observed. Furthermore, the brief peak near 5 CAD was preserved in the cycle-resolved results, indicating that this peak is also associated with small-scale, high-frequency turbulence, not simply cycle-to-cycle fluctuations in the phasing or location of the premixed combustion event.

The potential sources of the increased turbulent kinetic energy observed with fuel injection and combustion will be addressed in detail in the following section. However, at this point it is germane to examine the behavior of the turbulent kinetic energy under the same conditions as employed in Fig. 4.24, but with combustion inhibited. The corresponding results are shown in Fig. 4.25. Although the impact of fuel injection alone on the turbulence energy is less than the combined impact of fuel injection and combustion, the overall trends observed in both cases are remarkably similar: At fixed swirl ratio, increasing injection pressure results in a larger, more advanced increase in k, while at fixed injection pressure an increase in swirl ratio results in a larger increase in k. A notable difference does exist between Figs. 4.24 and 4.25, however. In the results obtained without combustion, the brief peak in k at 5 CAD is absent. Accordingly, this peak can be identified as turbulence generated by rapid gas expansion during the premixed burn. It is not caused directly by the fuel injection process or the flow structures it creates.

Overall, the similarity in the behavior of the late-cycle turbulent kinetic energy for fuel injection and combustion and for fuel injection alone suggests that the dominant mechanism by which the late-cycle turbulence is generated can be identified from measurements made with combustion inhibited. Without combustion, the complicating influence of rapid gas expansion is eliminated, as is the impact of large density variations on turbulence production. In this context, note that an order of magnitude analysis [69] suggests that buoyant production associated with density inhomogeneities in the strong, swirl-generated radial pressure gradient can be at least as large as turbulence production by the r-θ plane shear stresses.

4.4.4.3 Turbulence Production Mechanisms

When fuel injection occurs, high local levels of turbulence are generated by large velocity gradients in the vicinity of the sprays. Although this direct, spray-generated turbulence is expected to decay rapidly, the surviving turbulence can be transported to the measurement volume by the mean flow, suggesting the presence of a local source of turbulence production. Distinguishing surviving spray-generated turbulence from additional late-cycle turbulence generated by local flow structures is difficult. The central problem is that regions of intense spray-generated turbulence are spatially coincident with regions of high angular momentum fluid, which, as will be seen below, are essential to the formation of turbulence generating flow structures. Various approaches to separate these sources based on the gross anisotropy of the fluctuations, convective-flight times, and modeling studies have been proposed and examined [69]. However, the most convincing method is a close examination of the phasing, magnitude, and anisotropy of the normal stresses and correlation of these parameters with measured local production rates—as was done previously to identify the pre-TDC turbulence sources.

Figure 4.26 presents the individual component energies and turbulence production by the r-θ plane shear stress measured for $R_s = 2.5$ and $P_{inj} = 600$ bar. The general behavior of the normal stresses, a larger increase in $\left\langle u_r'^2 \right\rangle$ than in $\left\langle u_\theta'^2 \right\rangle$, is seen for all injection pressures at this swirl ratio. Production by the r-θ plane shear stress correlates well with the increasing component energies. However, the larger increase in the radial component energy is not supported by the approximately equal magnitude of the production of $\left\langle u_r'^2 \right\rangle$ and $\left\langle u_\theta'^2 \right\rangle$, suggesting the presence of an additional source of $\left\langle u_r'^2 \right\rangle$. $\langle S_{rr} \rangle$ exhibits a sharp minimum at this time [70], resulting in significant production of $\left\langle u_r'^2 \right\rangle$ by the isotropic normal stresses, which may explain this behavior.

Another feature of note in Fig. 4.26 is the sustained production of the radial fluctuations from 20–25 CAD. Concurrently, production of the tangential fluctuations has approached zero. This behavior is mirrored in the evolution of the component energies in the upper portion of the figure, wherein the radial energy reaches a plateau while the tangential energy continues to decrease. The close correspondence between the production and the component energies suggests that a significant fraction of the turbulence energy is associated with this production mechanism.

The peak magnitude of the net production by the r-θ plane shear stress, however, appears to contradict this suggestion. In Fig. 4.25, a peak dk/dt of $0.4 - 0.5$ S_p^2/CAD is seen, while the r-θ plane production shown in Fig. 4.26 only attains a value of approximately 0.1 S_p^2/CAD. The measurements indicate, however, that steep radial gradients in the production exist. A conservative estimate, at a location 1 mm inward, results in a production rate over twice as great—due to large radial gradients in both the angular momentum gradient and in $\left\langle u_r' u_\theta' \right\rangle$. At this time, the mean radial velocity is outwardly directed, such that turbulence produced at these inward locations will be convected past the measurement location. Thus, the

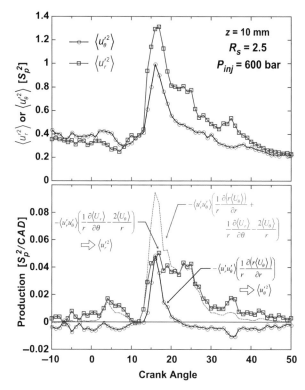

Fig. 4.26 Measured component energies and r-θ plane turbulence production for operation with fuel injection, but without combustion

magnitude of the r-θ plane production is largely in accord with the temporal evolution of k, although other production sources may also contribute.

Unlike the production of k by the r-θ plane shear stress observed prior to TDC, in this instance the production of $\langle u_r'^2 \rangle$ and $\langle u_\theta'^2 \rangle$ do not counteract each other, but are both positive during the period of greatest production, from about 13 to 20 CAD. Inspection of Eq. (4.11) indicates that, for modest departures from axisymmetry, this will only occur when the mean radial gradient in angular momentum is negative. The measured mean angular momentum gradient, shown in Fig. 4.27 with the other components of $\langle S_{r\theta} \rangle$, confirms this expectation.

Figure 4.28 illustrates the simulated flow fields that create the negative angular momentum gradient. At 16 CAD, when the minimum momentum gradient is measured, the lower vortex generated by the spray/swirl interaction has transported high angular momentum fluid below the measurement location to the pip, where it is deflected upwards and attempts to return to the bowl periphery. The measurement location is located near the outer edge of this high-momentum fluid, accounting for the negative radial gradient in angular momentum (and the increasingly negative momentum gradients observed at smaller radial locations). During this process, the remaining components of $\langle S_{r\theta} \rangle$ seen in Fig. 4.27 have changed little, and $\langle S_{r\theta} \rangle$

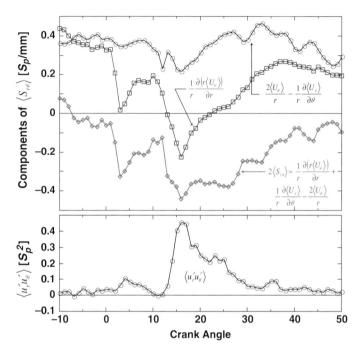

Fig. 4.27 The mean strain rate $\langle S_{r\theta} \rangle$ and the corresponding shear stress $\langle u'_r u'_\theta \rangle$, from which the r-θ plane turbulence production shown in Fig. 4.26 was obtained

reaches a local minimum. As the expansion stroke proceeds, the gradient in angular momentum at the measurement point again becomes positive, but remains small (cf. Figs. 4.17 and 4.20), and $\langle U_\theta \rangle$ changes little due to the slow radial movement of the high momentum fluid at this location. Accordingly, $\langle S_{r\theta} \rangle$ retains a significant negative magnitude and the net r-θ plane production remains important, though all of the energy is directed into the radial component fluctuations. Also shown in Fig. 4.27 is the evolution of the shear stress $\langle u'_r u'_\theta \rangle$. Clearly large magnitude shear stresses are observed concurrent with the minimum in $\langle S_{r\theta} \rangle$, as required to generate turbulence energy. The factors influencing the development of the shear stress will be discussed below.

Figure 4.28 merits one final point of discussion. Although the negative radial gradient in angular momentum is only briefly present at the measurement location, the flow structure responsible is both long-lived and spans a considerable portion of the bowl volume. As the cycle proceeds, the simulations suggest that similar negative momentum gradients (or at least negative gradients in $\langle U_\theta \rangle$) persist in the outer regions of the bowl until well into the expansion stroke, and likely represent a sustained source of increased late-cycle turbulence energy.

A similar analysis of the evolution of k, r-θ plane production, shear stresses and the angular momentum gradient at the lowest swirl ratio, $R_s = 1.5$, can be found elsewhere [68]. Note from Figs. 4.24 and 4.25, however, that no significant increase in late-cycle turbulence is observed for this swirl ratio, at the measurement location

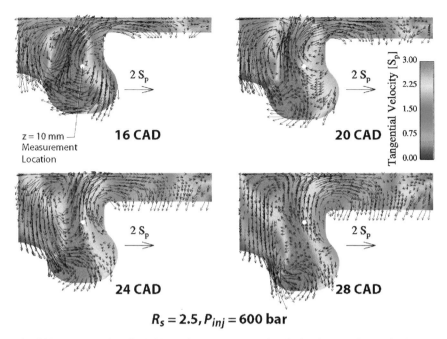

Fig. 4.28 Simulated flow fields illustrating the swirl velocity distribution associated with the negative radial gradient of angular momentum seen in Fig. 4.27

considered. Examination of the predicted flow fields in Fig. 4.22 for this swirl ratio suggests the reason for this observation: the 'high' momentum fluid is swept far in toward the bowl center, due to the smaller centrifugal forces acting on this fluid. The region of large negative angular momentum gradients, shown by the ellipse superimposed on the $P_{inj} = 800$ bar flow field, is thus found at considerably smaller radii than the measurement location. Furthermore, the mean flow streamlines are oriented such that turbulence generated at this interface is convected up toward the cylinder head, over the measurement location. Measurements of turbulent kinetic energy at an injection pressure of 600 bar, shown in Fig. 4.29, convincingly support this picture. As the radius of the measurement location is reduced, k increases by nearly an order of magnitude.

At the highest swirl ratio, $R_s = 3.5$, the largest increase in the late-cycle turbulent kinetic energy was observed both with and without combustion. The individual component energies measured in this case, with $P_{inj} = 600$ bar, are shown in Fig. 4.30. Like the results obtained at $R_s = 2.5$, the largest increase is observed in the radial component fluctuations, which attain a peak energy some 60% higher than the tangential component. The corresponding production terms associated with the r-θ plane shear stresses are shown in the lower part of the figure. As seen previously at $z = 12$ mm (Fig. 4.19), prior to TDC there is a significant net r-θ plane production, which preferentially enhances the radial fluctuations. Some small enhancement of the turbulence production after injection is observed, but the net production rapidly falls to near zero and remains small thereafter. During this time, the mean flow

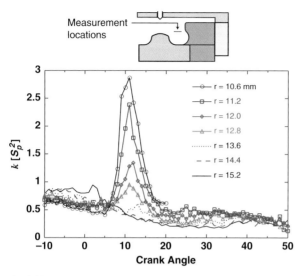

Fig. 4.29 Radial variation of the turbulent kinetic energy measured with $R_s = 1.5$ and $P_{inj} = 600$ bar

structure does not depart significantly from a solid-body-like structure, except for a brief period near 20 CAD when the shear stress $\langle u'_r u'_\theta \rangle$ is near zero.

Thus, although modest production of $\langle u'^2_r \rangle$ and $\langle u'^2_\theta \rangle$ is intermittently observed later in the cycle, the net contribution to k is insignificant. Clearly, the local production by the r-θ plane shear stress cannot account for the large observed late-cycle increase in turbulence energy. If r-θ plane production is important for this operating condition, it must occur elsewhere—and the turbulence generated convected to the measurement location by the mean swirl velocity and r-z plane flow structures. Measurements obtained closer to the fuel sprays, at several axial locations, indicate sizeable rates of net r-θ plane production (up to 0.24 S_p^2/CAD). Hence this may be a significant source of the increased k. Nevertheless, given the near 0.5 S_p^2/CAD dk/dt observed in Fig. 4.25, it is prudent to examine other potential sources of turbulence energy.

Of the production terms that can be directly estimated from the measurements, the production of $\langle u'^2_r \rangle$ by the anisotropic part of the radial normal stresses is the largest, and reaches a peak of approximately 0.04 S_p^2/CAD between 10 and 12 CAD [70]. Additionally, because $\langle S_{rr} \rangle$ is negative at this time, while $\langle S_{\theta\theta} \rangle$ and $\langle S_{zz} \rangle$ are positive, the turbulence production by the isotropic part of the normal stresses—though negative overall—is providing energy to the radial fluctuations at the expense of the tangential and axial fluctuations. These two sources undoubtedly contribute to the anisotropy in the normal stresses, but are not likely to account for the full increase observed in k.

The modest production of $\langle u'^2_r \rangle$ by the anisotropic part of the radial normal stresses is also observed in results obtained with numerical simulation—as is

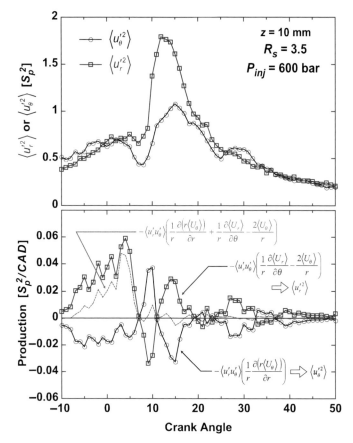

Fig. 4.30 Measured component energies and r-θ plane turbulence production for operation with fuel injection, but without combustion at $R_s = 3.5$

the more pronounced production of $\langle u_r'^2 \rangle$ associated with the isotropic stresses. More importantly, the simulation results provide insight into additional sources of k that were not accessible experimentally. In particular, substantial levels of local production associated with $\langle S_{rz} \rangle$ are predicted near 8 CAD and beyond 20 CAD, while production associated with $\langle S_{z\theta} \rangle$ is found to be significant near 13–14 CAD.

The simulated r-z plane production is shown in Fig. 4.31, along with the flow structures responsible for this production. Near 8 CAD, it is clear that the large predicted turbulence production is associated with the interface between the two counter-rotating vortices created by the spray/swirl interaction. The mean strain rate, superimposed on the figure and depicted in principal axes, shows that the lower vortex is attempting to pull a fluid element at the measurement location into the lower/outer region of the bowl, while the upper vortex attempts to bring it into the upper/inner region. The resulting high deformation rate results in the large predicted

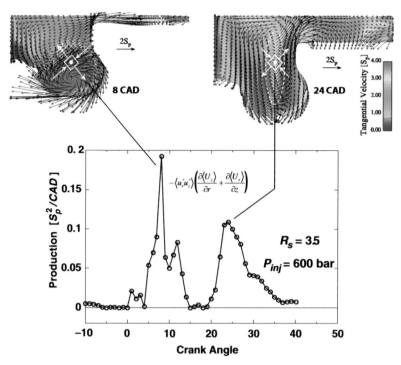

Fig. 4.31 Simulated turbulence production by the r-z plane shear stresses and the corresponding mean flow structures. The mean rate-of-strain on a fluid element at the measurement location is depicted in principal axes

turbulence production. Returning to the conventional coordinate system, the mean strain rate is dominated by $\frac{\partial \langle U_r \rangle}{\partial z}$, and thus the energy is provided predominantly into the radial fluctuations, in accordance with the behavior observed in Fig. 4.30. In the later portion of the cycle, the flow deformation corresponds to nearly pure shear (the principal axes are rotated approximately $45°$ from the r-z coordinate axes), and is dominated by $\frac{\partial \langle U_z \rangle}{\partial r}$. Accordingly, we expect to find more energetic axial component fluctuations at this time. Note that this shear region is quite large, extending over an area approximated by the dashed ellipse superimposed on the predicted flow field. Similar regions appear at other operating conditions along the inner edge of the region of high-$\langle U_\theta \rangle$ fluid, and are caused by the axial elongation as the piston descends of the inner, upper vortex formed by spray/swirl interaction. See, for example, the 24 and 28 CAD flow fields depicted in Fig. 4.28.

The z-θ plane production, and the corresponding flow structure are shown in Fig. 4.32. In this case, the production is dominated by $\frac{\partial \langle U_\theta \rangle}{\partial z}$. Still greater gradients in $\langle U_\theta \rangle$ clearly exist near the bowl pip, in the left-hand portion of the dashed rectangle superimposed on the figure. As the lower, outer vortex generated by the spray/swirl interaction is transporting high angular momentum fluid inward and upward, the fluid spins-up, increasing its tangential velocity. Simultaneously, the

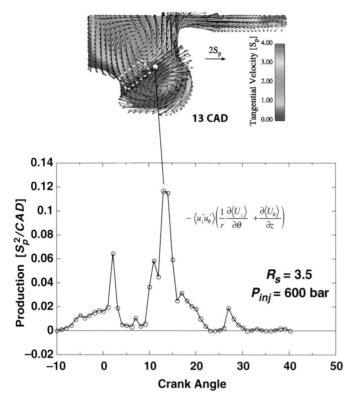

Fig. 4.32 Simulated turbulence production by the z-θ plane shear stresses. Regions of still higher shear are found within the dashed rectangle superimposed on the figure

upper, inner vortex is transporting low angular momentum downward and outward. This fluid spins-down, decreasing its tangential velocity. At the interface between the two vortices, in addition to the r-z plane deformation described above, the low and high angular momentum fluid meets. Accordingly, large radial and axial gradients in $\langle U_\theta \rangle$ are formed, with expected large turbulence production rates. Note that the mean flow streamlines in the r-z plane suggest that turbulence generated along this interface will subsequently be transported through the measurement location. Hence, a considerable fraction of the increased k seen at this swirl ratio may be due to additional production mechanisms that exist at the vortical interface.

Although measurements and calculations obtained without combustion have been the focus of this discussion, very similar results are obtained with combustion, albeit with increased statistical scatter due to the necessity of skip-firing the optical engine and, consequently, acquisition of data over fewer cycles. As seen with fuel injection alone, the late-cycle radial component fluctuations invariably exceed the tangential fluctuations at their peak, and further exhibit an increase that is sustained for a greater period of time. At $R_s = 2.5$, negative radial gradients in the mean flow angular momentum are observed that are coincident with this peak, along with large rates of turbulence energy production by the r-θ plane shear stresses. For higher swirl, the negative momentum gradients are no longer seen, and the late-cycle r-θ plane production at the measurement location is generally considerably smaller, in close accord with the results shown in Fig. 4.30.

4.4.5 The Role of the Bulk Flow Structures in Combustion

The discussion in the previous section has provided a detailed picture of how flow swirl and the fuel sprays interact to form large-scale, post-injection flow structures. The effect of changing swirl ratio and injection pressure has been examined, and the impact of changing speed and load follow inductively. Further, the role of these structures in generating post-TDC turbulence, and how the spatial distribution of the turbulent kinetic energy may vary, have been examined in detail. However, the complementary role of these bulk flow structures in transporting fuel and oxidant throughout the bowl has not yet been addressed.

Figure 4.33 depicts the evolution of the mean velocity field and the fuel distribution simultaneously, for a swirl ratio of 2.5 and an injection pressure of 600 bar. From the predicted fuel distributions, the bulk of the fuel is seen to remain in the lower portion of the bowl, and is spatially coincident with the high-angular momentum (high-$\langle U_\theta \rangle$) fluid. As the lower vortex transports high-momentum fluid inward and upward toward the vortical interface, the unburned fuel is also transported to the same location. Similarly, while the upper vortex transports low-momentum fluid to the interface, it simultaneously brings in fresh oxidant. Thus, a well-organized mixing and combustion system exists: this system employs bulk flow structures to not only transport fuel and air to the same region, but also to generate intense turbulence and enhance the small scale mixing at the fuel/air interface. Furthermore, a portion of the fuel will be transported along the interface to a region where large r-θ plane production, associated with negative radial gradients in angular momentum, exists.

The simulations on which Fig. 4.33 is based were performed without combustion. For late-injection, extended-ignition-delay combustion modes the main heat release is often delayed until approximately 15 CAD. Under these circumstances, the fuel distributions and flow structures depicted in Fig. 4.33 realistically depict the pre-combustion mixing process for fired engine operation as well. However, the flow structures that form after combustion are also of interest. As will be seen below, simulations performed with combustion show much the same behavior.

In light of the additional role played by the large scale structures in transporting fuel and air to a common interface, it is interesting to examine the effect of changing the relative sizes of the two vertical structures, as could be accomplished by varying injection pressure, swirl ratio, speed, or load. Accordingly, the variation in the fuel and oxidant distributions with swirl ratio at 17 CAD is shown in Fig. 4.34. In this figure, simulated flow fields obtained with combustion are presented, and the mass fraction of CO is employed as a marker for the location of partially burned fuel.

Fuel

Tangential Velocity

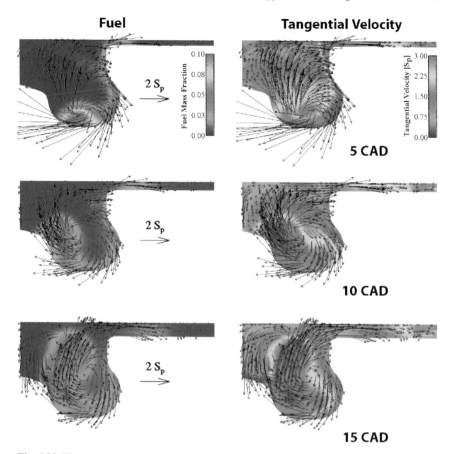

Fig. 4.33 The evolution of fuel and tangential velocity distributions at $R_s = 2.5$ and $P_{inj} = 600$ bar. In these simulations, combustion is inhibited

These simulations were performed for a load and speed of 3 bar imep (≈ 9 mg per injection) and 1500 rpm. Similar results are obtained when combustion is inhibited.

At the lowest swirl, $R_s = 1.5$, the location of the high CO regions and the r-z plane flow structures indicate that the unburned fuel will be largely confined to the inner region of the bowl. Although high local levels of turbulence will likely be found in this region (see Fig. 4.29), there is no opposing pair of vortical structures that actively bring the fuel and fresh air into close contact. A weak upper vortex forms later, but it does not encompass a significant volume of air with which to oxidize the fuel. Increasing the swirl ratio to $R_s = 2.5$ results in a considerably larger upper vortex, in which the bulk of the remaining O_2 is located. The two vortices clearly play the same role as discussed previously in the case without combustion— transporting fresh air and unburned fuel to a common interface where turbulence production can be expected to be large. One thus expects that mixing rates will be significantly enhanced at this swirl ratio. Further increasing the swirl ratio, the upper

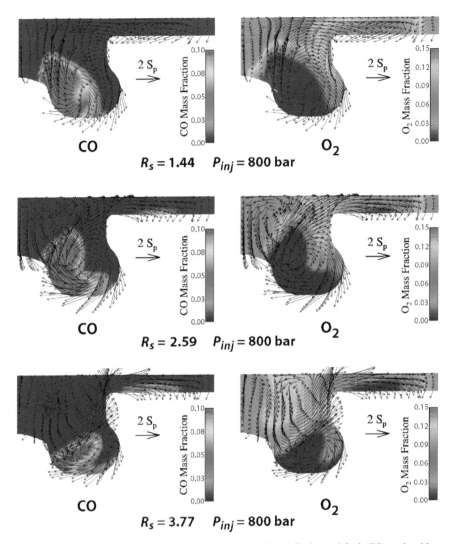

Fig. 4.34 The effect of swirl ratio on the distribution of partially burned fuel (CO) and oxidant at 17 CAD. In these simulations the premixed portion of combustion peaked at approximately 12 CAD, and these results correspond to the later, mixing-controlled burning. For these simulations, the flow was initialized with a solid-body-like mean velocity profile

vortex continues to expand at the expense of the lower, fuel-containing vortex. As the lower vortex shrinks, the fuel tends to recirculate and may become trapped deep within the bowl.

The relative sizes and positions of the two vortices generated by the spray/swirl interaction can thus play a key role in the efficacy of the mixing process, during both the early, mixture-preparation stage of combustion and in the later, mixing-controlled burning. Either too little or too much swirl appears to inhibit the mixing process. An ideal compromise might entail making the upper vortex as large as possible, while still producing mean flow streamlines at the interface between the two vortices that transport the combustible mixture out of the bowl and into the squish volume. Such a compromise would likely be similar to the $R_s = 2.59$ flow field shown in Fig. 4.34.

Finally, as discussed above, the relative sizes and positions of the two vortices will also be affected by engine speed. At higher speeds the lower vortex is expected to be smaller and lower in the bowl, in much the same manner as is seen with increased swirl ratio. The formation of an excessively small lower vortex, confining the fuel deep in the bowl, may be one factor responsible for the well-known need to reduce swirl level at high engine speeds.

4.5 Model Assessment

In the preceding sections the physical development of the in-cylinder mean flow structures has been described in detail, and the predictions of the numerical simulations have been shown to agree well with the measured mean flow properties. The simulation results have further allowed us to extract understanding about the flow development, potential turbulence production mechanisms, and the role of the bulk structures in combustion that would have been impossible based on the measurements alone. However, the accuracy of the simulations in predicting the turbulence properties is far less satisfactory. An example is given in Fig. 4.35, which illustrates the agreement achieved between the predicted and the measured turbulent kinetic energy for engine operation with fuel injection, but no combustion. Note that both during the latter part of the compression stroke and in the period of the increased late-cycle turbulence there are significant discrepancies between the measured and simulated turbulence energies. Another notable inconsistency is the spatial distribution of k predicted in the simulations. From Fig. 4.14 it is clear that, for $R_s = 2.5$, the highest pre-TDC turbulence energy is measured in the upper regions of the cylinder, near the bowl lip. Predicted results, however, exhibit the opposite trend: the highest turbulence levels are exhibited in the central region of the bowl, as seen in Fig. 4.36.

Due to the central role played by turbulence and mixing processes in diesel combustion, improvement of the turbulence modeling is a necessary prerequisite to the use of numerical simulations to design a truly optimized combustion system. RANS-based turbulence models, particularly variants of the two-equation k-ε model, are the current standard in use by industry. Although alternative modeling approaches can offer increased accuracy, or the possibility of resolving cycle-to-cycle variations in the flow (e.g., large eddy simulation techniques), the increased computational expense presents a formidable barrier to their widespread use for engine design and optimization. For that reason, in this section the fundamental assumptions inherent in the k-ε model are assessed, with a view toward improving the accuracy of these less expensive models. As will be seen below, the level of detail that can be captured using relatively simple models for the turbulent stresses in the extremely complex flows considered here is both surprising and encouraging.

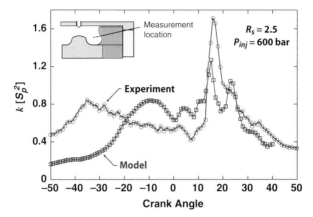

Fig. 4.35 Comparison of the measured and simulated turbulent kinetic energy. Significant differences are observed both before and after the fuel injection event

Fig. 4.36 Spatial distribution of the turbulent fluctuating velocity observed at $R_s = 2.5$. The maximum fluctuations are predicted to occur in the central region of the bowl, in contrast to the measurements reported in Fig. 4.14

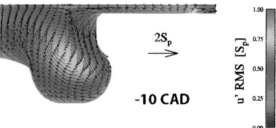

4.5.1 Turbulent Shear Stress Modeling

An evaluation of the modeling of the turbulent shear stress $\langle u'_r u'_\theta \rangle$ is provided in Fig. 4.37. In this figure, the measured stress is compared against the stress computed from Eq. (4.18), using Eqs. (4.19)–(4.21); i.e., no results from numerical simulations are employed. Results for four different operating conditions/locations are presented: the first three correspond to the three swirl ratios and spatial locations considered in the discussion of the pre-TDC flow evolution, and the last corresponds to operation with fuel injection. Recall that the evaluation of the turbulence production mechanisms for these conditions revealed four distinct flow structures:

1. $R_s = 1.5$, $z = 8$ mm: Spatially homogeneous turbulence energy, with production of k dominated by the isotropic stresses. Mean velocity gradients are generally small.
2. $R_s = 2.5$, $z = 4$ mm: Large positive radial gradient in angular momentum (greater than would be observed for solid-body-rotation) leading to large, positive $\langle S_{r\theta} \rangle$ (Fig. 4.17), coincident with significant local production of k by the r-θ plane shear stresses.

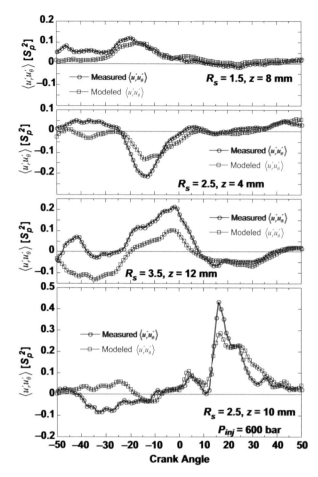

Fig. 4.37 Comparison of the measured shear stress to the modeled stress estimated from experimental data using Eqs.(4.18)-(4.21)

3. $R_s = 3.5$, $z = 12$ mm: Small positive radial gradient in angular momentum (less than would be observed for solid-body-rotation), with corresponding large, negative $\langle S_{r\theta} \rangle$ (Fig. 4.20) and significant local production of k by the r-θ plane shear stresses.

4. $R_s = 2.5$, $z = 10$ mm, $P_{inj} = 600$ bar: Negative radial gradient in angular momentum, with large negative $\langle S_{r\theta} \rangle$ (Fig. 4.27) and very large local production of k by the r-θ plane shear stresses.

For the first three operating conditions the engine was motored—hence the lack of pronounced activity during the expansion stroke. Note that although the size of the graph depicting the stress measured with fuel injection is larger, the vertical scale has been maintained the same.

There are several general observations to be made from Fig. 4.37:

- At the lowest swirl ratio, or in the expansion stroke (in the absence of fuel injection), the magnitude of the shear stress is small, and it is generally modeled with impressive accuracy by Eq. (4.18). This is consistent with the expectation of a small turbulent to mean flow time-scale ratio.
- When turbulence production by the r-θ plane shear stresses is significant, the model captures the trends in the measured stress reasonably well.
- Although the peak modeled shear stresses are typically less than the measured stresses, there is no evidence that the choice of the constant $A = 0.55$ for use in Eq. (4.21) is inappropriate. Larger values of A would cause larger errors in other portions of the cycle, e.g., prior to -20 CAD for $R_s = 3.5$ and $z = 12$ mm. Additionally, comparisons at other operating conditions or locations [72] occasionally exhibit over-predicted peak stresses.
- The evolution of the measured stress is often captured qualitatively by the model equations, even when the magnitude is incorrect (e.g., prior to TDC for $R_s = 3.5$ and $z = 12$ mm.

To obtain further insight into the possible causes for the discrepancies observed in Fig. 4.37, it is useful to examine the individual terms responsible for the production of $\langle u'_r u'_\theta \rangle$, as given in Eq. (4.16). All but the latter two of these terms can be estimated directly from measured velocities. Figure 4.38 shows the production of $\langle u'_r u'_\theta \rangle$ by the leading term of Eq. (4.16) [corresponding approximately to Eq. (4.18)], compared against the net production by the anisotropic part of the normal stresses and the production given by $-\langle u'_r u'_\theta \rangle (\langle S_{rr} \rangle + \langle S_{\theta\theta} \rangle)$. The latter term is generally near zero, and therefore is not visible at all crank angles.

In the upper plot shown in Fig. 4.38 ($R_s = 1.5$ and $z = 8$ mm), the net production by the anisotropic stresses is generally small, as would be expected from the near equal normal stresses seen in Fig. 4.12. At other locations or swirl ratios, however, production of $\langle u'_r u'_\theta \rangle$ by the anisotropic normal stresses can be significant, and rival the magnitude of the leading term in Eq. (4.16). Under some circumstances, these additional production terms are consistent with the discrepancies observed in Fig. 4.37. For example, at $R_s = 3.5$ and $z = 12$ mm, the near-zero measured stress prior to -20 CAD may be due to the approximate self-cancellation of the two dominant production terms seen in Fig. 4.38. Conversely, for other circumstances, the anticipated influence of the anisotropic production is to increase the discrepancies between the model and measurement—such as the difference observed between -20 and -10 CAD for $R_s = 2.5$ and $z = 4$ mm. Including the effects of production by the anisotropic stresses would only serve to increase this difference. Note from Fig. 4.16, however, that at this time axial gradients in both $\langle U_r \rangle$ and $\langle U_\theta \rangle$ are expected. Accordingly, production by the latter two terms in Eq. (4.16) cannot be neglected.

Overall, the conclusion to be drawn from Fig. 4.38 is that production of $\langle u'_r u'_\theta \rangle$ by the anisotropic portion of the normal stresses, and potentially by the remaining unevaluated terms in Eq. (4.16), can be significant in these flows. Including these

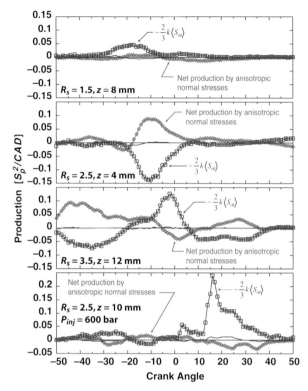

Fig. 4.38 Measured terms in the production of $\langle u'_r u'_\theta \rangle$. The solid line without symbols, not labeled in the figure, is the production by the shear stresses $-\langle u'_r u'_\theta \rangle (\langle S_{rr} \rangle + \langle S_{\theta\theta} \rangle)$

terms in the modeling of the shear stress could result in considerably enhanced model accuracy.

4.5.2 Turbulent Normal Stress Modeling

Recall that in the previous discussion of the pre-TDC flow structure evolution (Sect 4.4.3), significant anisotropy in the normal stresses was observed when production of turbulent kinetic energy by the shear stresses acting with $\langle S_{r\theta} \rangle$ became important. From this observation, it is clear that a model for the normal stress anisotropy of the form of Eq. (4.18) will fail when production associated with $\langle S_{r\theta} \rangle$ is significant. On the other hand, when the isotropic part of the normal stresses dominates production of the turbulence energy (and, hence, the normal stress anisotropy), Eq. (4.18) might be expected to perform well. Both of these expectations are largely fulfilled. Figure 4.39 compares the predictions of Eq. (4.18) against the difference in the measured normal stresses for the latter case, when production by the isotopic part of the normal stresses dominates. Note that when the unmeasured normal stress

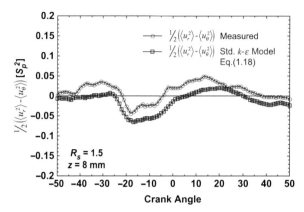

Fig. 4.39 Comparison of the difference in the measured normal stresses $1/2(\langle u_r'^2 \rangle - \langle u_\theta'^2 \rangle)$ with model estimates based on Eqs.(4.18)-(4.21)

$\langle u_z'^2 \rangle$ is approximated as $1/2 \left(\langle u_r'^2 \rangle + \langle u_\theta'^2 \rangle \right)$, the anisotropic part of $\langle u_r'^2 \rangle$ can be estimated as $1/2 \left(\langle u_r'^2 \rangle - \langle u_\theta'^2 \rangle \right)$, the quantity displayed in Fig. 4.39. Overall, very satisfactory agreement is achieved in modeling the temporal evolution of the normal stress anisotropy, although there is a general trend towards under-prediction of the measured anisotropy.

In contrast, Fig. 4.40 depicts a situation in which production by the shear stresses is dominant. Under these circumstances, with $R_s = 2.5$ and $z = 4$ mm, production of $\langle u_\theta'^2 \rangle$ and negative production of $\langle u_r'^2 \rangle$ by the r-θ plane shear stresses becomes important at approximately -20 CAD (v. Fig. 4.15). At this time, the measured difference in the normal stresses deviates radically from the predictions of Eq. (4.18). Likewise,

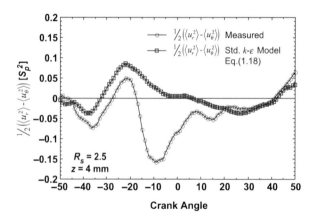

Fig. 4.40 Comparison of the difference in the measured normal stresses $1/2(\langle u_r'^2 \rangle - \langle u_\theta'^2 \rangle)$ with model estimates based on Eqs.(4.18)-(4.21). When turbulence production associated with shear stresses is important (near-10 CAD), the linear stress model performs poorly

differences observed between approximately -40 and -30 CAD and between 5 and 25 CAD might be alleviated if the model accounted for the non-trivial production associated with $\frac{\partial \langle U_\theta \rangle}{\partial z}$ shown at this time in Fig. 4.15. As will be seen below, adoption of more general stress models can significantly improve the model predictions during each of the periods identified above.

4.5.3 Higher Order Stress Models

Higher-order, non-linear stress models in the form of Eq. (4.23) have the potential to correct many of the short-comings of the linear stress model without adding undue computational expense. It can be argued [77] that for high Reynolds number, nearly homogeneous flows, the mean flow rate of strain and rotation tensors, along with the turbulence scales k and ε, contain all of the information necessary to model the turbulent stress tensor. Furthermore, when the coefficients of these non-linear models are found from the explicit solution of an algebraic stress formulation (e.g. [42, 43, 77, 99]), the various sources of production of the individual stress components are considered directly. Hence, these models can be expected to provide an improvement in situations where the latter terms on the RHS of Eq. (4.16) contribute substantially to $\langle u_r' u_\theta' \rangle$ (v. Fig. 4.38), or when production by the anisotropic part of the stress tensor constitutes a significant fraction of the production of the normal stresses.

Although the most general form of a non-linear stress model contains terms up to 5th order in the mean velocity gradients, attention is restricted here to the simpler, quadratic models. Preliminary assessments by the author of cubic [28] and quadratic [99] models indicate that the additional complexity of these models does not improve the accuracy of the modeled stresses. To evaluate these models, information on the unmeasured gradient $\frac{\partial \langle U_z \rangle}{\partial r}$ is provided from numerical simulations. The model predictions considered are found to be insensitive to the accuracy of this gradient.

The quadratic models of Gatski and Speziale [42] and of Girimaji [43] are compared against the shear stress measured with $R_s = 3.5$ and $z = 12$ mm in Fig. 4.41. Recall that under these conditions, the production of $\langle u_r' u_\theta' \rangle$ by the anisotropic normal stresses is comparable to the production by the isotropic part of the stress tensor, and a large benefit might be expected from the use of a more advanced stress model. The models shown were selected for comparison because they incorporate the same dependency on the mean flow strain and rotation rate tensors—only the dependency of the coefficients α_1–α_3 on the normalized invariants of these tensors differs. Both models avoid the contentious term found in some quadratic models that depends on the mean flow rotation alone [89]. Although the stress estimates from both models are improved over the estimates from Eq. (4.18), a large benefit is not found. Recall that at this swirl ratio the turbulence is inhomogeneous (v. Fig. 4.18), a fact which may be partly responsible for the remaining discrepancies. Because the models have the same form (only α_1–α_3 are non-zero), the differences in the modeled stresses are

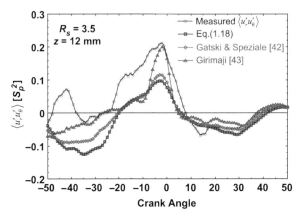

Fig. 4.41 Evaluation of quadratic models for the r-θ plane shear stress $\langle u'_r u'_\theta \rangle$. The linear model, Eq.(4.18), is also shown for comparison

due to differences in the variability of the model coefficients as the characteristic strain and rotation rates vary.

A more significant difference in the predictions of the two quadratic models as compared to the linear model is seen when the anisotropy in the normal stresses is considered. Fig. 4.42 shows that the large discrepancy seen in Fig. 4.40 between the difference in the measured normal stresses and the difference predicted by the linear model is substantially reduced when the quadratic models are employed. It is clear that both models capture the increased production of $\langle u'^2_\theta \rangle$ by the r-θ plane shear stresses between roughly −20 CAD and −5 CAD, although the change is under-predicted by the Gatski and Speziale model and somewhat over-predicted by the Girimaji model. The enhanced production of $\langle u'^2_\theta \rangle$ associated with $\frac{\partial \langle U_\theta \rangle}{\partial z}$ between

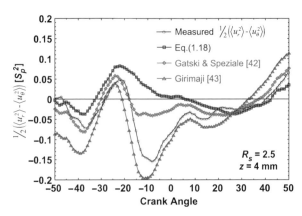

Fig. 4.42 Assessment of the quadratic model performance in predicting the difference in the normal stresses

approximately -40 and -30 CAD and between 5 and 25 CAD also appears to be reflected in the quadratic model predictions, but its influence on the difference in the normal stresses is over-predicted by the Girimaji model.

Figures 4.41 and 4.42 clearly indicate that more accurate modeling of the anisotropic stresses can be achieved through non-linear stress models of the form given by Eq. 1.23. However, both the form of the model (the number of terms retained) and the dependency of the coefficients on the characteristic strain and rotation rates influence the accuracy of the models. Several additional non-linear stress models are available and remain to be evaluated. A detailed assessment of the performance of these models over as wide a range of conditions as possible is an important step toward improving our ability to predict these flows.

4.5.4 Dissipation Modeling

Assessment of current engine flow modeling practice—based on the k-ε model—has focused until now on the modeling of the turbulent stresses. This is the fundamental assumption made in the modeling of the production term in the k-equation, a term which also figures in the dissipation (ε) equation. However, the dissipation equation is of at least equal importance, and its modeling merits evaluation as well. This evaluation is complicated by two factors: First, a specific modeling relationship, such as the turbulent stress models discussed above, that can be examined with experimental data is lacking. Second, a direct experimental measurement of the dissipation has not been made. From Eq. (4.21), however, it is seen that the evolution of the turbulent dissipation can be assessed—at least qualitatively—from the evolution of a measured turbulent length scale and the turbulent kinetic energy. A similar assessment has been performed previously by Ikegami and co-workers [53].

The evolution of the measured longitudinal and the lateral length scales, obtained with fuel injection, are compared with the results of the numerical simulations in Fig. 4.43. Length scales measured at various locations under motored conditions are similar [72]. The measured length scales have been scaled to be consistent with the simulation results as described in [70].

A number of features exhibited in Fig. 4.43 merit discussion:

- At the start of the period shown, near -50 CAD, the length scale predicted by the simulation is generally overestimated at all swirl ratios. At this time the simulated turbulence energy is significantly underestimated [70] for all three swirl ratios considered. The simulated turbulence energy increases rapidly through approximately -20 CAD, however, during which time the length scale decreases considerably. Despite simulated turbulence energy that is comparable to (or greater than) the measured energy as TDC is approached, the simulated length scale generally exceeds the measured length scales, suggesting that the simulated dissipation is too small during this period.
- Just after the injection event, a sharp decrease in the simulated length scale is observed, that is not seen consistently in either of the two measured scales.

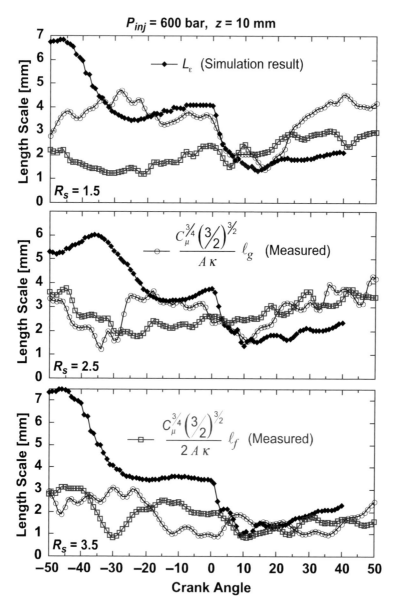

Fig. 4.43 Comparison of the measured longitudinal and lateral length scales with the results of numerical simulation

Turbulence energies are not changing significantly at this time. Accordingly, the simulation is predicting a large impact of the fuel injection event on the turbulence dissipation that is not supported by the measurements.

- Beyond approximately 10 CAD, the trends in the simulated length scale track well with the measured length scales, although the magnitude is underestimated (dissipation overestimated) at the lower swirl ratios.

The length scale comparison made in Fig. 4.43 is based on the validity of Eq. (4.21), which assumes that the turbulence is in equilibrium. Some authors (e.g., [49, 54]) argue that a linear relationship between the simulated length scale and the Taylor scale of the turbulence may be more appropriate than the relationship implied by Eq. (4.21). A comparison based on this alternative scaling does not provide better agreement [70], however.

Overall, the qualitative assessment of the dissipation modeling made here indicates that although the simulation captures the approximate magnitude and some of the trends in the dissipation rate, further improvement is needed.

4.6 Future Research Areas

As evidenced by the material reviewed in this chapter, considerable progress has been made in the last three decades toward advancing our knowledge of the fluid dynamics of direct-injection diesel engines. Nevertheless, it is also clear that there remain significant gaps in our understanding of the physics of these flows, in our knowledge of how the physics can best be utilized in practical combustion systems, and in our ability to quantitatively predict these flows in a computationally affordable manner. Although certainly not complete, a list of areas in which future research efforts may prove fruitful is given below.

1. The swirl velocity distribution has been shown to exert a dominant influence on the formation of flow structures and the generation of turbulence both prior to TDC (the squish/swirl interaction) and after injection (the spray/swirl interaction). Emissions measurements (e.g., [18, 56, 74]) show, however, that knowledge of the bulk flow swirl ratio alone is insufficient to predict emissions behavior—vastly different emissions are measured for different port geometries, though equal swirl ratios are attained. One likely explanation for this behavior is port-induced differences in the spatial distribution of the swirl velocity (or angular momentum). Although the effects of different spatial distributions of the swirl velocity on the development of flow structures has been investigated numerically in one study [33], this area remains largely unexplored. Additional understanding of the differences developed in flow structures and how they correlate with engine emissions would be a useful contribution.

2. A better understanding of the effects of the spatial distribution of swirl velocity on the flow structures, as discussed above, would likely lead to a desire to tailor the swirl velocity distribution generated during the intake process. Increased

insight into how such tailored distributions could be generated is therefore desirable. Studies incorporating the effects of port geometry, valve lift profiles, valve seat angles, etc. on the formation of a stratified swirl field, as well as the persistence of the stratification achieved as the compression stroke proceeds would be valuable.

3. The role of the large-scale structures in promoting mixing of fuel and air through both bulk transport and small scale mixing has been clarified from the results of numerical simulation. Experimental verification of the spatial distributions of unburned or partially-burned fuel, and their correlation with the large-scale structures is still required. Similarly, experimental assessment of the rate of fuel/air mixing, through measurement of scalar dissipation rates would be a useful contribution, both for validating model predictions as well as for clarifying the degree to which various flow features are influencing the mixing rate.

4. Little work has been performed to date to clarify how various features of the bowl geometry interact with the fuel spray and the development of large-scale flow structures. Small changes in bowl geometry can result in large changes in particulate emissions. Issues of interest are the shape of the bowl at the location of spray impact and the influence of the bowl pip on the size and location of the vertical plane vortical structures generated by the spray/swirl interaction.

5. Multiple injection events have been shown to have a sizeable impact on engine emissions. Assessment of the effect of multiple injections on the development of bulk flow structures and late-cycle mixing processes could be productive.

6. Both numerical and experimental studies (e.g. [66]) have shown that much of the late-cycle oxidation of particulates occurs well outside the bowl, above the piston top. Flow structures in this area and their dependency on combustion chamber geometry, swirl levels, and injection system parameters has not yet been explored.

7. The sensitivity of engine emissions to small changes in bowl geometry and to details of how the flow swirl is generated demonstrate that optimization of these combustion systems will not be achieved without the assistance of accurate multi-dimensional modeling efforts. Improved turbulence models have been available for many years, but there has been no experimental data with which to evaluate the potential benefits these improved models offer. The stress model assessment described above is a first step in this direction, but many additional experiments must be performed and models assessed before recommendations can be made with confidence.

8. Likewise, evaluation of the dissipation has been limited to indirect evaluation based on measured length scales. Planar or three-dimensional global measurement techniques offer the potential of direct evaluation of the turbulent energy dissipation. An example of recent progress in this area is given in reference [41]. These full-field techniques offer the additional advantage of providing single-cycle flow structures that can be compared with large-eddy simulations of the flow. Temporally-resolved point measurements, employing Taylor's hypothesis and assumptions of small scale isotropy can also yield information on the temporal evolution of the dissipation rate and should be pursued.

Acknowledgments The recent experimental data reviewed in this work were obtained at the Combustion Research Facility of Sandia National Laboratories. Sandia is a multi-program laboratory operated by Sandia Corporation, a Lockheed Martin Company, for the United States Department of Energy's National Nuclear Security Administration under contract DE-AC04-94AL85000.

The author would like to acknowledge the contributions of M. Megerle, D. Choi, J. Hammer, and M.-C. Lai (affiliated with Wayne State University) to the data acquisition and analysis. The numerical results reviewed were obtained at the University of Wisconsin Engine Research Center by M. Bergin, Y. Liu, Z. Nagel, B.H. RempelEwert and K. Richards, under the direction of R.D. Reitz. The efforts of B.H. RempelEwert in generating the numerical results depicted in Figs. 4.33 and 4.34 specifically for this contribution are gratefully acknowledged. Support for this work, at both Sandia National Laboratories and the University of Wisconsin, was provided by the US Department of Energy, Office of Vehicle Technologies under the guidance of Gurpreet Singh and Kevin Stork.

References

1. Ahmadi-Befrui A, Brandstätter W, Pitcher G, Troger C, Wigley G (1990) Simulationsmodell zur Berechnung der Luftbewegung in Zylindern von Verbrennungsmotoren. MTZ Motortechnische Zeitschrift 51: 440–447, Vieweg and Teubner
2. Amsden AA (1997) KIVA-3 V: a block structured KIVA program for engines with vertical or canted valves. Los Alamos National Laboratory Report No. LA-13313-MS
3. Arcoumanis C, Bicen AF, Vlachos NS, Whitelaw JH (1982) Effects of flow and geometry boundary conditions on fluid motion in a motored IC model engine. Proc Inst Mech Eng 196(4): 1–10
4. Arcoumanis C, Bicen AF, Whitelaw JH (1982) Measurements in a motored four-stroke reciprocating model engine. J Fluids Eng 104: 235–241
5. Arcoumanis C, Bicen AF, Whitelaw JH (1983) Squish and swirl-squish interaction in motored model engines. J Fluids Eng 105: 105–112
6. Arcoumanis C, Begleris P, Gosman AD, Whitelaw JH (1986) Measurements and calculations of the flow in a research diesel engine. SAE technical paper 861563, Society of Automotive Engineers, Warrendale, PA
7. Arcoumanis C, Hu Z, Vafidis C, Whitelaw JH (1990) Tumbling motion: a mechanism for turbulence enhancement in spark-ignition engines. SAE technical paper 900060, Society of Automotive Engineers, Warrendale, PA
8. Arcoumanis C, Hadjiapostolou A, Whitelaw JH (1991) Flow and combustion in a hydra direct-injection diesel engine. SAE technical paper 910177, Society of Automotive Engineers, Warrendale, PA
9. Arcoumanis C, Whitelaw JH, Hentschel W, Schindler K-P (1994) Flow and combustion in a transparent 1.9 liter direct injection diesel engine. Proc Inst Mech Eng D 208: 191–205
10. Auriemma M, Corcione FE, Macchioni R, Seccia G, Valentino G (1996) LDV measurements of integral length scales in an IC engine. SAE technical paper 961161, Society of Automotive Engineers, Warrendale, PA
11. Auriemma M, Corcione FE, Macchioni R, Valentino G (1998) Interpretation of air motion in reentrant bowl-in-piston engine by estimating Reynolds stresses. SAE technical paper 980482, Society of Automotive Engineers, Warrendale, PA
12. Ball WP, Pettifer HF, Waterhouse CNF (1983) Laser doppler velocimeter measurements of turbulence in a direct-injection diesel combustion chamber. In: Proc Instn Mech Engrs international conference on combustion in engineering, C52/83, Oxford, pp 163–174. Institution of Mechanical Engineers, London: http://www.imeche.org/

13. Béard P, Mokaddem K, Baritaud T (1998) Measurement and modeling of the flow-field in a DI diesel engine: effects of piston bowl shape and engine speed. SAE technical paper 982587, Society of Automotive Engineers, Warrendale, PA

14. Bianchi GM, Richards K, Reitz RD (1999) Effects of initial conditions in multidimensional combustion simulations of HSDI diesel engines. SAE technical paper 1999-01-1180, Society of Automotive Engineers, Warrendale, PA

15. Bianchi GM, Pelloni P, Corcione FE, Mattarelli E, Lupino Bertoni F (2000) Numerical study of the combustion chamber shape for common rail HSDI diesel engines. SAE technical paper 2000-01-1179, Society of Automotive Engineers, Warrendale, PA

16. Bopp S, Vafidis C, Whitelaw JH (1986) The effect of engine speed on the TDC flowfield in a motored reciprocating engine. SAE technical paper 860023, Society of Automotive Engineers, Warrendale, PA

17. Bradshaw P (1973) Effects of streamline curvature on turbulent flow. AGARDograph 169, AD-768316, Advisory Group for Aerospace Research and Development, North Atlantic Treaty Organization, Paris

18. Brandl F, Reverencic I, Cartellieri W, Dent JC (1979) Turbulent air flow in the combustion bowl of a DI diesel engine and its effect on engine performance. SAE technical paper 790040, Society of Automotive Engineers, Warrendale, PA

19. Catania AE, Spessa E (1996) Speed dependence of turbulence properties in a high-squish automotive engine combustion system. SAE technical paper 960268, Society of Automotive Engineers, Warrendale, PA

20. Celik I, Yavuz I, Smirnov A, Smith J, Amin E, Gel A (2000) Prediction of in-cylinder turbulence for IC engines. Combust Sci and Tech 153: 339–368

21. Chen K-H, Liu N-S (1998) Evaluation of a non-linear turbulence model using mixed volume unstructured grids. AIAA paper 98-0233, American Institute of Aeronautics and Astronautics, Reston, VA

22. Christensen M, Johansson B (2002) The effect of in-cylinder flow and turbulence on HCCI operation. SAE technical paper 2002-01-2864, Society of Automotive Engineers, Warrendale, PA

23. Cipolla G, Puglisi A, Vafidis C (1987) In-cylinder velocity field measurements in a motored diesel engine. SAE technical paper 870373, Society of Automotive Engineers, Warrendale, PA

24. Collins N, Roughton AW, Ma T (1987) Turbulence length scale measurements in a motored internal combustion engine. SAE technical paper 871692, Society of Automotive Engineers, Warrendale, PA

25. Comte-Bellot G, Corrsin S (1971) Simple-eulerian time correlation of full- and narrow-band velocity signals in grid-generated, 'isotropic' turbulence. J Fluid Mech 48: 273–337

26. Corcione FE, Valentino G (1994) Analysis of in-cylinder flow processes by LDA. Combust Flame 99: 387–394

27. Corcione FE, Valentino G (1994) Analysis of in-cylinder turbulent air motion dependence on engine speed. SAE technical paper 940284, Society of Automotive Engineers, Warrendale, PA

28. Craft TJ, Launder BE, Suga K (1996) Development and application of a cubic eddy-viscosity model of turbulence. Int J Heat Fluid Flow 17: 108–115

29. Dent JC, Salama NS (1975) The measurement of turbulence characteristics in an internal combustion engine cylinder. SAE technical paper 750886, Society of Automotive Engineers, Warrendale, PA

30. Deslandes W, Dupont A, Baby X, Charney G, Boree J (2003) PIV measurements of internal aerodynamic of diesel combustion chamber. SAE technical paper 2003-01-3083, Society of Automotive Engineers, Warrendale, PA

31. Dimopoulos P, Boulouchos K (1995) Reynolds stress components in the flow field of a motored reciprocating engine. SAE technical paper 950725, Society of Automotive Engineers, Warrendale, PA

32. Dimopoulos P, Boulouchos K (1997) Turbulent flow field characteristics a motored recip-rocating engine. SAE technical paper 972833, Society of Automotive Engineers, Warren-dale, PA
33. El Tahry SH (1982) A numerical study on the effects of fluid motion at inlet-valve closure on subsequent fluid motion in a motored engine. SAE technical paper 820035, Society of Automotive Engineers, Warrendale, PA
34. Fansler TD (1985) Laser velocimetry measurements of swirl and squish flows in an engine with a cylindrical bowl. SAE technical paper 850124, Society of Automotive Engineers, Warrendale, PA
35. Fansler TD, French DT (1987) Swirl, squish and turbulence in stratified-charge engines: laser velocimetry measurements and implications for combustion. SAE technical paper 870371, Society of Automotive Engineers, Warrendale, PA
36. Fansler TD, French DT (1988) Cycle-resolved laser-velocimetry measurements in a re-entrant-bowl-in-piston engine. SAE technical paper 880377, Society of Automotive Engineers, Warrendale, PA
37. Foster DE, Witze PO (1988) Two-component laser velocimetry measurements in a spark ignition engine. Combust Sci Tech 59: 85–105
38. Fraser RA, Bracco FV (1988) Cycle-resolved LDV integral length scale measurements in an IC engine. SAE technical paper 880381, Society of Automotive Engineers, Warrendale, PA
39. Fraser RA, Bracco FV (1989) Cycle-resolved LDV integral length scale measurements investigating clearance height scaling, isotropy, and homogeneity in an IC engine. SAE technical paper 890615, Society of Automotive Engineers, Warrendale, PA
40. Fujita H, Kovasznay LS (1968) Measurement of Reynolds stress by a single rotated hot wire anemometer. Rev Sci Instrum 39: 1351–1355
41. Funk C, Sick V, Reuss DL, Dahm WJA (2002) Turbulence properties of high and low swirl in-cylinder flows. SAE technical paper 2002-01-2841, Society of Automotive Engineers, Warrendale, PA
42. Gatski TB, Speziale CG (1993) On explicit algebraic stress models for complex turbulent flows. J Fluid Mech 254: 59–78
43. Girimaji SS (1995) A galilean invariant explicit algebraic stress model for turbulent curved flows. Phys Fluids 9: 1067–1077
44. Glover AR, Hundleby GE, Hadded O (1988) The development of scanning LDA for the measurement of turbulence in engines. SAE technical paper 880379, Society of Automotive Engineers, Warrendale, PA
45. Gosman AD, Johns RJR, Watkins AP (1980) Development of prediction methods for in-cylinder processes in reciprocating engines. In: Mattavi JD, Amann CA (eds) Combustion modeling in reciprocating engines. Plenum, New York, London
46. Gosman AD (1986) Flow processes in cylinders. In: Horlock JH, Winterbone D (eds) Ther-modynamics and gas dynamics of internal combustion engines, vol 2. Oxford University Press, Oxford, pp 616–772
47. Hadded O, Denbratt I (1991) Turbulence characteristics of tumbling air motion in four-valve SI engines and their correlation with combustion parameters. SAE technical paper 910478, Society of Automotive Engineers, Warrendale, PA
48. Han Z, Reitz RD (1995) Turbulence modeling of internal combustion engines using RNG k-ε models. Combust Sci Tech 106: 267–295
49. Han Z, Reitz RD, Corcione FE, Valentino G (1996) Interpretation of k-ε computed length scale predictions for turbulent flows. In: Proc 26th Intl Symp on Combustion, The Combus-tion Institute, Pittsburgh, PA, pp 2717–2723
50. Heywood JB (1988) Internal combustion engine fundamentals. McGraw-Hill, New York
51. Hinze JO (1959) Turbulence. McGraw-Hill, New York, Toronto, London

52. Hoult DP, Wong VW (1980) The generation of turbulence in an internal combustion engine. In: Mattavi JD, Amann CA (eds) Combustion modeling in reciprocating engines. Plenum, New York, London
53. Ikegami M, Shioji M, Nishimoto K (1987) Turbulence intensity and spatial integral scale during compression and expansion strokes in a four-cycle reciprocating engine. SAE technical paper 870372, Society of Automotive Engineers, Warrendale, PA
54. Kaario O, Larmi M, Tanner F (2002) Relating integral length scale to turbulent time scale and comparing k-ε and RNG k-ε turbulence models in combustion simulation. SAE technical paper 2002-01-1117, Society of Automotive Engineers, Warrendale, PA
55. Kido H, Wakuri Y, Murase E (1983) Measurements of spatial scales and a model for small-scale structure of turbulence in an internal combustion engine. Proc ASME-JSME Thermal Eng Joint Conf, Honolulu, vol 4, pp 191–198
56. Krieger RB, Siewert RM, Pinson JA, Gallopoulos NE, Hilden DL, Monroe DR, Rask RB, Solomon ASP, Zima P (1997) Diesel engines: one option to power future personal transportation vehicles. SAE technical paper 972683, Society of Automotive Engineers, Warrendale, PA
57. Lancaster DR (1976) Effects of engine variables on turbulence in a spark-ignition engine. SAE technical paper 760159, Society of Automotive Engineers, Warrendale, PA
58. Li Y, Zhao H, Peng Z, Ladommatos N (2002) Particle image velocimetry measurement of in-cylinder flow in internal combustion engines—experiment and flow structure analysis. Proc Inst Mech Eng D 216: 65–81
59. Lin L, Shulin D, Jin W, Jinxiang W, Xiaohong G (2000) Effects of combustion chamber geometry on in-cylinder air motion and performance in DI diesel engine. SAE technical paper 2000-01-0510, Society of Automotive Engineers, Warrendale, PA
60. Liou T-M, Santavicca DA (1983) Cycle resolved turbulence measurements in a ported engine with and without swirl. SAE technical paper 830419, Society of Automotive Engineers, Warrendale, PA
61. Liou T-M, Santavicca DA (1985) Cycle resolved LDV measurements in a motored IC engine. J Fluids Eng 107: 232–240
62. Lumley JL (1999) Engines—an introduction. Cambridge University Press, Cambridge
63. Marc D, Boree J, Bazile R, Charnay G (1997) Tumbling vortex flow in a model square piston compression machine: PIV and LDV measurements. SAE technical paper 972834, Society of Automotive Engineers, Warrendale, PA
64. Margary R, Nino E, Vafidis C (1990) The effect of intake duct length on the in-cylinder air motion in a motored diesel engine. SAE technical paper 900057, Society of Automotive Engineers, Warrendale, PA
65. Matsuoka S, Kamimoto T, Urushihara T, Mochimaru Y, Morita H (1985) LDA measurement and a theoretical analysis of the in-cylinder air motion in a DI diesel engine. SAE technical paper 850106, Society of Automotive Engineers, Warrendale, PA
66. Miles PC (2000) The influence of swirl on HSDI diesel combustion at moderate speed and load. SAE technical paper 2000-01-1829, Society of Automotive Engineers, Warrendale, PA
67. Miles P, Megerle M, Sick V, Richards K, Nagel Z, Reitz R (2001) The evolution of flow structures and turbulence in a fired HSDI diesel engine. SAE technical paper 2001-01-3501, Society of Automotive Engineers, Warrendale, PA
68. Miles PC, Megerle M, Nagel Z, Liu Y, Reitz RD, Lai M-C, Sick V (2002) The influence of swirl and injection pressure on post-combustion turbulence in a HSDI diesel engine. In: Proc thermo- and fluid dynamic processes in diesel engines, THIESEL2002, Sept. 10–13, Valencia, Spain, pp 415–433
69. Miles P, Megerle M, Hammer J, Nagel Z, Reitz RD, Sick V (2002) Late-cycle turbulence generation in swirl-supported, direct-injection diesel engines. SAE technical paper 2002-01-0891, Society of Automotive Engineers, Warrendale, PA

70. Miles P, Megerle M, Nagel Z, Reitz RD, Lai M-C, Sick V (2003) An experimental assessment of turbulence production, Reynolds stress, and length scale (dissipation) modeling in a swirl-supported DI diesel engine. SAE technical paper 2003-01-1072, Society of Automotive Engineers, Warrendale, PA

71. Miles PC, RempelEwert BH, Reitz RD (2003) Squish-swirl and injection-swirl interaction in direct-injection diesel engines. In: Sixth international conference on engines for automobile, ICE2003, Sept. 14–19, Capri/Naples, Italy

72. Miles PC, Choi D, Megerle M, RempelEwert BH, Reitz RD, Lai M-C, Sick V (2004) The influence of swirl ratio on turbulent flow structure in a motored HSDI diesel engine—a combined experimental and numerical study. SAE technical paper 2004-01-1678, Society of Automotive Engineers, Warrendale, PA

73. Mohammadi A, Kidoguchi Y, Miwa K (2002) Effect of injection parameters and wall-impingement on atomization and gas entrainment processes in diesel sprays. SAE technical paper 2002-01-0497, Society of Automotive Engineers, Warrendale, PA

74. Monaghan ML, Pettifer HF (1981) Air motion and its effect on diesel engine performance and emissions. SAE technical paper 810255, Society of Automotive Engineers, Warrendale, PA

75. Payri F, Desantes JM, Pastor JV (1996) LDV measurements of the flow inside the combustion chamber of a 4-valve DI diesel engine with axisymmetric piston-bowls. Exp Fluids 22: 118–128

76. Petersen U, MacGregor SA (1996) Jet mixing in a model direct injection diesel engine with swirl. Proc Inst Mech Eng C 210: 69–78

77. Pope SB (1975) A more general effective-viscosity hypothesis. J Fluid Mech 72: 331–340

78. Pope SB (2000) Turbulent flows. Cambridge University Press, Cambridge

79. Rask RB (1981) Comparison of window, smoothed-ensemble, and cycle-by-cycle data reduction techniques for laser doppler anemometer measurements of in-cylinder velocity. In: Proc ASME symposium on fluid mechanics of combustion systems, June 22–23, Boulder, CO, pp 11–20, (ASME, New York, 1981)

80. Rask RB, Saxena V (1985) Influence of geometry on flow in the combustion chamber of a direct-injection diesel engine. In: International symposium on flows in internal combustion engines III, ASME-FED, vol 28, pp 19–28, (ASME, New York, 1985) Library of Congress Cat Card No: 82-73181

81. RempelEwert BH (2003) Personal communication

82. Reynolds WC (1980) Modeling of fluid motions in engines—an introductory overview. In: Mattavi JD, Amann CA (eds) Combustion modeling in reciprocating engines. Plenum, New York, London

83. Rhim D-R, Farrell PV (2000) Characteristics of air flow surrounding non-evaporating transient diesel sprays. SAE technical paper 2000-01-2789, Society of Automotive Engineers, Warrendale, PA

84. Saito T, Daisho Y, Uchida N, Ikeya N (1986) Effects of combustion chamber geometry on diesel combustion. SAE technical paper 861186, Society of Automotive Engineers, Warrendale, PA

85. Sasaki S, Akagawa H, Tsujimura K (1998) A study on surrounding air flow induced by diesel sprays. SAE technical paper 980805, Society of Automotive Engineers, Warrendale, PA

86. Schäpertöns H, Thiele F (1986) Three dimensional computations for flowfields in DI piston bowls. SAE technical paper 860463, Society of Automotive Engineers, Warrendale, PA

87. Shih T-H (1997) Some developments in computational modeling of turbulent flows. Fluid Dyn Res 20: 67–96

88. Speziale CG (1987) On nonlinear k-ℓ and k-ε models of turbulence. J Fluid Mech 178: 459–475

89. Speziale CG (1998) A consistency condition for non-linear algebraic Reynolds stress models in turbulence. Int J Non-Linear Mech 33: 579–584

90. Spicher U, Velji A, Huynh NH, Kruse F (1987) An experimental study of combustion and fluid flow in diesel engines. SAE technical paper 872060, Society of Automotive Engineers, Warrendale, PA

91. Sugiyama K (1986) LDV measurement and simulation of air motion in a re-entrant combustion bowl for a DI diesel engine. SAE technical paper 865008, Society of Automotive Engineers, Warrendale, PA

92. Sun J, Dong Y, Xu Y, Tsao KC (1989) In cylinder gas motions via non-isotropic turbulent modeling and experiment. SAE technical paper 891915, Society of Automotive Engineers, Warrendale, PA

93. Tennekes H, Lumley JL (1972) A first course in turbulence. MIT Press, Cambridge, MA

94. Townsend AA (1976) The struct.ure of turbulent shear flows. Cambridge University Press, Cambridge

95. Tritton DJ (1977) Physical fluid dynamics. Van Nostrand Reinhold, Berkshire

96. Vafidis C (1984) Influence of induction swirl and piston configuration on air flow in a four-stroke model engine. Proc Inst Mech Eng C 198(8): 71–79

97. Valentino G, Auriemma M, Corcione FE, Macchioni R, Seccia G (1996) Evaluation of time and spatial turbulence scales in a DI diesel engine. In: Eighth international symposium on applications of laser anemometry to fluid mechanics, July 8–11, Lisbon, Paper 16.1, Instituto Superior Técnico, Lisbon, 1996

98. Wakisaka T, Shimamoto Y, Isshiki Y (1986) Three-dimensional numerical analysis of in-cylinder flows in reciprocating engines. SAE technical paper 860464, Society of Automotive Engineers, Warrendale, PA

99. Wallin S, Johansson AV (2000) An explicit algebraic stress model for incompressible and compressible turbulent flows. J Fluid Mech 403: 89–132

100. Wigley G, Patterson AC, Renshaw J (1981) Swirl velocity measurements in a firing production diesel engine by laser anemometry. In: Proc ASME symposium on fluid mechanics of combustion systems, June 22–23, Boulder, CO, pp 29–39, (ASME, New York, 1981)

101. Wilcox DC (1998) Turbulence modeling for CFD. DCW Industries, La Cañada, CA

102. Witze PO (1980) Influence of air motion variation on the performance of a direct injection stratified charge engine. In: Proc Instn Mech Engrs international conference on stratified charge automotive engines, C394/80, London, pp 25–31, Institution of Mechanical Engineers, London: http://www.imeche.org/

103. Witze PO, Foster DE (1988) Two-component velocity probability density measurements during premixed combustion in a spark ignition engine. In: Proc Instn Mech Engrs international conference on combustion in engines—technology and applications, May 10–2, London, C51/88, pp 225–233, Institution of Mechanical Engineers, London: http://www.imeche.org/

104. Yianneskis M, Tindal MJ, Suen KO (1988) Swirl and squish effects in diesel engine cylinders at high speeds. In: Fourth international symposium on applications of laser anemometry to fluid mechanics, July 11–14, Lisbon, Paper 2.13, Instituto Superior Técnico, Lisbon, 1988

105. Yoshikawa S, Nishida K, Arai M, Hiroyasu H (1988) Visualization of fuel-air mixing processes in a small DI diesel engine using the liquid injection technique. SAE technical paper 880296, Society of Automotive Engineers, Warrendale, PA

106. zurLoya AO, Siebers DL, Mckinley TL, Ng HK, Primus RJ (1989) Cycle-resolved LDV measurements in a motored diesel engine and comparison with k-ε model predictions. SAE technical paper 890618, Society of Automotive Engineers, Warrendale, PA

Chapter 5
Recent Developments on Diesel Fuel Jets
Under Quiescent Conditions

Dennis L. Siebers

5.1 Introduction

Fuel jets play a dominant role in fuel–air mixing, combustion, and emission forma-tion processes in direct-injection (DI) diesel engines. This chapter discusses several recent developments in the understanding of fuel jets injected into quiescent-type diesel engine conditions. The discussions are based on research by the author and coworkers. Although the discussions are focused on quiescent-type engine condi-tions (typical of heavy-duty, DI diesel engines), the results presented also provide a baseline for assessing the additional effects on fuel jet development caused by strong in-cylinder flows (typically of light-duty, DI diesel engines).

The topics that are discussed include: the transient penetration of the tip of a fuel jet, the penetration and vaporization of liquid-phase fuel within a fuel jet, and flame lift-off and its impact on soot formation. Also discussed are the scaling of many of these processes with engine and injector parameters, and new insights on fuel jet development, soot formation, and combustion in a diesel fuel jet that were derived from the results. Conservation of mass, momentum, and energy principles, applied to an idealized model of a fuel jet, are used to provide a framework for presenting and analyzing the results. The results and insights discussed provide guidance for the design of several aspects of an engine combustion system and for the development of computational models for simulating diesel combustion and emission processes. Moreover, the results provide well-characterized, simplified baselines for compar-ison with the predictions of the computational models being developed.

Experimental results presented in the following sections were acquired in constant volume combustion vessels [1, 2] over a very wide range of conditions relevant to DI diesel engines. The fuel injector used for the experiments was an electron-ically controlled, common-rail fuel injector [1, 2] that had a "top-hat" injection rate profile (i.e., a near instantaneous start of injection, a constant injection rate during the injection period, and a sharp end of injection). The well-controlled and well-characterized nature of the experimental conditions in the combustion vessel

Dennis. L. Siebers
Sandia National Laboratories, Livermore, CA, USA

C. Arcoumanis, T. Kamimoto (eds.), *Flow and Combustion in Reciprocating Engines*, 257
DOI: 10.1007/978-3-540-68901-0_5, © Springer-Verlag Berlin Heidelberg 2009

and the simplicity of the injection rate profile allowed the effects of several parameters on fuel jet processes to be more clearly understood than has been possible with data from engines. However, when applying the results to engines, the limitations imposed by the relatively simple conditions under which the experiments were conducted must be recognized. Factors in an engine such as time-varying in-cylinder conditions resulting from piston motion, complex injection rate profiles, multiple jets, and detailed injector tip/orifice flow effects, while not affecting the fundamental nature of the physical processes involved in free jet regions, can impact the magnitude of the effect of the various parameters considered.

5.2 Penetration of Diesel Fuels Jets

A diesel fuel jet parameter often measured is the transient tip penetration (e.g., [1, 3, 4, 5, 6, 7, 8]. Penetration of a fuel jet and its commensurate entrainment of air are needed to promote efficient air utilization in a DI diesel engine, which is important for optimizing engine performance. The first part of this section presents a scaling law for the penetration of a non-vaporizing fuel jet under quiescent conditions driven by a top-hat injection rate profile. The penetration scaling law was developed by Naber and Siebers [1] using an idealized model of a diesel fuel jet. The scaling law will be shown to capture the important features of fuel jet penetration and the effects of various parameters on penetration. The penetration scaling law and analysis are extended and utilized in later sections for analyzing liquid-phase fuel penetration and for helping interpret the impact of flame lift-off on soot formation in diesel fuel jets.

Following the development of the penetration scaling law is a discussion of the fuel jet spreading angle dependence that appears in the scaling law and a comparison of the scaling law with non-vaporizing fuel jet penetration data. Then comparisons of both the non-vaporizing penetration data and the scaling law are made with vaporizing (i.e., non-combusting) and combusting fuel jet penetration data to show the effects of vaporization and combustion on fuel jet penetration. In the final parts of the section, the scaling law is compared with other commonly cited penetration correlations in the literature and a discussion of the application of the penetration scaling law is presented.

5.2.1 Scaling Law for Fuel Jet Penetration

A scaling law for the penetration of a non-vaporizing (i.e., isothermal) fuel jet was recently developed by Naber and Siebers [1]. The penetration scaling law is valid in both the near-field and the far-field (i.e., short- and long-time limits) of a fuel jet with an instantaneous start of injection and a constant injection rate during the injection period. The derivation of the scaling law followed the fuel jet penetration analyses of Wakuri et al. [7] and Hays [9], but with important differences. One difference was the use of a non-dimensional method of analysis that both simplified the analysis and accounted for the effects of ambient gas density, fuel density, and various orifice parameters in a more complete manner. A second difference was the use of a clearly defined relationship between the "model" fuel jet used in the development of the penetration scaling law and a "real" fuel jet. The defined relationship between the real and model fuel jets (to be discussed later) allowed an arbitrary constant that appears in the penetration scaling law to be estimated [1].

The scaling law was developed by applying conservation of mass and momentum principles to the idealized, isothermal, incompressible "model" fuel jet shown in Fig. 5.1. The model fuel jet has a constant spreading angle (α), radially uniform velocity and fuel concentration profiles, an instantaneous start of injection, no significant azimuthal flow at the orifice exit, and dynamic equilibrium between the liquid and gas phases (i.e., no velocity slip between the injected fuel and the entrained air). The model fuel jet is defined such that it has the same cross-sectional average density and the same total mass and momentum fluxes at any axial location as a "real" fuel jet with the same initial and boundary conditions and a spreading angle of θ (The relationship between the model and real fuel jet spreading angles, α and θ, will be discussed in the next section). The defined relationship between the model and real fuel jets means that the tip penetration and air entrainment rate of the model and real fuel jets are the same, and that for a self-preserving real fuel jet, the spreading angles of the model and the corresponding real fuel jet can be approximately related by a constant.

The penetration scaling law derived by Naber and Siebers is given in dimensionless form by the following relationship (see Ref. [1] for details of the derivation):

$$\tilde{t} = \frac{\tilde{S}}{2} + \frac{\tilde{S}}{4} \cdot \sqrt{1 + 16 \cdot \tilde{S}^2} + \frac{1}{16} \cdot \ln\left(4 \cdot \tilde{S} + \sqrt{1 + 16 \cdot \tilde{S}^2}\right). \tag{5.1}$$

Equation (5.1) is an expression for the dimensionless penetration time, $\tilde{t}$, as a function the dimensionless penetration distance, $\tilde{S}$. The dimensionless penetration time and distance are defined by the following relationships:

$$\tilde{t} = t/t^+, \tag{5.2}$$

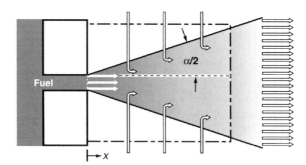

Fig. 5.1 Schematic of the idealized model fuel jet

$$\tilde{S} = S/x^+. \tag{5.3}$$

In Eqs. (5.2) and (5.3), the terms t and S are the penetration time and distance, respectively, and the terms t^+ and x^+ are characteristic time and length scales of the fuel jet defined as:

$$t^+ = \frac{d_f \cdot \sqrt{\tilde{\rho}}}{\tan(\alpha/2) \cdot U_f}, \tag{5.4}$$

$$x^+ = \frac{d_f \cdot \sqrt{\tilde{\rho}}}{\tan(\alpha/2)}. \tag{5.5}$$

The angle α in Eqs. (5.4) and (5.5) is the spreading angle of the model fuel jet in Fig. 5.1. The other terms in Eqs. (5.4) and (5.5) are the effective orifice diameter (d_f), the density ratio (ρ), and the fuel velocity at the orifice exit (U_f). The latter three terms are defined by the following relationships:

$$\tilde{\rho} = \rho_f/\rho_a, \tag{5.6}$$

$$d_f = \sqrt{C_a} \cdot d, \tag{5.7}$$

$$U_f = C_v \cdot \sqrt{\frac{2 \cdot (P_f - P_a)}{\rho_f}}. \tag{5.8}$$

In the definitions given by Eqs. (5.6) through (5.8), ρ_f and ρ_a are the fuel and ambient gas densities, respectively; d is the orifice diameter; C_a and C_v are the orifice area contraction and velocity coefficients, respectively; and P_f and P_a are the fuel pressure in the injector and the ambient gas pressure, respectively.

The product in the numerator in Eqs. (5.4) and (5.5) is the modified orifice diameter proposed by Thring and Newby [10] to account for a difference in density between the injected fluid and the ambient fluid. This modified orifice diameter has often been used in modeling as a length scale for diesel fuel jets (e.g., [11]). The spreading angle in Eqs. (5.4) and (5.5) has been found to account for orifice flow effects and secondary effects of the density difference between the injected fluid and the ambient gas on turbulent air entrainment into the fuel jet. The area contraction coefficient in Eq. (5.7), C_a, accounts for changes in the effective flow area through the orifice due to factors such as cavitation bubbles reaching the orifice exit (e.g., [12, 13, 14, 15, 16]) or "hydraulic flip" [17]. The velocity coefficient C_v in Eq. (5.8) accounts for the effects of factors such as friction loss and cavitation on the flow velocity through the orifice. By definition, the product of C_a and C_v equals the discharge coefficient, i.e., $C_d = C_a \cdot C_v$.

Two useful limits to the scaling law given by Eq. (5.1) are the near-field (or short-time) limit and far-field (or long-time) limit [1]. The near-field limit is:

$$\underset{\tilde{S}(or\tilde{t})\to 0}{\text{Limit}} \Bigg| \quad \tilde{S} = \tilde{t}, \tag{5.9}$$

which in dimensional form becomes:

$$S = C_v \cdot \sqrt{\frac{2 \cdot (P_f - P_a)}{\rho_f}} \cdot t. \tag{5.10}$$

The far-field (or long-time) limit is:

$$\underset{\tilde{S}(or\tilde{t})\to \infty}{\text{Limit}} \Bigg| \quad \tilde{S} = \tilde{t}^{1/2}, \tag{5.11}$$

which in dimensional form becomes:

$$S = \sqrt{\frac{C_v \cdot \sqrt{2 \cdot C_a}}{\tan(\alpha/2)}} \cdot \sqrt{\sqrt{\frac{(P_f - P_a)}{\rho_a}} \cdot d \cdot t}. \tag{5.12}$$

The limits show that there is a near-field region close to the injector where penetration is linearly dependent on time and a far-field region where penetration depends on the square-root of time, as noted by Hiroyasu and coworkers [5, 6]. These limits will be referred to later in comparisons with other penetration theory in the literature. The limits given by Eqs. (5.9) and (5.11) were also used in Ref. [1] to define an approximate inverse of Eq. (5.1), i.e., a relationship for penetration as a function of time:

$$\tilde{S} = \left(\left(\frac{1}{\tilde{t}}\right)^n + \left(\frac{1}{\tilde{t}^{1/2}}\right)^n \right)^{-1/n} = \tilde{t} / \left(1 + \tilde{t}^{n/2}\right)^{1/n}, \tag{5.13}$$

where $n = 2.2$. The approximation given by Eq. (5.13) agrees exactly with Eq. (5.1) in the near-field and far-field limits and to within 5% in the transition region between the near-field and far-field.

The transition between the near-field and far-field limits occurs at $\tilde{t} = 1$ (or $\tilde{S} = 0.71$). In physical coordinates, the transition time is given by:

$$t_r = \frac{\sqrt{C_a/2}}{C_v \cdot \tan(\alpha/2)} \cdot \frac{d \cdot \tilde{\rho}^{1/2}}{\sqrt{\frac{(P_f - P_a)}{\rho_f}}}. \tag{5.14}$$

The transition time (t_r) corresponds to the time when the fuel jet reaches a location where the mass in the fuel jet transitions from being dominated by the injected fuel to being dominated by entrained gas [1]. This transition occurs when the penetrating tip of the fuel jet reaches a distance of S_r, where:

$$S_r = \frac{\sqrt{C_a} \cdot d \cdot \tilde{\rho}^{1/2}}{\tan(\alpha/2)}.$$ (5.15)

Some other useful equations that can be derived following the analysis in Ref. [1] are an equation for the total air entrainment ($\dot{m}_a$) in a non-vaporizing fuel jet up to any axial location (x) relative to the amount of fuel injected ($\dot{m}_f$),

$$\frac{\dot{m}_a}{\dot{m}_f} = \frac{\sqrt{1 + 16 \cdot \tilde{x}^2} - 1}{2},$$ (5.16)

and the cross-sectional average axial velocity [$U(x)$] profile,

$$\frac{U(x)}{U_f} = \frac{2}{\sqrt{1 + 16 \cdot \tilde{x}^2} + 1}.$$ (5.17)

In Eqs. (5.16) and (5.17), the dimensionless axial distance x is x/x^+.

Equation (5.16) for air entrainment can be expressed in terms of a cross-sectional average equivalence ratio ($\bar{\phi}$) [1]:

$$\bar{\phi} = \frac{2 \cdot f_s}{\sqrt{1 + 16 \cdot \tilde{x}^2} - 1},$$ (5.18)

where f_s is the stoichiometric air/fuel mass ratio for a given fuel. In the far-field, Eq. (5.16) approaches (in dimensional form),

$$\frac{\dot{m}_a}{\dot{m}_f} = \frac{2 \cdot x \cdot \tan(\alpha/2)}{\sqrt{C_a} \cdot d \cdot \sqrt{\tilde{\rho}}}.$$ (5.19)

One importance of Eq. (5.19) is that it can be shown that it reduces to the entrainment relationship given by Ricou and Spalding [18] and others (e.g., [19]) for the far-field of a jet ($\tilde{x} \gg 1$) with a density ratio much closer to one than occurs for a diesel fuel jet (e.g., for an atmospheric pressure gas jet). Substituting into Eq. (5.19) a C_a equal to one (applicable for a gas jet), the relationship between "real" and "model" jet outer boundary spreading angles (θ and α) given later in Eq. (5.23), and an outer boundary spreading angle ($\theta/2$) of 12° (a value typically measured for a gas jet [20]) gives:

$$\frac{\dot{m}_a}{\dot{m}_f} \approx 0.32 \cdot \frac{x}{d \cdot \sqrt{\tilde{\rho}}},$$ (5.20)

i.e., the Ricou and Spalding jet entrainment relationship.

The area contraction coefficient and the spreading angle terms that appear in Eq. (5.19), and not in Eq. (5.20), are believed to capture orifice flow effects and effects of a large density difference between the fuel and the ambient gas on the turbulent entrainment processes in a diesel fuel jet. The orifice flow and the large density difference between the fuel and the ambient gas most likely influence entrainment by affecting the large scale turbulent structures that determine entrainment (e.g., [21, 22, 23]).

It is also interesting to note that Eq. (5.17) can be simplified to the centerline velocity decay profile typically cited for the far-field of gas jets (e.g., see [24]). In the far-field ($\tilde{x} \gg 1$), using the same assumptions that were used to develop Eq. (5.20) from Eq. (5.19), Eq. (5.17) simplifies to:

$$\frac{U_{cl}(x)}{U_f} \approx 6.0 \cdot \frac{d}{x}, \tag{5.21}$$

where the centerline velocity, $U_{cl}(x)$, is approximately twice the cross-sectional average axial velocity, $U(x)$, in Eq. (5.17).

The development of Eqs. (5.20) and (5.21) from Eqs. (5.16) and (5.17), respectively, highlights some of important differences that exist between the global features of atmospheric pressure gas jets and high injection pressure diesel fuel jets (i.e., highly atomized diesel fuel jets). However, the development also demonstrates that gas jets and diesel fuel jets are related.

5.2.2 Spreading Angle Dependence

The scaling law given by Eqs. (5.1) through (5.8) indicates that the penetration of an isothermal fuel jet is dependent on the fuel jet spreading angle, in addition to other parameters. The near-field and far-field limits given by Eqs. (5.10) and (5.12) indicate that this spreading angle dependence varies with distance from the orifice, changing from no dependence in the near-field to a significant dependence in the far-field. The dependence of penetration on spreading angle in the far-field, as it appears in Eq. (5.12), was first suggested by Wakuri et al. [7].

Based on Eq. (5.12), an increase in the spreading angle results in slower penetration of a fuel jet in the far-field. The slower penetration occurs because a larger spreading angle (for an isothermal jet) reflects greater air entrainment (i.e., a greater total mass in the fuel jet). From conservation of jet momentum, a greater mass of air in the fuel jet results in lower overall jet velocities at any axial location, and therefore, a slower tip penetration.

Research has shown that the spreading angle of a fuel jet is a function of the ratio of the fuel and the ambient gas densities, orifice geometry parameters (e.g., sharp *versus* smooth orifice edges, aspect ratio, orifice orientation, etc.) (e.g., [1, 5, 25, 26, 27]), and effects of the injector needle and tip design on flow through an orifice (e.g., [28, 29]). Based on discussion in the previous section, the spreading angle in the penetration scaling law accounts for orifice flow effects and the effects of a large density difference between the fuel and the ambient gas on penetration, both of which impact turbulent mixing and air entrainment processes in a jet. Their appearance means that penetration depends on more than just the total jet momentum at the orifice exit, as is commonly assumed in the development of fuel jet penetration correlations. Other than the penetration correlation developed by Wakuri et al. [7] (valid in the far-field) and that given by Eq. (5.1), most commonly cited penetration correlations (e.g., [3]) do not directly account for the effects of various parameters on turbulent mixing and air entrainment processes.

Unfortunately, the effects of various parameters on the spreading angle of a fuel jet through their effect on turbulent transport are complex and not understood as yet. As a result, experimentally measured spreading angles are needed for the penetration scaling law. However, neither a standard measurement technique nor a commonly accepted definition exists for spreading angle. For the orifices and the conditions discussed in this chapter, the mean outer boundary angles were measured for non-vaporizing and vaporizing fuel jets. The outer boundary angles were measured with a schlieren/extinction diagnostic [1, 30]. The spreading angles measured were the angles of the quasisteady portion of the fuel jet, i.e., the upstream 80–90% of the fuel jet. This region of the fuel jet is responsible for the majority of the air entrainment. Use of the spreading angle for the quasisteady region is therefore consistent with the development of the scaling law.

The tangent of the outer boundary angles measured for the orifices considered in this work were found to be correlated by the empirical relationship [30]:

$$\tan(\theta/2) = c_1 \cdot \left(\left(\frac{\rho_a}{\rho_f} \right)^{0.19} - c_2 \cdot \sqrt{\frac{\rho_f}{\rho_a}} \right). \tag{5.22}$$

The constant c_1 in Eq. (5.22) is orifice dependent. So far, values for c_1 have been found in the range from 0.26 to 0.40 for different orifices [1, 30]. The constant c_2 has a value of zero for non-vaporizing fuel jets and 0.0043 for vaporizing fuel jets.

The measured mean outer boundary spreading angle for a real fuel jet given by Eq. (5.22) is approximately related to the spreading angle of the model fuel jet (α) in Fig. 5.1 by a constant [1]:

$$\tan(\alpha/2) = a \cdot \tan(\theta/2) \tag{5.23}$$

The recommended value for the constant a in Eq. (5.23) is 0.75. (This value is higher than the previous recommendation for a of 0.66 in Ref. [1]. The reason for this change is discussed later).

The value for a in Eq. (5.23) can be derived by two methods. The first method is empirical and involves a best fit of Eq. (5.1) to the experimentally measured penetration and spreading angle data, which results in the value of 0.75 for a. The second method involves developing a relationship between the spreading angle of the model fuel jet and a real fuel jet by equating the density, the mass flux, and the momentum flux in the model and the real fuel jets [1]. The second method requires knowledge of typical radial velocity and fuel concentration profiles in a fuel jet, but no prior knowledge of fuel jet penetration or spreading angles. This later approach was presented in Ref. [1] and resulted in values of a in the far-field of a fuel jet (where the spreading angle is important for penetration) that varied from 0.75 to 0.71 as $\tilde{S}$ varied from of one to infinity. An average value for a based on the second method is 0.73, which is in close agreement with the empirically derived value of 0.75.

The variation in a with axial distance indicated by the second method of determining a in the previous paragraph suggests that the spreading angle of the model fuel jet and the real fuel jet are not exactly related by a constant as is assumed in Eq. (5.23). However, in the far-field region where the spreading angle is important to penetration, the variation in a noted (0.75 to 0.71) has less than a three percent effect on the penetration, making the use of a constant value for a reasonable. The magnitude of the effect of the variation in a on penetration is considerably less than the experimental repeatability noted in typical fuel jet penetration data. The importance of the second method for estimating a is that it shows that a scaling law for fuel jet penetration (for a top-hat injection rate profile) that agrees with measured penetration data to within a few percent can be developed from conservation of mass and momentum considerations, without any prior knowledge of experimental penetration and spreading angle data.

Use of spreading angles measured by others in the penetration scaling law must be approached cautiously in light of the discussion in the previous paragraphs. For example, spreading angles that include the transient tip of the fuel jet in their determination should not be used in the scaling law. Including the transient tip in the measurement of the spreading angle generally results in spreading angles that are dominated by changes in the width of the transient tip region that occur as the jet develops. These temporal changes in the transient tip have little relationship to the air entrainment occurring in the quasisteady region of the fuel jet, which accounts for most of the air entrainment at any instant in time.

In addition, spray dispersion angles determined from atomization theory should not be used in the penetration scaling law. Atomization theory gives the dispersion angle of liquid droplets formed during breakup of an initial intact liquid core. This theory predicts a square-root dependence of the liquid-phase fuel dispersion angle on the ratio of the ambient gas and the fuel densities for diesel-like conditions [3]. The atomization based dispersion angles, however, are not directly related to the amount of air entrained into a fuel jet. At most, these angles are relevant in the first few millimeters near the injector orifice for current high-pressure DI diesel injectors. For high injection pressure conditions, research suggests significant intact liquid core lengths do not exist (e.g., [3, 16, 31]) and that dynamic equilibrium between the liquid and gas phases in a fuel jet is approached even in the near injector region [30].

As previously mentioned in the discussion following Eq. (5.23), the new value of 0.75 recommended for a in Eq. (5.23) is higher than the previous value of 0.66 recommended in Ref. [1]. The new recommendation results from new measurements of C_a for the orifices used in Ref. [1] made with an improved technique [30]. The new values for C_a for the three orifices used in Ref. [1], which had minimum diameters of 185, 241, and 330 μm, are 0.76, 0.74, and 0.69, respectively.[1] These values apply for an injection pressure difference across the injector orifice of 140 MPa (see Ref. [30] for injection pressure effects on C_a). The corresponding values for C_d for each orifice based on the orifice minimum diameter for the three orifices in Ref. [1] are 0.65, 0.71, and 0.65, respectively.

5.2.3 Comparison with Non-Vaporizing Penetration Data

The fuel jet penetration scaling law is compared with penetration data measured for non-vaporizing fuel jets in terms of dimensionless time and distance in Fig. 5.2. Data are shown for a wide range of conditions that include: ambient gas densities from 3.6 to 196 kg/m³, injection pressures from 75 to 160 MPa, fuel temperatures from 300 to 450 K, and three orifice diameters: 185, 241, and 330 μm.

Orifice area contraction coefficients (C_a) and discharge coefficients (C_d) for the three orifices used in acquiring the data were given in the previous section, Section 5.2.2. The velocity coefficient (C_v) for each orifice is given by the relationship

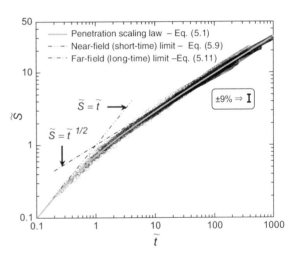

Fig. 5.2 A comparison of the measured non-vaporizing fuel jet penetration data (the symbols) and the penetration scaling law given by Eq. (5.1) (the light gray curve) in dimensionless coordinates. The broken lines are the short-time and long-time limits of the penetration scaling law given by Eqs. (5.9) and (5.11)

---

[1] Manufacturing techniques for orifices typically generate orifices with a slight taper in their diameter from inlet to outlet. The minimum diameter and the diameter at the orifice exit were reported for the three orifices in Ref. [1]. The orifice coefficients reported in Ref. [1] were based on the exit diameter, the larger of the two diameters. However, it does not matter which of the two diameters is used in the scaling law, as long as the measured orifice coefficients are defined based on the diameter that is used. The new C_a's reported here are based on the minimum diameter. The change to coefficients based on the minimum diameter was done for consistency with publications after Ref. [1] and all following discussions in this chapter. All orifice diameters cited in following discussions are minimum diameters, and the orifice coefficients are those based on the minimum diameters.

$C_d = C_a \cdot C_v$. Spreading angles needed in the scaling law were determined with Eqs. (5.22) and (5.23), where the value used for a in Eq. (5.23) was the new value of 0.75 and the constant c_1 in Eq. (5.22) for the 185, 241, and 330-μm orifices were 0.31, 0.31, and 0.4, respectively [1]. [The constant c_2 in Eq. (5.22) is 0 for the non-vaporizing conditions of Fig. 5.2].

Figure 5.2 shows that there is excellent agreement between the scaling law and non-vaporizing fuel jet penetration data. (Approximately 2000 data points are shown.) In general, the experimental data fall within $\pm 9\%$ ($\pm 2\sigma$) of the scaling law. The agreement is comparable to the $\pm 7\%$ ($\pm 2\sigma$) repeatability noted in the penetration data for any given set of conditions [1]. With the new value of 0.75 used for a, even the slight systematic 8% deviation of the data above the scaling law near a $\tilde{t} = 1$ noted in Ref. [1] no longer exists. Previously the deviation was thought to be caused by velocity slip between the liquid and gas phases in this near-injector region of the fuel jet. However, the revised fit shown in Fig. 5.2 suggests that this is not a significant factor for the conditions of the experiments.

5.2.4 Effects of Vaporization and Combustion on Penetration

Figure 5.3 compares fuel jet penetration data acquired under vaporizing, but non-combusting conditions, with the penetration predicted by the scaling law given by Eq. (5.1). Combustion was inhibited by having no oxygen in the ambient gas [1]. The orifice diameter, the pressure drop across the orifice, and the ambient-gas density were 241 μm, 137 MPa, and 1000 K for the data shown. The ambient temperature is representative of the top-dead-center temperature in a diesel engine.

The comparison in Fig. 5.3 shows that fuel vaporization slows penetration relative to non-vaporizing fuel jets, with the effect being most apparent at the lower

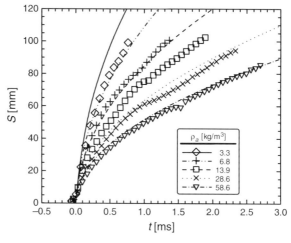

Fig. 5.3 Measured penetration data as a function of time for vaporizing fuel jets (the symbols). The curves give the penetration predicted by the non-vaporizing fuel jet penetration scaling law, Eq. (5.1), for each ambient-gas density

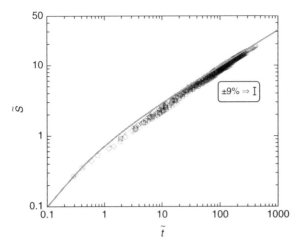

Fig. 5.4 A comparison between the vaporizing fuel jet penetration data (the symbols) and the penetration scaling law given by Eq. (5.1) (the gray curve) in dimensionless coordinates

density conditions. The reduction is as much as 20% at these lower density conditions. At higher gas densities the effect becomes significantly less noticeable.

Figure 5.4 shows a comparison of the vaporizing fuel jet penetration data with the scaling law in terms of dimensionless coordinates. Fuel jet spreading angles measured for vaporizing fuel jets were used in the scaling law [$c_2 = 0.0043$ in Eq. (5.22)]. The data shown in Fig. 5.4 were taken for ambient gas densities from 3.3 to 59 kg/m^3, injection pressures from 75 to 160 MPa, the three orifices discussed with Fig. 5.2, and an ambient gas temperature of 1000 K.

Figure 5.4 shows that the data are very well correlated in the dimensionless coordinates. The correlation of the data is as good as that noted for the non-vaporizing data in Fig. 5.3. The only significant difference between the data and the scaling law is that the data fall below the scaling law in the range of $\tilde{t}$ between 1 and 20, which corresponds to a range of $\tilde{S}$ between 0.7 and 5.0. The difference is the result of the slower penetration noted in Fig. 5.3. The peak deviation is about 15–20% and occurs around a $\tilde{t}$ of 3 or an $\tilde{S}$ of 1.5. The region where the vaporizing fuel jet deviates most from scaling law corresponds to the general region where fuel vaporization is complete and the liquid length establishes for typical engine conditions [30]. After the fuel jet penetrates beyond the liquid length, the impact of vaporization on the fuel jet tip penetration begins to fade.

The effect of vaporization on fuel jet penetration noted in Figs. 5.3 and 5.4 is most likely the result of a combination of competing effects caused by cooling of entrained air by the vaporization process. The cooling results in an overall higher density mixture in the fuel jet relative to a non-vaporizing fuel jet for the same ambient density. Based on conservation of jet momentum, a higher overall density in the fuel jet will result in a slower penetration velocity. A competing effect, however, is that air entrainment is likely to be reduced by vaporization, thus resulting in a slower deceleration of the fuel jet, i.e., faster penetration. Vaporization reduced air entrainment is suggested by the reduced spreading angles measured for vaporizing fuel jets [see Eq. (5.22)].[2] Since Figs. 5.3 and 5.4 show that the fuel jet penetration slows as a result of vaporization, the dominant effect appears to be the reduction in penetration speed due to an overall higher density in the fuel jet and conservation of momentum.

Figure 5.5 compares fuel jet penetration data (the symbols) acquired under combusting conditions with the data from Fig. 5.3 acquired under vaporizing, non-combusting conditions. The vaporizing data are represented by the curves which are best fits of a function of the form of Eq. (5.1) to the vaporizing data in Fig. 5.3. The combusting penetration data (previously unpublished) were analyzed in a manner similar to that described in Ref. [1] for non-vaporizing and vaporizing fuel jets. The ambient-gas temperature for all the data in Fig. 5.5 was 1000 K.

The comparison in Fig. 5.5 shows that the effects of combustion depend on ambient gas density. For the lowest two density conditions, the penetration data agree with the vaporizing only penetration data. The agreement results because ignition does not occur until after the fuel jet has penetrated to beyond 100 mm, the extent of the penetration data shown. For the higher three density conditions, however, the combusting fuel jets penetrate faster than the vaporizing fuel jets after an initial period during which they follow each other. For the 13.9, 28.6 and 58.6 kg/m^3 conditions, the deviation of the combusting fuel jet penetration data from the vaporizing fuel jet penetration data begins at about 1.2, 0.45 and 0.3 ms, respectively. The time at which the deviation from a vaporizing fuel jet begins occurs

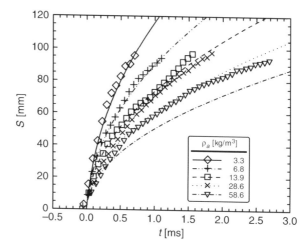

Fig. 5.5 Measured penetration data as a function of time for combusting fuel jets (the symbols). The curves are fits to the vaporizing fuel jet penetration data shown in Fig. 5.3 for each respective ambient-gas density. The orifice diameter, the pressure drop across the injector orifice, and the ambient-gas temperature are the same as for Fig. 5.3

---

[2] Note that some of the reduction in spreading angle for a vaporizing fuel jet may simply be due to a contraction of the fuel jet as a result of cooling of the entrained hot air by fuel vaporization. (The volume decrease resulting from contraction of entrained air as it cools is estimated to be significantly greater than the volume increase resulting from the vaporization of the fuel.)

progressively closer to the time of ignition as density increases. The times of ignition for the three densities were 0.6, 0.35 and 0.25 ms, respectively. Conditions at other temperatures, densities, and injection pressures showed similar trends. The results indicate that a fuel jet behaves like a vaporizing fuel jet until combustion begins. Once combustion begins, the fuel jet penetration in the region of the fuel jet that is burning decelerates less rapidly, and as a result penetrates further in the same time.

Figure 5.6 shows a comparison of the combusting fuel jet penetration data (the symbols) along with the scaling law for non-vaporizing fuel jets [Eq. (5.1) – the solid curve] in terms of dimensionless coordinates. The data shown in Fig. 5.6 were taken for the same range of conditions as the data in Fig. 5.4, except in an ambient gas with 21% oxygen (by volume). The spreading angle used in the length and time scales in Fig. 5.6 are those measured for the vaporizing fuel jets at the same ambient gas density and temperature. (Spreading angles for combusting fuel jets are not used because they are difficult to define and are less directly related to air entrainment. Much of the increase in spreading noted when combustion occurs is due to the dilatation from heat release. Research on atmospheric pressure gas jets shows that combustion actually decreases the entrainment of air into a jet by as much as a factor of 2.5. [19]).

The data in Fig. 5.6 confirm the observations from Fig. 5.5. Namely, the fuel jet penetration is initially slowed by vaporization, causing the penetration to be less than for a non-vaporizing fuel jet. However, after ignition occurs and combustion begins, the fuel jet decelerates less rapidly and ends up penetrating faster than a non-vaporizing fuel jet. Since ignition varies widely with conditions, the exact time

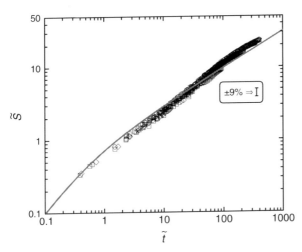

Fig. 5.6 A comparison of the combusting fuel jet penetration data (the symbols) and the penetration scaling law given by Eq. (5.1) (the gray curve) in dimensionless coordinates. The conditions are the same as for Fig. 5.4 except that the ambient gas has 21% (by volume) oxygen, allowing ignition and combustion to occur

at which combustion begins to play a role varies with conditions. For the data in Fig. 5.6, ignition typically occurs after a $\tilde{t}$ of approximately 20. As a result, as $\tilde{t}$ becomes larger than 30, the fuel jet begins to have a longer penetration distance than a non-vaporizing fuel jet on average for the data shown. Ultimately, the combusting fuel jet can penetrate as much as 20% further than a non-vaporizing fuel jet in the same time. The trends in Fig. 5.6 therefore suggest that the penetration scaling law derived for a non-vaporizing fuel jet is accurate to approximately ±20% for a burning fuel jet. (Note: the trends also suggest that an accuracy of ±20%, at best, applies for any fuel jet penetration correlation applied to a burning fuel jet if it does not directly account for combustion and vaporization effects.)

Based on the flame heights measured for burning fuel jets [32], combustion is expected to be completed by an $\tilde{S}$ of approximately 16, which corresponds to a $\tilde{t}$ of 270. For $\tilde{t}$ larger than 270, the percentage difference between the combusting fuel jet and the scaling law for a non-vaporizing fuel jet should begin to decrease since combustion is complete.

Vaporization effects on penetration observed in Figs. 5.3–5.6, which occur near the injector, will play a role in fuel jet penetration (up to the piston bowl wall) in engines. Exactly how significant the combustion effects shown in Figs. 5.5 and 5.6 are in an engine will depend on when ignition occurs, the parameters in the characteristic length and time scales [Eqs. (5.4) and (5.5)], and the distance from the injector to the piston bowl wall (i.e., the maximum penetration distance prior to wall impingement). For a typical moderate-density condition in an engine with a 50 mm distance between the injector and the piston bowl wall, estimates indicate that $\tilde{t}$ will be about 35 when the fuel jet hits the wall. Comparison of Figs. 5.4 and 5.6 indicates that at this time the combusting fuel jet will have penetrated as much as 10% further than a vaporizing fuel jet on average, but have about the same penetration as a non-vaporizing fuel jet. For larger piston bowls (i.e., larger engines) or higher density conditions, the effects of combustion on penetration up to the piston bowl wall should become more significant. For smaller piston bowls (i.e., smaller engines) or lower densities, the difference between the penetration of a combusting fuel jet and a vaporizing fuel jet up to the piston bowl wall should become less significant.

The effects of combustion on penetration noted in Figs. 5.5 and 5.6 are the result of at least two factors. First, the average density in the fuel jet should decrease when heat release occurs, resulting in a faster penetration rate based on conservation of jet momentum. Second, as previously mentioned, heat release has been found to significantly decrease air entrainment into atmospheric pressure, combusting gas jets [19]. A similar decrease in air entrainment at diesel-like conditions due to heat release will result in faster penetration, due to less mass in the fuel jet and conservation of jet momentum. A decrease in the ambient gas entrainment rate with heat release is supported by Eq. (5.19). Heat release in a diesel fuel jet occurs largely at the periphery of the jet [33, 34]. Heat release at the periphery will decrease the density at the periphery and effectively act like a reduction in ambient gas density, which Eq. (5.19) suggests will result in a reduction in air entrainment.

5.2.5 Comparisons with Other Penetration Correlations

A comparison of fuel jet penetration scaling law given by Eq. (5.1) to a large number of other penetration correlations in the literature was made in Ref. [1]. This was accomplished by comparing the penetration dependence on the major parameters noted by various researchers to those predicted by Eq. (5.1). The comparison suggested that a large portion of the differences in penetration predicted by various correlations (as noted by Hay and Jones [4] and others) can be traced to not accounting for the near-field and far-field (i.e., short-time and long-time) differences in the penetration characteristics of fuel jets reflected in Eqs. (5.9) and (5.11). Extrapolation of correlations that do not account for the near-field and far-field differences in penetration to conditions outside the data range on which they are based can easily result in the large differences noted in penetration predicted by various correlations.

Detailed comparisons to two specific fuel jet penetration correlations widely cited in text books (e.g., [3, 35, 36, 37]), one by Hiroyasu and Arai [5] and one by Dent [8], were also made in Ref. [1]. As mentioned previously, Hiroyasu and Arai recognized the need to take into account differences in fuel jet penetration in the near-field and the far-field. Their recommended penetration correlations for the near-field and far-field are similar to the near-field and far-field limits of Eq. (5.1), given by Eqs. (5.10) and (5.12), respectively. In the near-field, their correlation is the same as Eq. (5.10), except for the appearance of a constant with a value of 0.39 in place of the orifice dependent velocity coefficient (C_v), which typically has a value greater than 0.85 [30]. The difference between typical values for C_v and the constant of 0.39 cited by Hiroyasu and Arai is most likely the result of an injection rate profile with a more gradual initial increase than the top-hat profile used in Ref. [1]. Penetration data acquired with a more gradual increase in the initial injection rate would effectively give the appearance of a smaller C_v.

In the far-field, the penetration correlation developed by Hiroyasu and Arai has the same dependencies on orifice diameter, fuel pressure drop across the orifice, and time as shown by the far-field limit of Eq. (5.1), i.e., Eq. (5.12). However, their correlation lacks the spreading angle and the orifice coefficient terms present in Eq. (5.12). Given the spreading angle dependence on the ambient gas density and the fuel density shown in Eq. (5.22), the net result is that the far-field correlation of Hiroyasu and Arai has different overall dependencies on the ambient gas and fuel densities than shown by Eq. (5.12).

Dent's [8] penetration correlation, which does not account for differences in the near-field and far-field penetration, is similar to Hiroyasu and Arai's far-field correlation, and therefore, has similar differences with Eq. (5.12) as noted for Hiroyasu and Arai's correlation. Dent's correlation does have a temperature term, but this term does not account for the effects of fuel vaporization or combustion on penetration noted in the results presented in Figs. 5.3–5.6.

An additional difference between the penetration scaling law given by Eq. (5.1) and the penetration correlation of Hiroyasu and Arai is the interpretation placed on the transition between the near-field and the far-field. The development of the scaling law suggests the transition occurs when the entrained air mass in the fuel jet becomes greater than the injected fuel mass, as discussed in conjunction with Eqs. (5.14) and (5.15). Hiroyasu and Arai interpret the transition as a "breakup time" corresponding to the time when the intact core is completely atomized. However, research suggests that for current DI diesel injection pressures, a fuel jet will breakup very near the injector orifice (e.g., [3, 16]), and possibly even start breaking up within the orifice as a result of cavitation [31].

5.2.6 Application of the Penetration Scaling Law

The penetration scaling law given by Eq. (5.1) is most useful (a) for the understanding it provides regarding the physical processes and the parameters that control fuel penetration, (b) for assessing the relative effects of various parameters on penetration, and (c) for the simplified, well characterized baseline it provides for comparing with computational models being developed to simulate diesel combustion processes. The penetration scaling law can also be used to approximate fuel jet penetration in an engine if the limitations imposed by the development of the scaling law and their potential impact on penetration predictions are considered. This later statement is true of any penetration correlation in the literature. Not taking into account the limitations often leads to confusion when comparing correlations developed with data acquired under significantly different conditions, for example, with significantly different injection rate profiles or orifice flows.

The fuel jet penetration scaling law given by Eq. (5.1) is most directly applicable to a non-vaporizing fuel jet generated by an injector with a top-hat or near top-hat injection profile, and an orifice flow that does not have a significant radial asymmetry or azimuthal flow (i.e. the jet momentum is predominantly axial at the orifice exit). When applying the scaling law to combusting and vaporizing fuel jets, Figs. 5.3–5.6 show that the accuracy of the scaling law will be limited to about $\pm 20\%$. For a vaporizing fuel jet the penetration will be over-predicted by as much as 20%. For a combusting fuel jet, the initial penetration will be over-predicted by as much as 20%, but after ignition, the fuel jet penetration in the far-field will accelerate and eventually be under-predicted by as much as 20%. For injections with a non-top-hat injection rate profile, the accuracy of any prediction will depend on the degree of rate-shaping in the injection rate profile and the average pressure used to represent the actual time-varying injection pressure. As the rate-shape begins to diverge significantly from a top-hat profile, the transient penetration profile will begin to differ from that predicted by Eq. (5.1).

Use of the scaling law to predict penetration also requires knowledge of orifice flow coefficients and fuel jet spreading angles. When only a discharge coefficient, C_d, is known, it is recommended that C_a and C_v be replaced by the square-root of C_d. This approximation equally apportions the discharge coefficient effects to C_a and C_v and is consistent with the fact that C_d is the product C_a and C_v. For conditions where the spreading angle is unknown, the average spreading angle dependence on the ambient gas to fuel density ratio measured in Refs. [1, 30] is recommended [i.e., $c_1 \approx 0.3$ in Eq. (5.22)]. If some penetration data is available for a particular injector, but not the spreading angle data, it is also possible to "best fit" the scaling law to the available penetration data by adjusting the constant c_1 in Eq. (5.22).

5.3 Liquid-Phase Fuel Penetration

Liquid-phase fuel penetration and vaporization are important factors in the DI diesel engine combustion processes. Penetration of the fuel is needed to promote fuel–air mixing as mentioned in the previous section, however, over-penetration of the liquid-phase fuel can lead to higher emissions if the liquid fuel impinges and collects on piston bowl walls. The effects of a wide range of parameters on the maximum extent of liquid-phase fuel penetration in diesel fuel jets have been examined with the goal of identifying the parameters and the processes that control fuel vaporization and liquid-phase fuel penetration [2, 30, 38]. Parameters varied in the investigations included: injection pressure, orifice diameter and aspect ratio, ambient gas temperature and density, and fuel temperature and volatility. The fuels considered included cetane, heptamethylnonane (HMN), and a #2 diesel fuel (DF2) [2, 30], as well as several alternative diesel fuels and gasolines [38].

The maximum penetration distance of the liquid-phase fuel is referred to as the "liquid length." The liquid length occurs at a location where the total fuel vaporization rate (between the orifice and the liquid length) equals the fuel injection rate. Time-averaged images of Mie-scattered light from the liquid-phase fuel in diesel fuel jets, such as those shown in Fig. 5.7, were used to determine the quasi-steady mean liquid lengths. Each image in Fig. 5.7 was acquired at a different ambient gas density with all other parameters held fixed. The light region in each image is Mie-scattered light from the liquid-phase fuel in the diesel fuel jet. The fuel was

Fig. 5.7 Time-averaged, Mie-scattered light images of three vaporizing fuel jets injected from left to right into the ambient gas density given in the upper left corner of each image. The liquid lengths determined from each image are given in the lower-right corner. The orifice pressure drop and diameter, the ambient-gas temperature, the fuel temperature, and the fuel were 135 MPa, 246 μm, 1000 K, 438 K and DF2

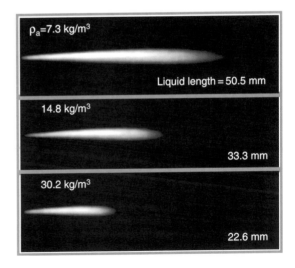

injected from left to right from an injector orifice located at the left edge of each image. The liquid length is determined from the images of Mie-scattered light by defining a threshold light intensity and then determining the maximum axial extent of the fuel jet with a scattered light intensity above the threshold intensity [2]. The liquid lengths for the three images in Fig. 5.7 are given in the lower right corner of each image.

The first part of this section presents trends observed in the liquid length data with respect to various engine parameters. In the second part, a scaling law developed by Siebers [30] for liquid length is presented and compared with the liquid length data.

5.3.1 Effects of In-cylinder and Injector Parameters

Two of the major trends observed in the liquid length data are shown in Figs. 5.8 and 5.9. Figure 5.8 shows that the liquid length has a linear dependence on orifice diameter, decreasing as the orifice diameter decreases for all conditions examined. Figure 5.9 shows that the liquid length is independent of the pressure drop across the injector orifice for all conditions. (The measured data are the solid gray symbols. The open symbols and lines through them in the figures are derived from a liquid length scaling law discussed in the next section.) The lack any dependence on the injection pressure drop across the orifice was previously noted by Kamimoto et al. [39] and Yeh et al. [40].

The results in Fig. 5.8 are important for the small-bore engines currently being developed for automotive applications, since liquid fuel impingement on piston

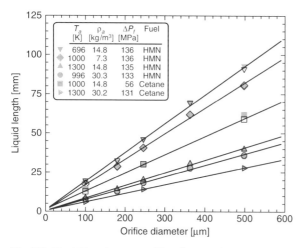

Fig. 5.8 Liquid length *versus* orifice diameter for a wide range of conditions. Measured data are given by the solid gray symbols. The open symbols are predictions made using the liquid length scaling law, Eq. (5.24). The lines are least-squares fits to the scaling law predictions for each set of conditions given in the legend. The aspect ratio of the orifices and the fuel temperature were 4.2 and 438 K

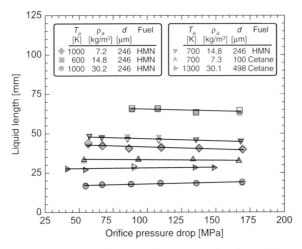

Fig. 5.9 Liquid length *versus* orifice pressure drop for a wide range of conditions. Measured data are given by the solid gray symbols. The open symbols are predictions made using the liquid length scaling law, Eq. (5.24). The lines are least-squares fits to the scaling law predictions for each set of conditions given in the legend. The aspect ratio of the orifices and the fuel temperature were 4.2 and 438 K

bowl walls is a concern in these engines. Figure 5.8 indicates that the smaller orifice sizes used in small-bore engines will help mitigate liquid impingement on piston bowl walls. Also, Fig. 5.9 indicates that the current trend toward higher injection pressures in diesel engines for emissions control does not by itself increase the likelihood of liquid impingement on piston bowl walls.

Figures 5.10 and 5.11 show the effects of ambient gas temperature and density on liquid length for two single-component, diesel-like fuels. The data points in Fig. 5.10 represent liquid lengths measured for heptamethylnonane (HMN), while in Fig. 5.11 they represent liquid lengths measured for cetane. The gray region in each figure shows the range of liquid lengths expected in engines for each fuel. (As in Figs. 5.8 and 5.9, the curves through the data points are derived from a liquid length scaling law discussed in the next section.) The results show that the liquid length decreases in a strong non-linear manner with increasing density or temperature for each fuel, similar to trends observed in an engine over a narrower range of conditions [41]. The decrease in liquid length with increasing density and temperature is due to factors such as increased air entrainment with increasing density, the increased energy content of higher temperature air, and the temperature and pressure dependent thermodynamic properties of the fuels [30].

Cross-comparison of the liquid lengths for each of the fuels in Figs. 5.10 and 5.11 also shows that liquid length is longer for cetane, the lower volatility fuel of the two fuels. (Cetane has a 40 K higher atmospheric pressure boiling point than HMN.) The differences due to fuel volatility are most pronounced at the lower temperature conditions (i.e., cold start and light load engine conditions). The effects of fuel volatility on liquid length were explored for a wider range of fuels by Higgins et al.

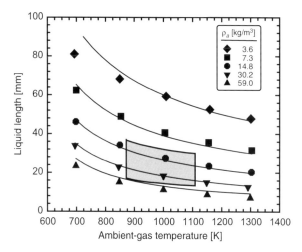

Fig. 5.10 Liquid lengths for HMN as a function of ambient-gas temperature for five ambient-gas densities. The symbols are measured data. The curves are predictions made with the liquid length scaling law, Eq. (5.24). The orifice pressure drop, the orifice diameter, and the fuel temperature were 136 MPa, 246 μm, and 438 K. The light gray region represents the range of liquid lengths expected for representative of TDC conditions in light- and heavy-duty DI diesels

[38] and Canaan et al. [42]. Both of these works showed that for diesel-like fuels there is a linear decrease in liquid length with decreasing boiling point temperature for a single component fuel or decreasing temperature at the 90% point on the distillation curve for a multi-component fuel. However, when a much wider range

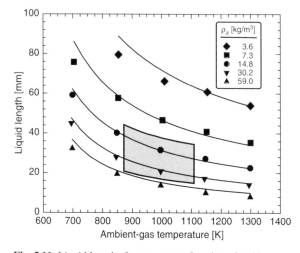

Fig. 5.11 Liquid lengths for cetane as a function of ambient-gas temperature for five ambient gas densities. The symbols are measured data. The curves are predictions made with the liquid length scaling law, Eq. (5.24). See Fig. 5.10 for the experimental conditions and the definition of the light gray region

of fuels that included methanol, gasolines, and biodiesel was considered, the linear relationship with volatility (as measured by boiling point or distillation temperatures) no longer held [38]. Methanol, for example was found to have liquid lengths close to those measured for a diesel fuel. The relatively long liquid lengths for the high volatility methanol resulted from the high heat of vaporization of methanol. Because of methanol's high heat of vaporization, almost as much energy is required to vaporize methanol as is required to vaporize DF2 when starting from the same initial fuel temperature. This means nearly the same length of the fuel jet will be required to entrain enough energy to vaporize either DF2 or methanol (i.e., the two fuels will have nearly the same liquid length).

5.3.2 Liquid Length Scaling Law

The liquid length trends with respect to orifice diameter in Fig. 5.8 and orifice pressure drop in Fig. 5.9 suggest that the vaporization process in a DI diesel fuel jet is mixing-limited, i.e., that fuel jet entrainment and mixing processes, and not transport processes at droplet surfaces, control or limit vaporization [2, 30]. Assuming mixing-limited vaporization, the liquid-phase fuel will be completely vaporized at an axial distance where the total hot air entrained contains enough energy to vaporize the injected fuel. The quantity of fuel injected is proportional to the injection velocity (i.e., the square-root of the orifice pressure drop) and the orifice diameter squared, while the total air entrained by a fuel jet up to any axial location is directly proportional to the axial distance, the orifice diameter, and the injection velocity [30]. These dependencies indicate that to vaporize the additional fuel injected through a larger orifice diameter, the axial distance needed to entrain enough hot air to vaporize the fuel must increase in proportion to an increase in orifice diameter, thus resulting in the linear increase in liquid length with increasing orifice diameter noted in Fig. 5.8. With respect to injection velocity, the dependencies mean that equal increases in the total amount of air entrained up to any axial location and the fuel injection rate occur with an increase in injection velocity. As a result, no net change in the axial distance required to entrain enough hot air to vaporize the fuel will occur with a change in injection velocity. The result is no change in the liquid length with a change in injection velocity (i.e., orifice pressure drop), as noted in Fig. 5.9, just faster fuel vaporization that compensates for a higher fuel injection rate. If on the other hand transport processes at droplet surfaces are assumed to limit vaporization, instead of the overall fuel jet entrainment and mixing processes, no similar simple explanation for the trends in Figs. 5.8 and 5.9 can be developed [2].

A scaling law for liquid-phase fuel penetration can be derived based on the concept of mixing-limited vaporization by considering an energy balance for the model fuel jet in Fig. 5.1 in addition to the mass and momentum balances used to determine the penetration scaling law for the fuel jet tip, Eq. (5.1). The mass, momentum, and energy balances are used to solve for the axial location at which enough energy has been entrained into the fuel jet to vaporize the fuel, i.e., the liquid length. The development is presented in detail in Ref. [30]. Three important approximating assumptions made in addition to those already discussed with the development of the fuel jet tip penetration scaling law, Eq. (5.1), are that saturated conditions exist at the liquid length, that idealized phase equilibrium can be applied, and that the form of the equation for air entrainment rate given by Eq. (5.16) for non-vaporizing fuel jets applies for vaporizing fuel jets. The agreement that will be shown between the form of the liquid length scaling law developed and the measured liquid length data will suggest that the additional assumptions are reasonable, or at least closely approached.

The resulting liquid-phase fuel penetration scaling law is given in dimensionless form by [30]:

$$\tilde{L} = 0.47 \cdot \sqrt{\left(\frac{2}{B(T_a, P_a, T_f)} + 1 \right)^2 - 1}. \tag{5.24}$$

The term $\tilde{L}$ in Eq. (5.24) is the liquid length L normalized by x^+ defined by Eq. (5.5). The constant, 0.47, in Eq. (5.24) was derived from a best fit of the liquid length scaling law to all the liquid length data acquired.[3]

The term B in Eq. (5.24) is derived from the energy balance on the fuel jet and an equation of state. For single component fuels, B is given by:

$$B(T_a, P_a, T_f) = \frac{Z_a(T_s, P_a - P_s) \cdot P_s M_f}{Z_f(T_s, P_s) \cdot [P_a - P_s] \cdot M_a} = \frac{h_a(T_a, P_a) - h_a(T_s, P_a - P_s)}{h_f(T_s) - h_f(T_f, P_a)}. \tag{5.25}$$

Equation (5.25) is an implicit equation for the saturated fuel/ambient-gas mixture temperature, T_s, at the liquid length in terms of the fuel and ambient gas properties and the initial fuel and ambient gas conditions. The temperature T_a in Eq. (5.25) is the ambient gas temperature. The pressures P_a and P_s are the ambient gas pressure and the saturation (i.e., partial) pressure of the fuel at the liquid length, respectively. This implies that the partial pressure of the entrained ambient gas in the fuel jet is given by $P_a - P_s$. Since the fuel is at saturated conditions at the liquid length, the saturation temperature T_s is related to the partial pressure of the vapor fuel, P_s, through the saturation pressure-temperature relationship for a given fuel. The terms h_f and h_a and M_f and M_a refer to the specific enthalpies and the molecular weights of the fuel and ambient gas, respectively, while Z_f and Z_a are the compressibilities of the vapor-phase fuel and the ambient gas at saturated conditions, respectively. The

---

[3] Note that the value of the constant in Eq. (5.24), 0.47, is different than the value of 0.41 originally recommended in Ref. [30]. The change is required by the change in the constant a in Eq. (5.23), which appears in Eq. (5.24) through the normalization of L by x^+. The change in the value of the constant in Eq. (5.24) is proportional to the change recommended in a in Eq. (5.23), and therefore, cancels out the effect of the change in a in Eq. (5.23) on the liquid length scaling law. The net result is that there is no change in liquid lengths predicted with Eq. (5.24) and the same agreement between Eq. (5.24) and the experimental data previously noted in Ref. [30] is obtained.

fuel at the orifice exit is assumed to be at the injector tip temperature, T_f, and the ambient pressure, P_a.

The enthalpy difference in the numerator on the right side of Eq. (5.25) is the specific enthalpy transferred from the entrained ambient gas to heat and vaporize the fuel. The enthalpy difference in the denominator is the specific enthalpy required to heat and vaporize the liquid fuel.

The method for determining a liquid length from the scaling law is to iteratively solve Eq. (5.25) for T_s, which defines B for a given set of conditions. The values for B derived from Eq. (5.25) and x^+ from Eq. (5.5) can then be used into Eq. (5.24) to determine the liquid length, L. The detailed temperature and pressure dependent thermodynamic properties for single component fuels, such as HMN and cetane, and the ambient gas were obtained from the API [43] and DIPPR [44] databases as discussed in Ref. [30].

Figure 5.12 shows the comparison between Eq. (5.24) and all the measured liquid length data for cetane and HMN from Ref. [1] for all conditions considered. The data points in the figure are the measured liquid lengths normalized by x^+ (symbols) plotted *versus* the value for B determined from Eq. (5.25) for each set of conditions examined for HMN and cetane. The conditions included in the figure cover ambient gas temperatures and densities from 700 to 1300 K and 3.3 to 60 kg/m³, injection pressures from 40 to 190 MPa, orifice diameters from 100 to 500 μm, orifice aspect ratios from 2 to 8, and fuel temperatures from 375 to 440 K. A total of seventy data

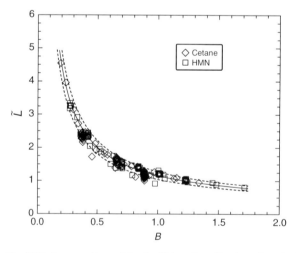

Fig. 5.12 A comparison of the liquid length scaling law given by Eq. (5.24) (the solid curve) and all the liquid lengths measured for cetane and HMN in dimensionless coordinates ($\tilde{L}$ versus B). The conditions included in the figure cover ambient-gas temperatures and densities from 700 to 1300 K and 3.3 to 60 kg/m³, injection pressures from 40 to 190 MPa, orifice diameters from 100 to 500 μm, orifice aspect ratios from 2 to 8, and fuel temperatures from 375 to 440 K. The band between the dashed lines represents $\pm 2\sigma$ in the difference between the scaling law and the experimental data

points are included in the figure (each the average of several measurements [2, 30]). The solid curve in the figure is the liquid length scaling law given by Eq. (5.24).

Figure 5.12 shows that there is close agreement between the scaling law and the measured data over the range of conditions considered. The standard deviation between the scaling law and all the measured liquid lengths is ±4%. The dashed lines in Fig. 5.12 represent a band of ±8% (or ±2σ). The data points falling outside of the ±8% band (generally on the low side) are for the extreme low temperature-density conditions and the highest density conditions of the experiment [30]. The close agreement between the trends predicted by the scaling law and those observed in the data is also visible in physical coordinates in Figs. 5.8–5.11. The lines and curves in those figures give the liquid length trends predicted by the scaling law. Agreement similar to that shown in Figs. 5.10 and 5.11 for cetane and HMN was also found for methanol, a very high volatility fuel.

5.4 Flame Lift-Off

After autoignition, the free portion of a DI diesel fuel jet (i.e., the portion between the injector and the piston bowl wall) burns in a lifted, turbulent diffusion flame mode. The lifted, turbulent diffusion flame mode continues until the end of injection or until the tip of the penetrating fuel jet is deflected by the piston bowl wall and adjacent jets back to the upstreammost location of the turbulent diffusion flame, i.e., the lift-off length. (The lift-off length is defined as the distance between the fuel injector tip and the upstreammost extent of combustion in the fuel jet.)

Two types of combustion are believed to be occurring in the lifted flame: a turbulent diffusion flame and a rich, partially-premixed flame [33]. The turbulent diffusion flame surrounds the periphery of the fuel jet [33, 34], with its upstream extent defining the lift-off length. The rich, partially premixed combustion is believed to occur in the central region of the fuel jet just downstream of the lift-off length [33, 45]. Any air entrained upstream of the lift-off length reacts with fuel in this central region, generating a significant local heat release and a high-temperature product gas, rich in unburned and/or partially reacted fuel. The product gas of the rich, central reaction zone becomes the "fuel" for the diffusion flame at the jet periphery. Moreover, soot formation has been hypothesized to begin in the product gas, with soot concentration and particle size growing as the product gas is transported downstream in the fuel jet [33].

For typical quiescent DI diesel conditions, the majority of the fuel burned in the free portion of a fuel jet is consumed in the turbulent diffusion flame surrounding periphery of the fuel jet. The amount of fuel and air that react in the rich, central reaction zone will depend on the amount of air entrained upstream of the lift-off length, and therefore, on the parameters that affect the lift-off length and the entrainment of air into a fuel jet. For a moderate-load condition in a heavy-duty engine, fuel concentration measurements [33] and air entrainment estimates [46] indicate that as much as 20% of the air required to burn the injected fuel is entrained upstream of the lift-off length.

The discussion above implies that lift-off length plays an important role in diesel combustion and emission formation processes. The potential importance and an almost complete lack of understanding of flame lift-off on diesel fuel jets has prompted recent investigations of flame lift-off and its impact on fuel jet combustion and soot formation processes [32, 46, 47, 48, 49, 50, 51, 52]. The following sections summarize the trends in lift-off observed in the investigations, examine the amount of fuel–air mixing that occurs upstream of the lift-off length, and investigate the link between soot formation within a fuel jet and the amount of fuel–air mixing upstream of the lift-off length. Measurements of lift-off length and soot concentration in fuel jets are presented. Parameters varied in the investigations included: injection pressure, orifice diameter, and ambient gas temperature, density and oxygen concentration. All of the results presented were acquired using DF2 as the fuel.

5.4.1 Effects of In-Cylinder and Injector Parameters on Lift-Off

Time-averaged, line-of-sight images of light emitted at 310 nm from burning fuel jets, were used to determine lift-off length. Example images, acquired with an intensified CCD camera, are shown in Fig. 5.13. Each image in Fig. 5.13 was acquired for a different ambient gas temperature (given in the upper-left corner of an image) with all other parameters held fixed. The fuel was injected from left to right from an injector orifice located at the left edge of each image.

The light emitted in the vicinity of the flame lift-off length at a wavelength of 310 nm is dominated by chemiluminescence from excited-state OH (OH*) [47]. Unlike ground state OH which exists as an equilibrium product in regions of high temperature, OH* is short-lived and results from chemical reactions in near-stoichiometric, high heat release regions [53]. Since stoichiometric combustion is expected at the flame base at the lift-off length [54], OH chemiluminescence provides a good marker of the lift-off length. The distance between the injector orifice and the upstreammost location of OH chemiluminescence in the images is defined as the lift-off length [46, 47]. The lift-off length determined from each image in Fig. 5.13 is given in the lower-left corner of each respective image. It is also indicated by the length of a horizontal white line in each image.

Figures 5.14 through 5.18 show the effects of several in-cylinder and injector parameters on the flame lift-off length on DI diesel fuel jets under quiescent conditions determined from images such as those in Fig. 5.13. The effects of ambient gas temperature and density on lift-length are shown in Fig. 5.14. The figure is a plot of lift-off length *versus* ambient gas temperature for four ambient gas densities ranging from 7.3 to 58.5 kg/m^3. The ambient gas oxygen concentration, orifice diameter and pressure drop across the orifice were 21%, 180 μm and 138 MPa, respectively. The curves through the data are trend lines included to help visualize the effects of temperature at each density.

Figure 5.14 shows that both ambient gas temperature and density have strong effects on the lift-off length. As either parameter increases, the lift-off length

Fig. 5.13 Time-averaged, line-of-sight images of the relative light intensity emitted at 310 nm from three burning diesel fuel jets injected into ambient gas with 21% oxygen and the temperatures given in the upper-left corner of each image. Fuel was injected from left to right. The lift-off length determined from each image is marked by a horizontal white line and is given in the lower-left corner of each image. The ambient-gas density, injection pressure, and orifice diameter were 14.8 kg/m³, 138 MPa, and 180 μm

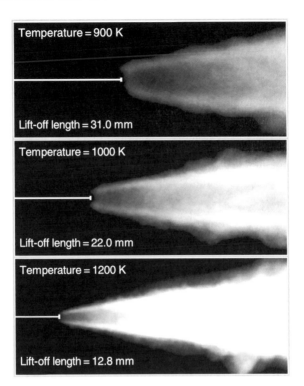

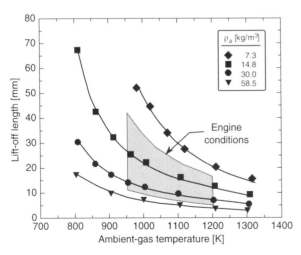

Fig. 5.14 Lift-off length *versus* ambient-gas temperature for a range of ambient-gas densities. The ambient-gas oxygen concentration, the pressure drop across the injector orifice and the orifice diameter were 21%, 138 MPa and 180 μm. The gray region represents the range of lift-off lengths expected in (quiescent) engines. The curves through the data represent the trends along lines of constant density. The lift-off length repeatability is ±7%

decreases. The effects of both parameters are non-linear, with the sensitivity of lift-off length to both parameters decreasing as they increase.

The temperature and density trends observed in Fig. 5.14 indicate that compression heating of in-cylinder gases as combustion proceeds during the injection period in an engine will cause the lift-off length to shorten with time. This combustion/lift-off length interaction must be accounted for in analyzing lift-off length data acquired in engines.

The gray region in Fig. 5.14 represents the range of lift-off lengths expected in a quiescent-type diesel engine for an orifice diameter of 180 μm and an injection pressure of 138 MPa. The range is about 7 to 30 mm, with a moderate-load, heavy-duty engine condition falling in the 10–15 mm range.

Figures 5.15 and 5.16 show the effects of ambient gas oxygen concentration. Figure 5.15 is a plot of lift-off length *versus* ambient gas temperature for four ambient gas oxygen concentrations ranging from 21% by volume (the concentration in air) down to 15% and ambient gas densities of 7.3 and 30.0 kg/m³. Figure 5.16 extends the data set in Fig. 5.15 to include ambient gas densities of 14.8 to 58.5 kg/m³. Figures 5.15 and 5.16 show that for any ambient gas temperature and density, the lift-off length increases as the ambient gas oxygen concentration decreases. The increase in lift-off length is approximately proportional to the decrease in oxygen concentration. Because of the proportional effect of oxygen concentration on lift-off length, the trends in lift-off length with respect to temperature and density for each oxygen concentration are the same as those noted in Fig. 5.14 for 21% oxygen. A similar effect of oxygen concentration on lift-off length was found for a wide range of other parameters [49].

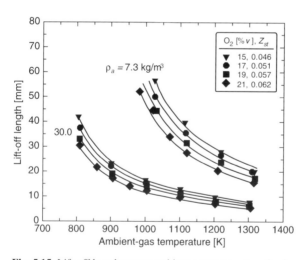

Fig. 5.15 Lift-off length *versus* ambient-gas temperature for four ambient oxygen concentrations and ambient-gas densities of 7.3 and 30.0 kg/m³. The pressure drop across the injector orifice and the orifice diameter were 138 MPa and 180 μm. The curves through the data represent the trends along lines of constant density for each oxygen concentration

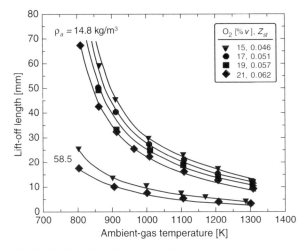

Fig. 5.16 Lift-off length *versus* ambient-gas temperature for four ambient oxygen concentrations at an ambient-gas density of 14.8 kg/m³ and for two oxygen concentrations at a density of 58.5 kg/m³. The other conditions are the same as for Fig. 5.15. The curves through the data represent the trends along lines of constant density for each oxygen concentration

Figures 5.17 and 5.18 show the effects of orifice diameter and the pressure drop across the injector orifice on lift-off length for different sets of conditions given in the legend of each figure. Data for orifice diameters from 45 to 363 μm and orifice pressure drops from 40 to 190 MPa are included in the respective figures. The curves through the data sets in the figures are trend lines. The figures show that

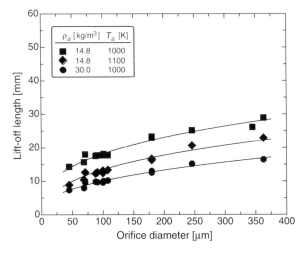

Fig. 5.17 Lift-off length *versus* orifice diameter for three different sets of conditions listed in the legend. The curves through the data represent the trend for each set of conditions. The ambient-gas oxygen concentration and the pressure drop across the injector orifice were 21% and 138 MPa

Fig. 5.18 Lift-off length
versus pressure drop across
the injector orifice for three
different sets of conditions
listed in the legend. The
curves through the data
represent the trend for each
set of conditions. The
ambient-gas oxygen
concentration and the orifice
diameter were 21% and
180 μm

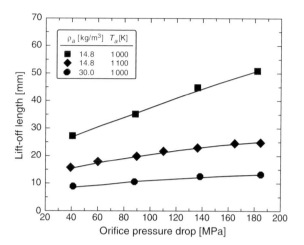

as either orifice diameter or injection pressure increase, the lift-off length increases.
The dependence of lift-off length on the pressure drop across the injector orifice
noted in Fig. 5.18 is linear with respect to the square-root of the orifice pressure
drop, i.e., linear with respect to injection velocity [46].

The lift-off length data in Figs. 5.14 through 5.18 provide a comprehensive
picture of the effects of many parameters on flame lift-off on DI diesel fuel jets.
However, further research is needed to fully understand flame lift-off and its impact
on diesel combustion and emission processes. Several areas that need to be investigated include determining the effects of fuel parameters (e.g., cetane number),
adjacent fuel jets (on a multi-orifice injector tip), piston bowl and fire-deck walls,
in-cylinder flows, and orifice and injector tip details on lift-off length.

5.4.2 Fuel–Air Mixing Upstream of the Lift-Off Length

The change in lift-off length observed in the previous section with respect to various
parameters can change the amount of air entrained into the fuel jet upstream of the
lift-off length, and therefore, the amount of fuel–air mixing that occurs prior to the
initial combustion zone. The quantity of air entrained prior to the lift-off length
can be estimated using the expression for the axial variation of the cross-sectional
average equivalence ratio in a fuel jet, $\bar{\phi}$, given by Eq. (5.18). The reciprocal of that
equivalence ratio relationship ($\times$ 100), when applied at the lift-off length, gives an
expression for the air entrained up to the lift-off location as a percentage of the total
air required to burn the fuel being injected. This percentage will be referred to as
the percentage of stoichiometric air, ζ_{st}, where

$$\zeta_{st}(\%) = \frac{100}{\bar{\phi}} = 100 \cdot \left(\frac{\sqrt{1 + 16 \cdot \tilde{H}^2} - 1}{2 \cdot f_s} \right) \qquad (5.26)$$

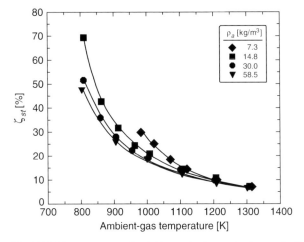

Fig. 5.19 The percentage of stoichiometric air entrained up to the lift-off length *versus* ambient-gas temperature for the conditions of Fig. 5.14. The curves through the data represent the trends along lines of constant density

In Eq. (5.26), $\tilde{H}$ is lift-off length normalized by x^+ and f_s is the stoichiometric air/fuel mass ratio. Note that f_s will change for different ambient gas oxygen concentrations and that ζ_{st} can also be viewed as the percentage of stoichiometric oxygen entrained upstream of the lift-off length. The latter meaning of ζ_{st} will be used when dealing with ambient gas with less than 21% oxygen.

Values of ζ_{st} determined with Eq. (5.26) and the lift-off length data in Figs. 5.14–5.18, are plotted in Figs. 5.19–5.23, respectively. Figure 5.19 shows that as ambient

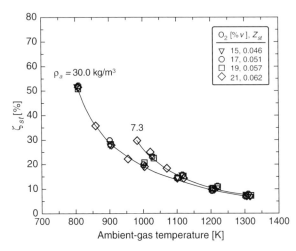

Fig. 5.20 The percent of stoichiometric oxygen entrained upstream of the lift-off length for the conditions of Fig. 5.15. The curves through the data represent the trends along lines of constant density

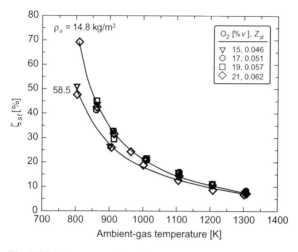

Fig. 5.21 The percent of stoichiometric oxygen entrained upstream of the lift-off length for the conditions of Fig. 5.16. The curves through the data represent the trends along lines of constant density

gas temperature increases for a constant ambient gas density, the total amount of air entrained upstream of the lift-off length decreases. This decrease occurs as a result of the decrease in lift-off length noted in Fig. 5.14 with increasing temperature.

With respect to ambient gas density, Fig. 5.19 shows that there is very little change in ζ_{st} with a change in density for a constant temperature. This trend is especially true over the range of conditions expected at the lift-off length in a

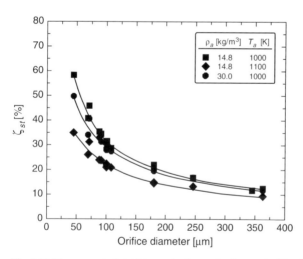

Fig. 5.22 The percent of stoichiometric air entrained up to the lift-off length *versus* orifice diameter for the conditions of Fig. 5.17. The curves through the data represent the trend for each set of conditions in the legend

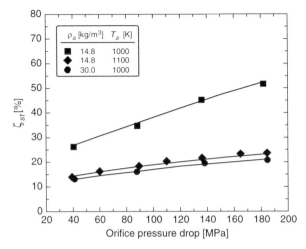

Fig. 5.23 The percent of stoichiometric air entrained up to the lift-off length *versus* the pressure drop across the injector orifice for the conditions of Fig. 5.18. The curves through the data represent the trend for the conditions in the legend

quiescent diesel ($10–45\,\text{kg/m}^3$ and 950 to 1200 K). The lack of a significant effect of ambient gas density must occur because the changes in air entrainment induced by changes in the lift-off length as density changes are largely compensated for by the direct effects of density on air entrainment observed in Eq. (5.16) [also see Eq. (5.19)].

The trends observed in Fig. 5.19 have two implications for an engine. The first is that changes in the operation of an engine that affect in-cylinder temperature near top-dead-center, such as changes in the turbocharging, intercooling, and wall heat transfer, can have a significant effect on the amount of fuel–air mixing that occurs upstream of the lift-off length. Second, changes in the operation of an engine that affect the in-cylinder density near top-dead-center will have a relatively small effect on the amount of fuel–air mixing prior to combustion.

Figures 5.20 and 5.21 show that as ambient gas oxygen concentration decreases for any set of conditions, ζ_{st} remains unchanged to within the repeatability of the data. In other words, the ζ_{st} estimated for each ambient gas oxygen concentration for any set of conditions is the same. This lack of an effect of a decrease in ambient gas oxygen concentration on ζ_{st} means that the decrease in oxygen entrainment due to less oxygen in the ambient gas is exactly compensated for by the increase in the total ambient gas entrained as a result of a longer lift-off length with reduced oxygen concentration (see Figs. 5.15 and 5.16) for the conditions considered in the experiment. The lack of an effect of oxygen concentration on ζ_{st} was noted for a wide range of orifice diameters and injection pressures as well [49]. Only for extremely low oxygen concentrations ($<10\%$) has this trend been noted to breakdown [52].

The trends in Figs. 5.20 and 5.21 suggest that the reduction in oxygen concentration in the in-cylinder gas in an engine resulting from the use of EGR will have little effect on the amount of fuel burned in the rich reaction zone just downstream of the lift-off length. Combustion in the rich reaction zone, however, will be more dilute due to the additional inerts entrained upstream of the lift-off length.

The effects of orifice diameter and the pressure drop across the injector orifice on the amount of fuel–air mixing prior to combustion are shown in Figs. 5.22 and 5.23, respectively. Figure 5.22 shows that although the lift-off length decreases with decreasing orifice diameter (see Fig. 5.17), ζ_{st} increases as the orifice diameter decreases. The increase in ζ_{st} is due to the strong increase in the amount of air entrained relative to the amount of fuel injected that occurs with decreasing orifice diameter [see Eqs. (5.16) and (5.19)]. The impact on ζ_{st} of this strong increase in the amount of air entrained relative to the amount of fuel injected is only partially counteracted by the impact of the decrease in lift-off length with decreasing orifice diameter.

Figure 5.23 shows that an increase in pressure drop across the injector orifice causes greater fuel–air mixing upstream of the lift-off length, i.e., an increase in ζ_{st}. This effect is due directly to the increase in lift-off length observed in Fig. 5.18 with increasing pressure drop across the injector orifice, since unlike orifice diameter, the orifice pressure drop has no additional affect on the amount of air entrained relative to the amount of fuel injected [see Eqs. (5.16) and (5.19)].

5.4.3 Lift-Off Length and Soot Formation

Based on the current picture of quiescent-chamber DI diesel combustion discussed in the introduction to Section 5.4, air entrained upstream of the lift-off length reacts with fuel in a rich reaction zone located just downstream of the lift-off length during the lifted, turbulent diffusion phase of combustion [33, 45]. The product gas of this reaction zone is ideal for forming soot and has been hypothesized to play an important role in the formation of soot in a diesel fuel jet [33]. Estimates of air entrainment upstream of the lift-off length, discussed in the previous section, indicate that the strength of the central reaction zone (i.e., the amount of fuel burned), and therefore the composition of the product gas of the central reaction zone will be dependent on engine parameters such as: load, intercooling, turbocharging, injection pressure, orifice size, etc. This in turn implies that the amount of soot formed within a fuel jet will depend on the same parameters.

Measurements of total soot incandescence from the burning diesel fuel jets were made simultaneously with the lift-off measurements shown in Section 5.4.1. The measurements were used to conduct initial explorations of the link between soot formation and fuel–air mixing upstream of the lift-off length [46, 49]. Although total soot incandescence is related in a complex manner to soot concentration (as discussed in Ref. [46]), it nevertheless is strongly dependent on soot concentration and can be used to provide a relative indication of significant changes in the amount of soot formed in a fuel jet. Figure 5.24 is a plot of the measured total soot incandescence normalized by the fuel flow rate as a function of ζ_{st} estimated with Eq. (5.26) for all conditions considered. These conditions included ambient

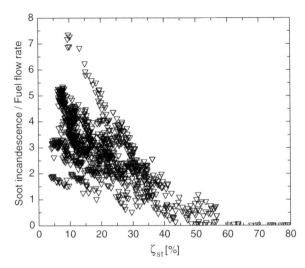

Fig. 5.24 The relative total soot incandescence normalized by the fuel flow rate *versus* the percent of stoichiometric air entrained up to the lift-off length. Data are shown for ambient-gas densities from 7.3 to $58.5\,kg/m^3$, ambient-gas temperatures from 800 to 1300 K, ambient-gas oxygen concentration from 15% to 21%, injection pressures from 40 to 190 MPa, orifice diameters from 100 to 363 μm, and ambient-gas oxygen concentrations from 15% to 21% (Each data point is for an individual injection)

gas densities from 7.3 to $58.5\,kg/m^3$, ambient gas temperatures from 800 to 1300 K, ambient gas oxygen concentrations from 15% to 21%, injection pressures from 40 to 190 MPa, and orifice diameters from 100 to 363 μm. Each data point in the figure is for an individual injection event. All data were acquired with DF2. Normalizing the soot incandescence measurements by the fuel flow rate helped eliminate differences in the total soot incandescence due to differences in the amount of soot formed as the quantity of fuel injected changed with orifice diameter and injection pressure.

The most visible and significant trend in Fig. 5.24 is the trend toward lower soot incandescence (i.e., reduced soot formation) as ζ_{st} increases. As ζ_{st} approaches values between 45% and 55% (independent of other conditions), the soot incandescence falls below the detectable limit of the photodiode used to measure the incandescence [46]. Values of ζ_{st} between 45% and 55% correspond to cross-sectional average equivalence ratios between 2.2 and 1.8 at the lift-off length, respectively [see Eq. (5.26)]. The disappearance of soot incandescence at an equivalence ratio of approximately two is in general agreement with studies of soot formation under premixed conditions. Under premixed conditions, typical hydrocarbon fuels do not produce soot for equivalence ratios less than approximately two [3].

The dominant trend of decreasing soot incandescence with increasing air entrainment upstream of the lift-off length noted in Fig. 5.24 clearly supports a link between the amount of soot formed within a fuel jet and the amount of air entrained (i.e., oxygen) upstream of the lift-off length. This link may help explain some of the soot emission changes noted with changes in various engine-operating conditions. For example, a portion of the reduction in soot emissions noted with increased injection pressure in production engines in all likelihood results from the additional air entrainment that occurs as a result of the longer lift-off length.[4] Reductions in soot in fuel jet plumes with increased injection pressure have also been observed by Kamimoto et al. [39] and by Dec and Espey [55].

The trends observed in Fig. 5.24 are also in agreement with recent detailed kinetic modeling studies by Flynn et al. [56]. Flynn et al. investigated the impact of the addition of an oxygenated fuel to a hydrocarbon fuel on soot precursor formation under rich premixed combustion conditions - the type of combustion conditions believed to occur in the central region of a diesel fuel jet downstream of the lift-off length [33]. They showed that as more oxygenated fuel was mixed with heptane under rich ($\phi = 4$) premixed conditions, less carbon went into the formation of soot precursors. When the atomic oxygen to carbon ratio (O/C) in the reactants reached values between 1.3 and 1.6, soot precursors were no longer formed. O/C ratios between 1.3 and 1.6 correspond to values of ζ_{st} between 45% and 55% in Fig. 5.24, since the O/C ratio is equal to $2.9\,\phi^{-1}$ (or equivalently, $0.029\,\zeta_{st}$) for typical diesel fuel with an H/C ratio of 1.8 [3]. The link between soot and air entrainment and the agreement with the analysis of Flynn et al. also suggests that oxygen, whether introduced *via* entrained air or bound in oxygenated fuels, will reduce the amount of soot formed in a diesel fuel jet.

Observations made using the qualitative soot incandescence data in Fig. 5.24 are substantiated in Figs. 5.25 and 5.26. Figures 5.25 and 5.26 present axial profiles of soot measured in diesel fuel jets that show detailed effects of injection pressure and orifice diameter on soot within a fuel jet [50]. The soot data are from a continuing series of investigations of soot in diesel fuel jets [50, 51, 52]. The figures show plots of the soot volume fraction as a function of axial distance. The soot measurements were made using a line-of-sight extinction technique [50]. Each data point represents the average soot volume fraction, $\bar{f}_v$, along a line-of-sight orthogonal to and passing through the fuel jet centerline [50]. The curves through the data show the axial trends in the measured soot data. The solid symbol on the x-axis in each figure near 20 mm gives the location of the lift-off length determined for each condition from an OH chemiluminescence image. Data are shown for four orifice pressure drops (43, 89, 138, and 184 MPa) and an orifice diameter of 100 μm in Fig. 5.25 and for four orifice diameters and an orifice pressure drop of 138 MPa in Fig. 5.26. Also presented in the legend of Figs. 5.25 and 5.26 are the ζ_{st} estimated with Eq. (5.26).

With respect to axial distance, Fig. 5.25 shows that the amount of soot present in the cross-section of the fuel jet starts increasing after the lift-off length location, reaches a peak near 50 mm, then decreases toward zero as the flame length is

---

[4] Soot emission reductions with increased injection pressure will also occur as a result of enhanced oxidation caused by enhanced mixing at the higher injection pressures, both during and after injection.

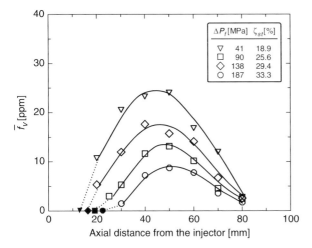

Fig. 5.25 The average soot volume fraction along the path length of the soot measurement as a function of axial distance from the injector for four orifice pressure drops. The curves show the trends. The solid symbols on the x-axis mark the lift-off length for each orifice pressure drop. Also, the ζ_{st} at the lift-off length is given in the legend for each orifice pressure drop. The orifice diameter was 100 μm

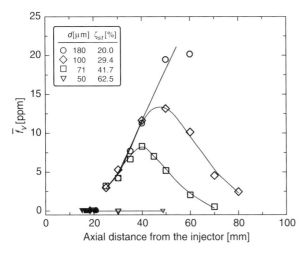

Fig. 5.26 The average soot volume fraction along the path length of the soot measurement as a function of axial distance from the injector for four orifice diameters. (Note the soot volume fraction axis extends slightly below zero to better show the no-soot 50-μm orifice data.) The curves show the trends. The solid symbols on the x-axis mark the lift-off length for each orifice diameter. Also, the ζ_{st} at the lift-off length for each orifice diameter is given in the legend. The orifice pressure drop was 138 MPa

approached. (The flame length for the conditions of Fig. 5.25 is slightly greater than the distance between the injector and the opposite wall of the combustion vessel in which the measurements were made, or approximately 100 mm [32]). The general soot trends noted in Fig. 5.25 are similar to trends noted in atmospheric pressure gas jets (e.g., [57, 58]), except that the peak soot volume fractions are considerably higher.

The initial rise in the cross-sectional average soot volume fraction downstream of the lift-off length with increasing axial distance observed in Fig. 5.25 is most likely due to the dominance of soot formation over oxidation in this upstream region of the jet. The rate of soot formation, however, must decrease with increasing axial distance as the fuel is consumed by combustion. The peak in total soot at 50 mm occurs at the location where soot formation is balanced by soot oxidation. Downstream of the peak, the total amount of soot decreases as soot oxidation begins to dominate over soot formation as the flame length is approached and the fuel is completely consumed. The soot data in Fig. 5.25 indicate that the soot is largely oxidized before the flame length is reached for the conditions considered in the figure.

With respect to injection pressure, Fig. 5.25 shows that two important changes in the axial soot distribution occur with changing injection pressure. The first and most important change is that the soot level in the fuel jet decreases with increasing pressure drop across the injector orifice. Analysis of the peak soot volume fraction in each profile indicates that the peak soot decreases linearly with increasing injection velocity (i.e., the square-root of the pressure drop across the injector orifice). The second change is that the axial location of the first soot is pushed downstream with increasing orifice pressure drop. This latter trend is due to the increase in lift-off length with increasing orifice pressure drop.

A likely reason for the decrease in the level of soot with increasing orifice pressure drop is the linear increase in the total air entrainment that occurs upstream of the lift-off length as a result of the linear increase in lift-off length with increasing injection velocity. The increase in air entrainment with increasing injection velocity is reflected in the increasing ζ_{st} at the lift-off length with increasing orifice pressure drop shown in the legend of the figure. The increased air entrainment leads to less soot formation as discussed in conjunction with Fig. 5.24.

An additional potential factor contributing to the decrease in soot with increasing pressure drop across the injector orifice (i.e., increasing injection velocity) is the decrease in residence time in the combusting region of the fuel jet (i.e., the distance between the lift-off length and the flame length) that occurs with increased injection pressure. This decrease in residence time limits the time for soot formation. The decrease in residence time occurs for two reasons. The most direct reason is that residence time in the combusting region of the fuel jet is inversely related to injection velocity. The second reason is that the length of the combustion region is shortened as the orifice pressure drop increases due to the increase in lift-off that occurs with increasing orifice pressure drop. The importance of residence time on soot formation has been shown in near-atmospheric turbulent gas jets for which soot levels increase as the velocity of the fuel jet decreases [57, 58]. However, the relative importance of residence time on soot formation at diesel conditions is not clear and needs to be investigated further.

The effects of orifice diameter on the axial distribution of soot in a diesel fuel jet are shown in Fig. 5.26. The data in the figure for the 180-μm orifice end at an axial distance of 60 mm because the fuel jet became too optically thick downstream of this location to make soot measurements by an extinction technique [50]. Figure 5.26 shows that as orifice diameter decreases, the level of soot in the fuel jet decreases. The magnitude of the soot reduction is such that while the sooting region for the 180-μm orifice is optically thick, the 50-μm orifice produces no detectable soot. Because of this strong trend in soot with orifice diameter, however, only two orifice diameters have complete axial profiles of soot, making it difficult to assess the exact dependence of soot on orifice diameter.

Considering the data for the two orifices with complete axial soot profiles, the 100-μm and 71-μm orifices, the soot trends with respect to axial distance from the injector appear the same as those observed in Fig. 5.25. However, unlike for a change in injection pressure in Fig. 5.25, the axial extent of the soot changes significantly with a change in orifice diameter, decreasing as orifice diameter decreases. The decrease in the axial extent of the soot with decreasing orifice diameter is largely the result of the near linear decrease in flame length that occurs with decreasing orifice diameter [32].

The reasons for the decrease in the soot levels within a fuel jet as orifice diameter decreases are similar to those mentioned when discussing the effects of injection pressure on soot formation. A major contributor to the decrease is the enhanced mixing of fuel and air that occurs as orifice diameter decreases. Enhanced fuel–air mixing upstream of the lift-off length with decreasing orifice diameter is reflected in the increasing ζ_{st} at the lift-off length (given in the legend of Fig. 5.26) with decreasing orifice diameter. This trend means that the rich combustion zone, hypothesized to exist just down-stream of the lift-off length [33], will occur under less rich conditions as orifice diameter decreases, resulting in less soot formation [46]. Studies of soot formation under rich, premixed fuel–air conditions also support this conclusion [59]. The increase in fuel–air premixing upstream of the lift-off length with decreasing orifice diameter is such that ζ_{st} at the lift-off length increases to greater than 60% for the 50-μm orifice, resulting in no soot formation. In addition to enhanced mixing upstream of the lift-off length, enhanced fuel–air mixing occurs along the entire fuel jet length with decreasing orifice diameter, allowing completion of the combustion and soot oxidation processes in a shorter axial distance, thus explaining the reductions in flame length and the length of the sooting region that occur as orifice diameter decreases.

Decreasing residence time for soot formation may also play a role in the soot decrease with decreasing orifice diameter, the same as suggested for an increasing orifice pressure drop. Residence times in the combusting region of a fuel jet will decrease approximately linearly with orifice diameter, resulting in less time for soot formation. However, as mentioned in previous discussions, the impact of residence time in diesel fuel jets is not understood yet and needs further investigation.

Fig. 5.27 Images of
laser-induced incandescence
(LII) for the conditions of
Fig. 5.26. The orifice
diameter and the camera gain
relative to the gain for the
100-μm orifice are shown in
the upper left and right
corners, respectively. Each
image is the average of at
least twelve individual
images. The LII laser sheet
penetrates the fuel jet from
the bottom in these images
and the gamma for each
image was set to 0.5 (Note:
The laser sheet did not cover
the full width of the image
field in the top image,
resulting in the dark region at
far right of the top image)

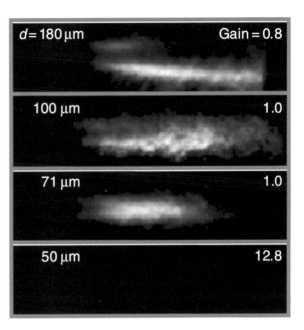

Images of soot across the cross-section of a fuel jet, acquired using laser-induced incandescence (LII) soot imaging [50], are shown in Fig. 5.27 for the conditions of Fig. 5.26. The LII images in Fig. 5.27 are not quantitative, but they do provide independent measurements that support the trends noted in Fig. 5.26, especially the no soot condition for the 50-μm orifice. Each image is the average of at least twelve individual LII images for each condition [50]. The orifice diameter and camera gain (relative to the camera gain for the 100-μm orifice) are given in the upper left and right corners of each image, respectively. The camera gain changes must be factored in when comparing intensities between images.

The trends in the LII intensities along the upper and lower edges of the fuel jets in Fig. 5.27 are the most revealing. Along the lower edge of the fuel jets, the signal intensity decreases with decreasing orifice diameter. Since the LII intensities along the lower edges are not as strongly affected by attenuation by soot in the fuel jet, and therefore are more directly related to the amount of soot, the lower edge trend clearly indicates a reduction in soot with decreasing orifice diameter. The reduction in soot with decreasing orifice diameter is such that for the 50-μm orifice no soot is detectable even though the camera gain is almost thirteen times higher than for the 71-μm orifice condition, supporting the laser extinction measurements of soot presented in Fig. 5.27.

In the top three images, the increase in LII intensity with decreasing orifice diameter along the upper edge of the fuel jets also indicates a decrease in soot. The trend along the upper edge with decreasing orifice diameter is due to a decrease in the soot optical thickness (i.e., a decrease in soot concentration) in each fuel jet with decreasing orifice diameter from top to bottom in the figure. The decrease in optical thickness with decreasing orifice diameter allows an increase in the transmission of the Nd:YAG laser pulse used for the LII imaging to the upper edge of the fuel jet and an increase in the transmission of the LII signal through the fuel jet to the camera. (Note: The soot levels are so high for the largest orifice diameter that the soot along the upper edge cannot be seen for most of the length of the jet due to extinction of the Nd:YAG laser pulse. By an orifice diameter of 71 μm, soot levels are low enough that the soot present is visible from the lower edge to the upper edge of the fuel jet.)

5.5 Implications for DI Diesel Fuel Jets Under Quiescent Conditions

The trends observed in the experimental data in the previous sections and the close agreement found between the data and the scaling laws developed from mass, momentum, and energy balances applied to an idealized fuel jet have significant implications regarding the processes that control the development of DI diesel fuel jets, and therefore, the modeling of these jets. This section discussions several implications and insights drawn from the data and the comparisons with the scaling laws. Included are discussions on the mixing-limited nature of DI diesel fuel jets, the relationship between fuel vaporization and combustion, and the changes in the structure of the DI diesel fuel jet that are occurring in response to changes being made in engine and injector parameters to meet new emissions regulations and improve engine performance.

5.5.1 The Mixing-Limited Nature of DI Diesel Fuel Jets

The close agreement between the liquid length scaling law, Eq. (5.24), and the experimental data shown in Figs. 5.8–5.12 over a very wide range of conditions (much wider than are possible to achieve in engines) strongly implies that mixing-limited fuel vaporization (the premise on which the liquid length scaling law is based) is closely approached in current-technology DI diesel fuel jets. These are high injection pressure, well-atomized fuel jets in what Hiroyasu and Arai [5] have called the "complete spray" regime. Mixing-limited fuel vaporization in turn implies that the local mass, momentum, and energy transport processes occurring at droplet surfaces do not control the overall rate of vaporization. The atomization process produces droplets small enough in size with enough total surface area such that the droplet surface transport processes are fast relative to the gas-phase fuel jet mixing processes. If this were not true, the liquid length scaling law, which contains no physics regarding droplet atomization or vaporization, would not agree with the experimental data as closely as it does over such a wide range of conditions [30]. This discussion means that better atomization (i.e., smaller droplets) alone will not promote increased fuel vaporization in a DI diesel fuel jet. Parameters such as injection pressure and orifice diameter affect fuel vaporization through their effects on jet mixing processes, not through their effects on droplet size via the atomization process.

The close agreement noted between the liquid length scaling law and the data also implies that other assumptions made in developing the scaling law are reasonable with respect to the overall fuel vaporization process. One critical assumption is that the fuel and ambient gas mixture in the vaporizing region is saturated. Achieving or closely approaching a saturated condition suggests a mechanism for control of vaporization by mixing. When a saturated conditions exists, the only means for further vaporizing fuel is to heat the liquid fuel to a higher temperature and/or transport fuel vapor away from the inner, vaporizing region of the fuel jet. A higher liquid fuel temperature would allow further vaporization by raising the fuel vapor pressure. Transport of fuel away from the inner region of the fuel jet (or equivalently, dilution of the fuel jet by entrained ambient gas) would allow further fuel vaporization by lowering the vapor fuel partial pressure below the fuel saturation pressure. Since both the energy entrainment (i.e., high-temperature ambient gas entrainment) needed to heat the fuel and the transport of fuel and ambient gas in the fuel jet are controlled by fuel jet mixing processes, vaporization will be controlled by mixing processes.

Significant deviation from mixing-limited fuel vaporization only begins to appear at the lower gas temperature and density conditions considered (see Figs. 5.10 and 5.11). At these conditions, which are well below those expected in an engine, the measured liquid lengths are shorter than those predicted with the scaling law. The reason for this deviation is most likely that local transport processes at droplet surfaces begin controlling fuel vaporization at the lower gas temperature and density conditions. Consider moving along the curve for the smallest ambient gas density considered in Fig. 5.10 toward lower gas temperatures, with all other conditions fixed. As the gas temperature is lowered, the mass and energy transport rates at droplet surfaces will decrease, but the turbulent transport of ambient gas to the inner region of the fuel jet and fuel transport away from the inner region will not. Therefore, as the ambient temperature is decreased, the overall droplet surface transport rates are likely at some temperature to become slow relative to the fuel jet mixing processes. When this transition occurs, the droplet transport processes will be unable to maintain saturated equilibrium conditions in the fuel jet inner region, and control of vaporization should shift to the droplet surface transport processes. Under these conditions liquid lengths should be shorter than those predicted by the scaling law, since vaporization would no longer be limited by gas-phase mixing in the jet and saturation.

Previous research has shown that combustion in a DI diesel fuel jet (after the ignition and premixed burn phases are completed) is also controlled by gas-phase mixing processes, i.e., by the rate at which fuel vapor and air mix [3]. Combustion controlled by mixing implies that the development of the combusting region of the fuel jet downstream of the lift-off length is controlled by gas-phase mixing processes. However, fuel vaporization occurs in the dense inner region of the fuel jet between injector and the liquid length [2, 33] (which overlaps with the combusting region of the fuel jet for some conditions, as discussed in the next section). Mixing-limited fuel vaporization therefore indicates that the mixing-limited nature of a DI diesel fuel jet does not just include the combusting region downstream, but extends essentially up to the injector orifice.

Control of the jet development in the near injector region by mixing processes is also supported by the agreement noted in Fig. 5.2 between the transient tip penetration scaling law and the experimental transient tip penetration data. The fuel jet tip penetration scaling law (also based on the premise that entrainment and mixing control the development of the fuel jet) correctly predicts penetration, and most importantly, the transition between the near-field and far-field penetration regimes. The transition, which occurs in the middle of the vaporization region of a fuel jet for typical diesel engine conditions, marks the location where the mass in the jet becomes dominated by entrained air, as discussed previously. In other words, the transition from the near-field to far-field tip penetration regimes in a DI diesel fuel jet can be explained by entrainment and mixing processes, as opposed to the breakup of an "intact liquid core" as has been previously proposed [5, 6].

The mixing-controlled nature of the near injector region of a DI diesel fuel jet has important implications for the modeling of DI diesel fuel jets as well. One implication is that the current "spray equation" model used in multidimensional CFD engine calculations for the near injector region may not be the most efficient or appropriate approach for modeling diesel fuel jets for current technology conditions. The "spray equation" model typically used is based on a stochastic particle method [60, 61, 62]. Solution of the spray equation model gives the full distribution function of droplet properties needed to predict non-equilibrium effects between the liquid and gas-phases. The liquid length results, however, suggest that dynamic and thermodynamic equilibrium are closely approached and that this information this may not be necessary. Also, O'Rourke et al. [63] recently extended the simple mixing-controlled model used to develop the liquid length scaling law to solve for the axial profile of the liquid volume fraction in the near injector region. Based on this broader application of mixing control, they found that the liquid volume fraction was greater than 0.1 over 20–50% of the liquid length. This means that the dilute spray assumption made when using the spray equation model does not apply over a significant fraction of the region with liquid-phase fuel, i.e., the region where the spray equation model is used. The observations discussed above indicate that for most diesel conditions, assuming equilibrium between the gas and liquid phases may be a better limit to consider when developing multidimensional CFD models of diesel fuel jets and may significantly simplify modeling of diesel fuel jets in the near injector region.

Another implication of the mixing-controlled nature of the near injector region is that modeling of gas entrainment and mixing processes (i.e., turbulence) in the near injector region needs more attention in future DI diesel fuel jet modeling activities. Accurate modeling of gas entrainment and mixing processes in the near injector region are critical to modeling the fuel vaporization process and the ensuing combustion and emission processes downstream. Modeling of the near injector region is inhibited, however, by the reliability of two-phase flow turbulence models and by grid resolution constraints. Grid resolution constraints arise as a result of limitations imposed by computational resources and by the spray equation model currently used (e.g., [64]). The spray equation model requires that the grid size for the Eulerian gas-phase flow portion of the model be larger than the smallest liquid fuel parcel, which is typically too coarse to resolve the jet features needed to model the mixing processes in the near injector region. The grid resolution problem is a major factor contributing to the grid dependent solutions documented by many authors when using "spray equation" based CFD engine models (e.g., [64, 65, 66, 67]).

5.5.2 Relationship Between Vaporization and Combustion

Although determining the exact nature of the relationship between the fuel vaporization and the combustion regions in a DI diesel fuel jet under quiescent conditions requires further research, comparisons of measured liquid lengths (Section 5.3) and lift-off lengths (Section 5.4) provide some insight. Figure 5.28 compares the liquid and lift-off lengths for DF2 as a function of orifice diameter. The pressure drop across the injector orifice and the ambient gas temperature and density for the results in Fig. 5.28 were 138 MPa, 1000 K, and 14.8 kg/m^3, respectively. The liquid lengths shown in the figure were determined with the liquid length scaling law for non-combusting, vaporizing fuel jets given in Section 5.3.2. The single component fuel used in the liquid length scaling law to model DF2 was n-heptadecane, which has been shown to accurately model the liquid lengths measured for the DF2 used in Ref. [30].

Figure 5.28 shows that as orifice diameter decreases, the liquid length shortens relative to the lift-off length, becoming shorter than the lift-off length at an orifice diameter of about 160 μm for the conditions of the figure. The relative trend between liquid and lift-off length is the result of the differing dependencies of liquid and lift-

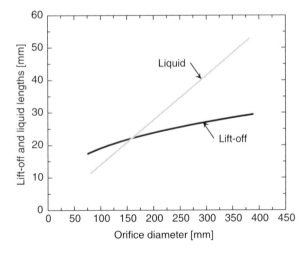

Fig. 5.28 Comparison of lift-off (black curve) and liquid lengths (gray line) for DF2 as a function of orifice diameter. The ambient-gas temperature and density were 1000 K and 14.8 kg/m^3. The pressure drop across the injector orifice was 138 MPa. The liquid lengths were determined with the liquid length scaling law, Eq. (5.24), using heptadecane as a surrogate for DF2 [30]

off length on orifice diameter. As discussed previously, liquid length has a strong linear dependence on orifice diameter, while lift-off length has a much weaker dependence on orifice diameter.

The relative trend between liquid and lift-off length in Fig. 5.28 suggests that orifice diameter has a strong effect on the relationship between fuel vaporization and combustion processes in a DI diesel fuel jet. For smaller orifice diameters, with liquid lengths shorter than lift-off lengths, the results in Fig. 5.28 indicate that fuel vaporization will be completed prior to any combustion. This means that there will be no direct interaction between the vaporization and combustion regions in the fuel jet. Combustion will only indirectly affect vaporization by compression heating in-cylinder gases, raising the temperature and density of the air entrained into the fuel jet, and thus shortening the liquid length [46].

For the larger orifice diameters, with liquid lengths longer than the lift-off lengths, Fig. 5.28 suggests that there will most likely be a more direct and strong interaction between vaporization and combustion processes. For this latter condition, vaporization must be occurring in a relatively cool, central region surrounded by combustion (on a time-averaged basis). Combustion is unlikely to occur in the central fuel vaporization region, since the high velocities, the low temperatures (<700 K [30]), and the rich mixtures expected in this region will not support combustion. The heat source from the combustion surrounding the fuel vaporization region will enhance the rate of fuel vaporization, shortening the liquid length from that measured in a vaporizing, non-combusting fuel jet (e.g., those shown in Figs. 5.8–5.12 and 5.28). As a result, liquid lengths measured in vaporizing, non-combusting fuel jets can only be interpreted as the maximum possible penetration distance of the liquid-phase fuel in a combusting fuel jet. (Evidence of such combustion shortening of the liquid length was found by Canaan et al. [42].)

The combustion region surrounding the fuel vaporization region for conditions with liquid lengths much longer than lift-off lengths is most likely composed of an inner, rich reaction layer similar to that previously hypothesized to occur downstream of the lift-off length and the liquid length for conditions with similar lift-off and liquid lengths [33]. In this rich reaction layer, the oxygen entrained upstream of the lift-off length will be consumed. The rich reaction layer in turn will be surrounded by the stoichiometric diffusion flame extending from the lift-off length downstream along the periphery of the burning fuel jet [33, 34].

Similar to the trends in Fig. 5.28 with decreasing orifice diameter, increasing pressure drop across the injector orifice and decreasing ambient (i.e., in-cylinder) gas temperature and density also cause a relative decrease in the liquid length with respect to the lift-off length. These trends suggest that changes in engine operation being made to meet new emission regulations or improve engine performance, such as increased injection pressure, reduced orifice diameter, or greater intake charge cooling, are leading to more gas jet-like diesel combustion, i.e., the two-phase, spray region of the fuel jet is having a decreasing direct impact on the combustion region. The opposite is true, however, for increased turbocharging, i.e., increased in-cylinder gas densities.

5.5.3 Evolution of DI Diesel Fuel Jets Under Quiescent Conditions

The lift-off length, liquid length, air entrainment, and soot incandescence information presented in this paper suggest that for different conditions, DI diesel fuel jets
can have very different characteristics in the quasisteady portion of the fuel jet. For
a given fuel, these characteristics depend on ambient gas conditions, orifice parameters, and injection pressure. Figure 5.29 shows two schematics that show possible
extremes in the time-averaged, fuel jet characteristics for a typical diesel fuel. The
schematics suggest how the relative spatial relationship between various regions in
the fuel jet change for the two extremes considered. The segment of the fuel jet in
each schematic represents the quasisteady, post-ignition, free region of the fuel jet.
The schematics are highly simplified and intended only for illustrating changes in
the time-averaged location and the characteristics of various regions in a fuel jet
that can be interpreted from the data discussed in this paper. (At the end of this
section, these extreme cases will be related to the schematic of diesel combustion
presented by Dec [33] for conditions typical of a moderate-load condition in current
heavy-duty DI diesel engines.)

   In each schematic, the dark gray region emanating from the injector orifice at
the bottom represents the maximum possible extent of the region containing fuel
droplets mixed with vaporized fuel and entrained air (i.e., the liquid length). This
region should not be interpreted as an "intact liquid core." Very little if any intact
liquid core is believed to exist in high-pressure DI diesel fuel jets (e.g., [3, 16, 31])
The lighter gray region outside the fuel droplet-containing region represents the
region with vaporized fuel mixed with entrained air. The black curve emanating

Fig. 5.29 Schematics
showing how the relative
spatial relationship between
fuel vaporization and
combustion zones and the
percent of stoichiometric air
entrained up to the lift-off
length can change with
conditions in a DI-type diesel
fuel jet under quiescent
conditions. The schematic at
the left is for an ambient-gas
temperature and density of
1100 K and 23 kg/m^3, and an
orifice pressure drop and
orifice diameter of 40 MPa
and 250 μm. The schematic at
the right is for an ambient-gas
temperature and density of
1000 K and 20 kg/m^3, and an
orifice pressure drop and
orifice diameter of 200 MPa
and 100 μm

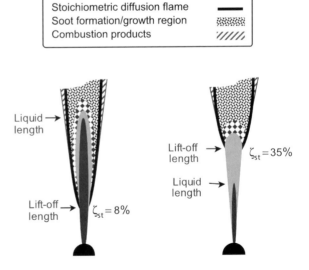

from the lift-off location in the downstream direction is the mixing-controlled combustion zone (i.e., diffusion flame zone), which has been shown to exist near the periphery of the fuel jet [33, 34]. The checkered region in the central region of the fuel jet downstream of the lift-off length represents the rich, reaction zone suggested by the works of Dec [33] and Chomiak and Karlsson [45]. Since no direct measurements of combustion in the rich central reaction zone have ever been made, the location and shape that is shown in the schematic for this region is purely speculative and intended only to indicate that such a zone exists. The dotted region in the figure downstream of the central reaction zone is the soot formation and growth region. Soot begins to form just downstream of this central reaction zone and evolves as the gases are transported downstream [33]. The crosshatched band shown at the very downstream periphery of the fuel jet, outside of the diffusion flame zone, represents a region of high temperature combustion products mixed with entrained air. The product zone becomes wider in the downstream direction as the diffusion flame consumes fuel and moves inward.

The conditions for the two schematics in Fig. 5.29 were selected to show potential extremes. The schematic on the left is for an orifice pressure drop of 40 MPa, an orifice diameter of 250 μm, and an ambient gas temperature and density of 1100 K and 23 kg/m^3. The temperature and density are indicative of those expected for a moderate-load, heavy-duty diesel engine condition after ambient gas compression heating from autoignition has occurred. An injection pressure of 40 MPa represents much older injector technology conditions.

The schematic on the right in Fig. 5.29 is for an orifice pressure drop of 200 MPa, an orifice diameter of 100 μm, and an ambient gas temperature and density of 1000 K and 20 kg/m^3. The temperature and density for this schematic are indicative of those expected for a heavy-duty diesel engine condition similar to the conditions for the schematic on the left, but with more intake charge cooling for NOx control. The orifice diameter for this set of conditions is much smaller and the injection pressure is greater than are typical in current engines.

The schematic on the left indicates that there is a significant quantity of unvaporized fuel at the lift-off location, and that only about 8% of the air required to burn the injected fuel is entrained by the lift-off length. The unvaporized fuel is confined to the central region of the fuel jet based on images of the gas and liquid phases of fuel jets (e.g., [2, 42]). Combustion downstream of the lift-off length surrounds this central region, most likely resulting in a significant length of the burning portion of the fuel jet having a vaporizing, relatively cool core. As described in the previous section, there is likely to be significant interaction between combustion and fuel vaporization processes for these conditions. Combustion will enhance the vaporization of any fuel not vaporized by the lift-off length, as well as inhibit air entrainment [19]. Vaporization will cool the gases, and potentially interact with the flame stabilization process at the lift-off length. As also mentioned in the previous section, the interaction implies that the vaporizing, non-combusting fuel jet liquid length shown in the schematic is the maximum liquid length that could exist for the ambient gas conditions. The actual liquid length with combustion occurring is likely to be shorter than indicated in the schematic.

The schematic on the right in Fig. 5.29 shows that the reduction in orifice diameter, increase in the orifice pressure drop, and changes in the ambient gas conditions from the schematic on the left, results in a dramatic change in the relationship between the liquid-phase fuel region and the combustion zones. In comparison to the schematic on the left, the schematic on the right indicates that the fuel will be completely vaporized long before the lift-off length. This effect is due to the significantly greater amount of hot air entrainment relative to the amount of fuel injected that occurs with the smaller orifice diameter, as can be observed in Fig. 5.22. There will therefore be no direct interaction between the fuel vaporization and combustion processes for the conditions on the right. The lift-off length also occurs much further downstream for the conditions on the right, with significantly more fuel–air mixing occurring by the lift-off length in comparison to the conditions on the left. As a result, the central reaction zone will be much more intense, and based on Figs. 5.24 through 5.27, there will likely be less soot formed within the fuel jet per unit of fuel injected for the conditions on the right than for the conditions on the left.

For a moderate load condition in current-technology, quiescent, heavy-duty DI diesel engine, the results presented in this paper indicate that the picture of diesel combustion falls in between the two schematics shown in Fig. 5.29. Under these conditions, the liquid and lift-off lengths have about a 20–30% overlap and approximately 15% of the air required to burn the fuel is entrained by the lift-off length. The small overlap indicates that hot air entrainment dominates the fuel vaporization process, but that some shortening of the liquid length will occur as a result of combustion. The picture for a moderate-load condition is in general agreement with the conceptual model for DI diesel combustion presented by Dec [33]. However, the changes between the left and the right schematics in Fig. 5.29 indicate that DI diesel combustion is evolving. Changes made over time to meet new emissions regulations and to maintain and improve diesel engine performance (such as injection pressure increases, retarded injection timing, and intake charge cooling) have led to an evolution of the diesel combustion process from the schematic on the right toward the schematic on the left. Moreover, this evolution of the diesel combustion process is likely to continue in the future given the need to meet ever more stringent emissions regulations.

Acknowledgments The author would like to acknowledge the contributions of Jeff Naber, Brian Higgins, and Lyle Pickett to the research discussed. They were coauthors on several of the supporting papers.

Support for this research was provided by the U.S. Department of Energy, Office of Vehicle Technologies. The research was performed at the Combustion Research Facility, Sandia National Laboratories, Livermore, California.

Nomenclature

a a constant with a value of 0.75 in Eq. (5.23)
B defined by Eq. (5.25)
c_1 an orifice dependent constant in Eq. (5.22)

c_2 a constant in Eq. (5.22) with a value of zero for non-vaporizing fuel jets and a value of 0.0043 for vaporizing fuel jets

C_a orifice area-contraction coefficient

C_d orifice discharge coefficient

C_v orifice velocity coefficient

d orifice diameter

f_s stoichiometric air/fuel mass ratio

$\bar{f}_v$ average soot volume fraction along the measurement line-of-sight

h enthalpy

H lift-off length

L liquid length (i.e., the maximum penetration distance of liquid-phase fuel)

m mass flow rate

M molecular weight

n the exponent in Eq. (5.13) with a value of 2.2

P pressure

S penetration distance

t time

T temperature

U axial velocity

U_f fuel velocity at orifice exit given by Bernoulli's equation, Eq. (5.8)

x axial distance from the orifice

Z compressibility factor

Z_{st} stoichiometric mixture fraction, $(1+f_s)^{-1}$

Greek

α full spreading angle of the model fuel jet

Δ incremental change

$\bar{\phi}$ cross-sectional average equivalence ratio

θ full spreading angle of a real fuel jet

ρ density

σ standard deviation

ζ_{st} percent of stoichiometric air

Subscripts

a ambient gas

cl centerline value

f fuel or effective orifice diameter

s saturated fuel vapor condition at the liquid length

r transition between the near- and far-field limits of a fuel jet

Superscripts

+   characteristic length or time
~   non-dimensional coordinates

References

1. Naber JD, Siebers DL (1996) Effects of Gas Density and Vaporization on Penetration and Dispersion of Diesel Sprays. SAE Trans., vol. 105, no. 3, pp 82–111
2. Siebers DL (1998) Liquid-Phase Fuel Penetration in Diesel Sprays. SAE Trans., vol. 107, no. 3, pp 1205–1227
3. Heywood JB (1988) *Internal Combustion Engine Fundamentals*. McGraw-Hill, New York
4. Hay N, Jones JL (1972) Comparison of the Various Correlations for Spray Penetration. SAE Paper 720776
5. Hiroyasu H, Arai M (1990) Structure of Fuel Sprays in Diesel Engines. SAE Trans., vol. 99, no. 3, pp 1050–1061
6. Hiroyasu H, Kadota T, Arai M (1980) Supplementary Comments: Fuel Spray Characterization in Diesel Engines. In: James N. Maltavi and Charles A. Amann (eds) *Combustion Modeling in Reciprocating Engines*. New york, Plenum Press, pp 369–408
7. Wakuri Y, Fujii M, Amitani T, Tsuneya R (1960) Studies of the Penetration of Fuel Spray in a Diesel Engine. Bull. JSME, vol. 3, no. 9, pp 123–130
8. Dent JC (1971) A Basis for the Comparison of Various Experimental Methods for Studying Spray Penetration. SAE Trans., vol. 80, pp 1881–1884
9. Hays WJ (1995) Personal Communication
10. Thring MW, Newby MP (1953) Combustion Length of Enclosed Turbulent Jets. Fourth Symposium (International) on Combustion, The Combustion Institute, Pittsburgh, pp 789–796
11. Bracco FV (1985) Modeling of Engine Sprays. SAE Trans., vol. 94, no. 7, pp 144–167
12. Chavez H, Knapp M, Kubitzek A, Obermeier F, Schneider T (1995) Experimental Study of Cavitation in the Nozzle Hole of Diesel Injectors Using Transparent Nozzles. SAE Paper 950290
13. Arcoumanis C, Flora H, Gavaises M, Kampanis N, Horrocks R (1999) Investigation of Cavitation in a Vertical Multi-Hole Injector. SAE Trans., vol 107, no. 3, pp 661–678
14. Arcoumanis C, Badami M, Flora H, Gavaises M (2000) Cavitation in Real-Size Multi-Hole Diesel Injector Nozzles. SAE Paper 2000-01-1249
15. Schmidt DP, Rutland CJ, Corradini ML (1999) Cavitation in Two-Dimensional Asymmetric Nozzles. SAE Trans., vol. 107, no. 3, pp 613–629
16. Fath A, Fettes C, Leipertz A (1998) Investigation of Diesel Spray Break-UP Close to the Nozzle at Different Injection Conditions. Fourth International Symposium on Diagnostics and Modeling of Combustion in Internal Combustion Engines, Kyoto, Japan, JSME, pp 429–434
17. Soteriou C, Andrews R, Smith M (1995) Direct Injection Diesel Sprays and the Effect of Cavitation and Hydraulic Flip on Atomization. SAE Trans., vol, 104, no. 3, pp 128–153
18. Ricou FP, Spalding DB (1961) Measurements of Entrainment by Axisymmetrical Turbulent Jets. J. Fluid Mech., vol. 11, pp 21–32
19. Han D, Mungal MG (2001) Direct Measurement of Entrainment in Reacting/Non-reacting Turbulent Jets. Combust. Flame, vol. 124, pp 370–386
20. White FM (1974) *Viscous Fluid Flow*. McGraw-Hill, New York
21. Broadwell JE, Mungal MG (1991) Large-Scale Structures and Molecular Mixing. Phys. Fluids A, vol. 3, no. 5, pp 1193–1206
22. Dahm WJ, Dimaotakis PE (1990) Mixing at Large Schmidt Number in the Self-Similar Far Field of Turbulent Jets. J. Fluid Mech., vol. 217, pp 299–330

23. Mungal MG, O'Neil JM (1989) Visual Observations of a Turbulent Diffusion Flame. Combust. Flame, vol. 78, pp 377–389
24. Hinze JO (1975) *Turbulence*. McGraw-Hill, New York
25. Varde K, Popa D, Varde L (1984) Spray Angle and Atomization in Diesel Sprays. SAE Trans., vol. 93, no. 4, pp 779–787
26. Reitz RD, Bracco F (1979) On the Dependence of Spray Angle and Other Spray Parameters on Nozzle Design and Operating Conditions. SAE Paper 790494
27. Wu KJ, Su CC, Steinberger RL, Santavicca DA, Bracco FV (1983) Measurements of the Spray Angle of Atomizing Jets. J. Fluids Eng., vol. 105, pp 406–413
28. Kim JH, Nishida K, Yoshizaki T, Hiroyasu H (1997) Characterization of Flows in the Sac Chamber and the Discharge Hole of a D.I. Diesel Injection Nozzle by Using a Transparent Model Nozzle. SAE Trans., vol. 106, no. 3, pp 2372–2384
29. Potz D, Christ W, Dittus B (2000) Diesel Nozzle – The Determining Interface Between Injection System and Combustion Chamber. In: Whitelaw JH, Payri F, Desantes JM (eds) Thermo- and Fluid-Dynamic Processes in Diesel Engines, Springer, Berlin, pp 134–143
30. Siebers DL (1999) Scaling Liquid-Phase Fuel Penetration in Diesel Sprays Based on Mixing-Limited Vaporization. SAE Trans, vol. 108, no. 3, pp 703–728
31. Soteriou C, Andrews R, Torres N, Smith M, Kunkulagunta R (2001) Through the Diesel Nozzle Hole – A Journey of Discovery. ILASS Americas, 14th Conference on Liquid Atomization and Spray Systems, Dearborn, MI
32. Pickett LM, Siebers DL (2001) Orifice Diameter Effects on Diesel Fuel Jet Flame Structure. Internal Combustion Engine Division of the ASME 2001 Fall Technical Conference, Chicago, IL, September
33. Dec JE (1997) A Conceptual Model of DI Diesel Combustion Based on Laser-Sheet Imaging. SAE Trans, vol. 106, no. 3, pp 1319–1348
34. Kosaka H, Nishigaki T, Kamimoto T, Sano T, Matsutani A, Harada S (1996) Simultaneous 2-D Imaging of OH Radicals and Soot in a Diesel Flame by Laser Sheet Techniques. SAE Trans., vol. 105, no. 3, pp 1184–1195
35. Bayvel L, Orzechowski Z (1993) *Liquid Atomization*. Taylor and Francis, Washington, DC
36. Lefebvre A (1989) *Atomization and Sprays*. Hemisphere Publishing Company, New York
37. Ramos J (1989) *Internal Combustion Engine Modeling*. Hemisphere Publishing Company, New York
38. Higgins BS, Mueller CJ, Siebers DL (1999) Measurements of Fuels Effects on Liquid-Phase Penetration in DI Sprays. SAE Trans., vol. 103, no. 3, pp 630–643
39. Kamimoto T, Yokota H, Kobayashi H (1987) Effect of High Pressure Injection on Soot Formation Processes in a Rapid Compression Machine to Simulate Diesel Flames. SAE Trans., vol. 96, no. 4, pp 783–791
40. Yeh C-N, Kamimoto T, Kobori S, Kosaka H (1993) 2-D Imaging of Fuel Vapor Concentration in a Diesel Spray via Exciplex-Based Fluorescence Technique. SAE Paper 932652
41. Espey C, Dec JE (1995) The Effect of TDC Temperature and Density on the Liquid-Phase Fuel Penetration in a DI Diesel Engine. SAE Trans., vol. 104, no. 4, pp 1400–1414
42. Canaan RE, Dec JE, Green RM, Daly DT (1998) The Influence of Fuel Volatility on the Liquid-Phase Fuel Penetration in a Heavy-Duty DI Diesel Engine. SAE Trans., vol. 107, no.3, pp 583–602
43. API (1997) *Technical Data Book-Petroleum Refining*, Twelfth Revision. American Petroleum Institute (API), Washington, DC
44. Daubert TE, Danner RP (eds) (1998) *Physical and Thermodynamic Properties of Pure Compounds: Data Compilation*. Design Institute for Physical Property Data (DIPPR), American Institute of Chemical Engineers, Taylor and Francis, Washington, DC
45. Chomiak J, Karlsson A (1996) Flame Liftoff in Diesel Sprays. Twenty-Sixth Symposium (International) on Combustion. The Combustion Institute, Pittsburgh, pp 2557–2504
46. Siebers DL, Higgins BS (2001) Flame Lift-Off on Direct-Injection Diesel Sprays Under Quiescent Conditions. SAE Trans, vol. 110, no. 3, pp 400–421

47. Higgins BS, Siebers DL (2001) Measurement of the Flame Lift-off Location on DI Diesel Sprays Using OH Chemiluminescence. SAE Trans., vol. 110, no. 3, pp 739–753

48. Siebers DL, Higgins BS (2000) Effects of Injector Conditions on the Flame Lift-Off Length of DI Diesel Sprays. In: Whitelaw JH, Payri F, Desantes JM (eds) Thermo- and Fluid-Dynamic Processes in Diesel Engines, Springer, Berlin, pp 253–277

49. Siebers DL, Higgins BS, Pickett LM (2002) Flame Lift-Off on Direct-Injection Diesel Fuel Jets: Oxygen Concentration Effects. SAE Trans., vol. 111, no. 3, pp 1490–1509

50. Siebers DL, Pickett LM (2002) Injection Pressure and Orifice Diameter Effects on Soot in DI Diesel Fuel Jets. In: Whitelaw JH, Payri F, Desantes JM, Arcoumanis C (eds) Thermo- and Fluid-Dynamic Processes in Diesel Engines, Springer, Berlin, pp 109–132

51. Pickett LM, Siebers DL (2002) An Investigation of Diesel Soot Formation Processes Using Micro-Orifices. Proc. Combust. Inst., vol. 29, pp 655–662

52. Pickett LM, Siebers DL (2004) Non-Sooting, Low Flame Temperature Mixing-Controlled DI Diesel Combustion. SAE Paper 2004-01-1399

53. Gaydon AG (1974) *The Spectroscopy of Flames*. Chapman & Hall, London

54. Peters N (2000) *Turbulent Combustion*. Cambridge University Press, New York

55. Dec JE, Espey C (1992) Soot and Fuel Distributions in a D.I. Diesel Engine via 2-D Imaging. SAE Trans., vol. 101, no. 4, pp 1642–1651

56. Flynn PF, Durrett RP, Hunter GL, zur Loye AO, Akinyemi OC, Dec JE, Westbrook CK (1999) Diesel Combustion: An Integrated View Combining Laser Diagnostics, Chemical Kinetics, and Empirical Validation. SAE Trans., vol. 108, no. 3, pp 587–600

57. Kent JH, Honnery D (1987) Soot and Mixture Fraction in Turbulent Diffusion Flames. Combust. Sci. Technol., vol. 54, pp 383–398

58. Magnussen BF, Hjertager BH, Olsen JG, Bhaduri D (1978) Effects of Turbulent Structure and Local Concentration on Soot Formation and Combustion in C2H2 Diffusion Flames. Proc. Combust. Inst., vol. 17, pp 1383–1393

59. Böhm H, Hesse D, Hander H, Lüers B, Pietscher J, Wagner HGG, Weiss M (1988) The Influence of Pressure and Temperature on Soot Formation in Premixed Flames. Proc. Combust. Inst, vol. 22, pp 403–411

60. Dukowicz JK (1980) A Particle-Fluid Numerical Model for Liquid Sprays. J. Comput. Phys., vol. 35, pp 229–253

61. O'Rourke PJ (1981) Collective Drop Effects in Vaporizing Liquid Sprays. Ph.D. Thesis 1532-T, Princeton University

62. Amsden AA, O'Rourke PJ, Butler TD (1989) KIVA-II: A Computer program for Chemically Reactive Flows with Sprays. Los Alamos National Laboratory Report LA-11560-MS

63. O'Rourke PJ, Siebers DL, Subramaniam S (2001) Some Implications of a Mixing-Controlled Vaporization Model for Multidimensional Modeling of Diesel Sprays. Los Alamos National Laboratory Report LA-UR-01-1716

64. Wan YP, Peters, N (1997) Application of the Cross-Sectional Average Method to Calculations of the Dense Spray Region in a Diesel Engine. SAE Trans., vol. 106, no. 3, pp 2243–2252

65. Gonzalez MA, Lian ZW, Reitz RD (1992) Modeling Diesel Engine Spray Vaporization and Combustion. SAE Trans., vol. 101, no. 3, pp 1064–1076

66. Abraham J (1997) What is Adequate Resolution in the Numerical Computations of Transient Jets. SAE Trans., vol. 106, no. 3, pp 141–155

67. Aneja R, Abraham J (1998) How Far Does the Liquid Penetrate in a Diesel Engine: Computed Results vs Measurements. Combust. Sci. Technol., vol. 138, pp 233–255

Chapter 6
Conventional Diesel Combustion

Makoto Ikegami and Takeyuki Kamimoto

6.1 Introduction

Current diesel engines employ high boost pressures, high injection pressures and high exhaust gas recirculation (EGR) rates than ever used before to pursue better fuel economy and meet stringent emissions standards. However, the combustion processes themselves are principally governed by mixture formation, auto-ignition and turbulent diffusion in the same manner as in the conventional diesel engines. Figure 6.1 shows the temporal variations of in-cylinder pressure, flame temperature and concentrations of chemical species measured at a location in the combustion chamber of a naturally aspirated single cylinder direct injection engine which has a bore $\times$ stroke $= 95\,\text{mm} \times 110\,\text{mm}$ and a compression ratio of 14.6 [3]. The data was acquired at an engine speed of 1,250 rpm, excess air ratio of 2.07 and an injection timing of $15°$ BTDC. Although the data is relatively old, it still provides useful information to understand the basic processes of conventional diesel combustion and pollutant formation.

As seen in the time record of soot concentration, soot starts its formation immediately after the start of combustion, followed by a rapid rise in concentration until the end of fuel injection. In this particular condition, approximately 14% of carbon molecules contained in the original fuel is converted into soot by a crank angle near TDC when the highest concentration is observed. After showing a peak, the soot concentration decreases due to oxidation to approximately 1–2% of the maximum concentration and remains nearly unchanged during the rest of the expansion stroke by the time of exhaust valve opening. Meanwhile, NO formation is very slow as compared to soot formation and the maximum NO concentration occurs $15°$ behind that of soot. As well known, thermal NO freezes after completing formation and the gradual decrease in NO concentration in the latter half of the combustion period can be indebted to the dilution by air entrained in the flame.

Makoto Ikegami
Fukui University of Technology, 3-6-1 Gakuen, Fukui-shi, Fukui, Japan,
e-mail: ikegami-m@m2.gyao.ne.jp

Takeyuki Kamimoto
Tokai University 1117 Kitakaname Hiratsuka-shi, Kanagawa Japan,
e-mail: takekamimoto@aol.com

C. Arcoumanis, T. Kamimoto (eds.), *Flow and Combustion in Reciprocating Engines*,
DOI: 10.1007/978-3-540-68901-0_6, © Springer-Verlag Berlin Heidelberg 2009

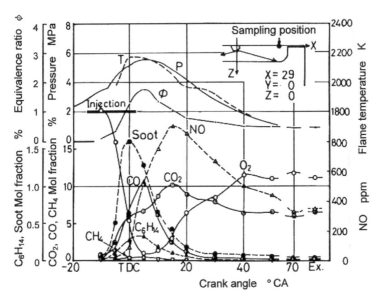

Fig. 6.1 Temporal variations of cylinder pressure, flame temperature, equivalence ratio, and concentrations of soot and NO in the engine cylinder of a DI diesel engine (1,250 rpm, excess air ratio = 2.07, injection timing = 15° BTDC) "Reprinted with permission from SAE Paper 800254 ©1980 SAE International"

In addition to the time based concentrations of chemical species, local equivalence ratio, ϕ, calculated from the concentrations of major combustion products and local flame temperature, T, measured by the two-color method are also plotted. The two series of data plots of ϕ and T permit to draw a time trajectory of cycle averaged local condition in the flame, represented by ϕ and T, on ϕ - T diagram as shown in Fig. 6.2 [19]. Although the trajectory is given by a single curve, the real condition in the flame at an instant should be expressed by a belt with a width indicating the extent of in-homogeneity because both ϕ and T in the figure have spatial distributions, respectively. The figure includes also two other trajectories for injection timings of 25° BTDC and 5° BTDC. The soot formation peninsula on the diagram was determined based on a constant volume combustion bomb study, while the NO formation peninsula was theoretically calculated using the Zeldovich reaction scheme. Inspecting the position of a trajectory relative to the two peninsulas allows a qualitative explanation of the trend of soot and NO emissions under the condition where the trajectory was obtained. It is seen that when the injection timing is retarded the trajectory shifts closer to the soot formation peninsula and farer from the NO formation peninsula, suggesting higher soot and lower NO emissions. An opposite trend is seen when the injection timing is advanced to 25° BTDC, suggesting higher NO and lower soot emissions. The trend observed here agrees well with the change of the tradeoff curve between NO and soot emissions when the injection timing is changed. The narrow region between the two peninsulas is

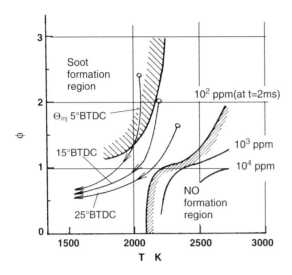

Fig. 6.2 Soot and NO formation regions on ϕ - T diagram. Three lines show temporal trajectories of local ϕ and T values in the combustion chamber at injection timings of 25°, 15° and 5° BTDC "Reprinted with permission from SAE Paper 880423 ©1988 SAE International"

"desirable path" that ϕ and T in the flame should follow to achieve simultaneous reduction of soot and NO emissions in conventional diesel combustion.

Figure 6.3 illustrates the new combustion concept, "mixed mode combustion", which is under development. In the mixed mode operation, conventional combustion mode dominates in medium and high load range, while advanced diesel combustion

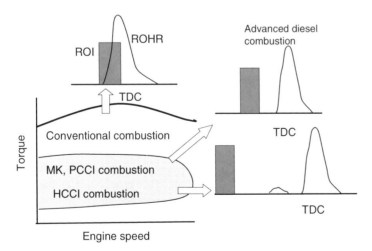

Fig. 6.3 Operation regions of conventional and new diesel combustion regimes

is applied under light load. The advanced diesel combustion includes modulated kinetics (MK) combustion, premixed charge compression ignition (PCCI) combustion, and homogeneous charge compression ignition (HCCI) combustion. These combustion regimes are all based on partially premixed or fully pre-mixed mixture combustion, featuring very low soot and NO emissions. As illustrated in the figure, ignition in these combustion regimes is controlled to occur after the end of injection, allowing injected fuel to sufficiently mix with air. The details of the advanced diesel combustion will be described in the next chapter.

This chapter deals with physical and chemical processes in conventional diesel combustion, starting with auto-ignition followed by the effects of flow and turbulence on combustion and pollutants formation. The control of conventional combustion by means of high boosting and multiple fuel injection will be also discussed in the latter part of the chapter.

6.2 Spray Evaporation and Thermal Decomposition

6.2.1 Spray Evaporation

The fuel injected into the engine cylinder through orifices of an injector undergoes breakup, atomization and evaporation, and simultaneously mixes with air entrained into the spray plume. The initial droplet size depends on the orifice diameter, injection pressure and air density, and ranges generally from 10 to 20 μm in diameter [21]. The cavitation bubbles generated in the nozzle and orifice flow collapse instantly when released in the high pressure ambient air. The effect of cavitation bubbles on atomization is discussed in ref [7, 32]. When the injection pressure is over some 200 MPa, the injection velocity exceeds the sound velocity in the in-cylinder air, and shockwaves originating from the orifice exit is observed.

The spray droplets transfer their momentum to the entrained air and decrease rapidly their relative velocities, and simultaneously receive heat from the entrained air. The increased vapor pressure on the hot surface of the droplets drives molecular mass transport, i.e., evaporation. With the progress of droplets evaporation inside the spray plume, both local mixture temperature and vapor pressure approach to their adiabatic-saturation conditions which depend on the local fuel–air ratio and initial air temperature. The increased vapor pressure as well as the decreased temperature within the spray plume slows the heat transfer and evaporation, respectively.

To know the thermal equilibrium temperatures under various conditions will be useful when considering auto-ignition process in reacting diesel sprays. The equilibrium temperature was calculated as a function of equivalence ratio with surrounding air temperature as a parameter, and the results for n-dodecane and n-heptane are plotted in Fig. 6.4 [20]. As seen in the figure, the equilibrium temperature, T_e, decreases with equivalence ratio at a given air temperature, but the corresponding curve becomes less dependent on the equivalence ratio at higher equivalence ratios because a part of fuel remains unvaporized under these conditions. The ratio of fuel

Fig. 6.4 Effect of equivalence ratio on thermal equilibrium temperature of fuel–air mixture with surrounding air temperature as a parameter (Initial fuel temperature is 300 K, surrounding air density = 6.5 kg/m³) "Reprinted with permission from SAE Paper 770413 ©1977 SAE International"

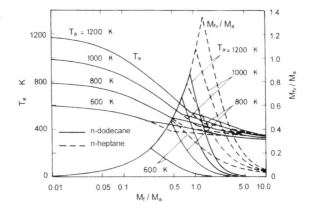

vapor to air mass at these equilibrium conditions is also shown. To the left of the peaks in the m_{fv}/m_a curves, only fuel vapor is present. To the right of these peaks, liquid fuel is also present because the vapor phase is saturated.

Although not shown here, when the ambient air pressure exceeds twice the fuel critical pressure and the ambient air temperature is higher than around 1.5 times the fuel critical temperature, the fuel droplet reaches critical temperature during evaporation and turns instantly into gas phase. This phenomena is considered to occur under high boost pressure conditions.

Evaporation process of an evaporating spray in a quiescent atmosphere was investigated using a rapid compression machine (RCM), which has a combustion chamber large enough to avoid spray impingement. The entire field of the combustion chamber is optically accessible to imaging diagnostics of various spray phenomena. The charged air was compressed to a condition equivalent to that in actual engines in 100 ms. The spatial distributions of fuel vapor concentration in an evaporating spray was measured using the laser induced fluorescence (LIF) imaging by taking into account the effect of temperature and pressure on LIF intensity [40]. Figure 6.5 shows the two-dimensional distributions of equivalence ratio in a transient evaporating spray at different timings after the start of injection. Three series of images at different runs under the same condition are presented to show the repeatability of the evaporation event. It is seen that fuel vapor–air mixtures with high equivalence ratios over 1.5 are distributed unevenly in the center and near the nozzle. The thermal equilibrium temperatures in these fuel-rich regions are about 50 K lower than that of surrounding air.

6.2.2 Thermal Decomposition During Ignition Delay

Pertaining to the ignition process, high-speed gas-sampling and analysis were performed to clarify how far thermal cracking proceeds during the ignition delay period [13]. Experiments were made on a two-stroke cycle engine that might afford to simulate basic behavior of a single spray under quiescent air condition. In Fig. 6.6

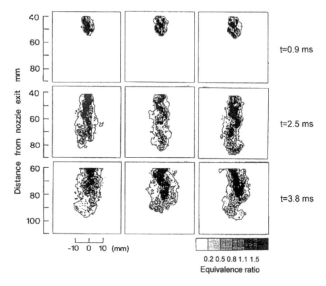

Fig. 6.5 Two-dimensional distributions of equivalence ratio in a transient evaporating spray (orifice diameter = 0.2 mm, orifice pressure drop = 30 MPa, ambient gas temperature = 718 K, ambient gas density = 14 Kg/m^3)

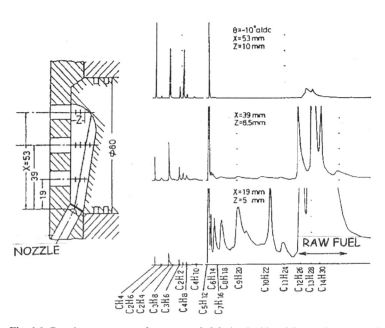

Fig. 6.6 Gas-chromatograms of gases sampled during ignition delay on the spray path

the results are given, showing gas-chromatograms of gas sampled in the spray at different sampling positions during ignition delay time. It is shown that even during the ignition delay period (1.2 ms in this case) when no appreciable pressure rise takes place, there are considerable amounts of split hydrocarbons having carbon numbers from five to eleven on the spray path at a position of 19 mm from the nozzle tip (the ratio of the position x to nozzle orifice diameter d being 63) and lighter saturated and unsaturated hydrocarbons having carbon numbers below six at the position of 39 mm, or at x/d being 130. At a position far downstream of the nozzle orifice at 53 mm, x/d being 180, raw fuel is no longer detected and instead there are certain amounts of inorganic products.

These lighter hydrocarbons are definitely decomposed products from raw fuel. According to different sampling experiments of traversing perpendicular to the spray at a fixed position of $x = 39$ mm, a pretty high concentration of carbon monoxide and acetylene is reached in the vicinity of the spray within the same delay period. This suggests that hot gas formed in the spray mantle might have been entrained into the spray body, thereby bringing about higher temperatures in the spray than without heat liberation in the mantle. These evidences might lead to a picture that even during the ignition delay fuel droplets are heated up quickly, changed into gas phase, and thermally decomposed into split hydrocarbon segments due to local exothermic reactions. Probably because of the endothermic processes during this period, such as heating-up, gasification and molecular splitting of raw fuel, the entire process may proceed almost adiabatically until the onset of ignition, which might be noticeable on the pressure-time trace. In summary, the ignition delay is such a period during which the spray is likely a gaseous jet containing a lot of unburned hydrocarbons due to gasification and thermal decomposition with assistance of partial oxidation of the spray mantle.

6.3 Auto-ignition

Fuel air mixtures formed during ignition delay period burn explosively when combustion starts and therefore, ignition delay together with the fuel injection rate and air motion plays a key role in determining the initial heat release rate. As the extent of homogeneity of fuel air mixtures formed during ignition delay is responsible for the spatial and temporal distributions of ϕ and T in the flame during the early stage of combustion, ignition delay affects largely the formation of NO and soot.

6.3.1 Factors Affecting Auto-Ignition

Auto-ignition of diesel sprays depends essentially on two processes that progress simultaneously, the physical process governing mixture formation and the chemical process leading to exothermic reactions. Therefore, all factors involved in both processes affect auto-ignition of diesel sprays. The physical properties of a fuel such as density, surface tension, viscosity, and volatility concern closely atomization, evaporation and mixture formation in the spray of this particular fuel. The cetane number is also an important index that relates closely chemical process. The in-cylinder air conditions such as pressure, temperature and oxygen concentration are all involved in both processes. The specifications of an injection system employed are also important as they play primary role in spray formation. Among them, the effect of orifice diameter on ignition delay attracts attention because it tends to become smaller with the increase in injection pressure. Also we have to know the effect of ambient air pressure as the boost pressure is continuously increasing.

Effects of orifice diameter and air pressure on ignition delay were investigated using a small rapid compression machine with a combustion chamber of 60 mm bore [23]. A blended fuel composed of n-cetane and heptamethylnonane with a cetane number of 55 was injected at a pressure of 100 MPa.

In order to see the effect of physical delay, five injection nozzles having different orifice numbers and diameters; 1 mm $\times$ 0.2 mm, 1 mm $\times$ 0.1 mm, 4 mm $\times$ 0.05 mm and 16 mm $\times$ 0.025 mm were prepared. The total cross sectional areas for the last four nozzles were kept constant by changing the number of orifices. The ignition delay, $\tau_1 + \tau_2$, was determined as the pressure recovery time from the pressure-time record as shown in Fig. 6.7. Figure 6.8 is an Arrhenius plot of ignition delay for different nozzles at an ambient air pressure of 4.2 MPa. Temperature, T_m, in the figure is not the gas law temperature but the temperature measured by a thin thermo-couple in the center of the combustion chamber.

It is worth noting that ignition delay decreases with the decrease in orifice diam-eter but remains unchanged when the orifice diameter is smaller than 0.05 mm. On the curves for small orifices, a step is observed in a temperature range between 750 and 850 K. This step obviously corresponds to the so-called negative temperature effect range, which is usually observed in shock tube experiments using homo-geneous mixtures [30]. This fact suggests that the ignition delay for 0.025 and 0.05 mm diameter orifices under the temperature range studied is dominated by the chemical delay because of the fast mixture formation with these small orifices. At temperatures above 1000 K, the difference between ignition delays for large and small orifices is bigger than that at lower temperatures, because the physical delay occupies a major portion in the total ignition delay at high ambient temperatures.

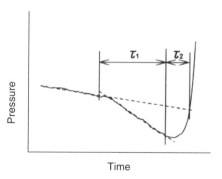

Fig. 6.7 Definition of ignition delay $\tau_1 + \tau_2$

Fig. 6.8 Effect of orifice diameter on ignition delay [23] "Permission is granted by the Council of IMechE"

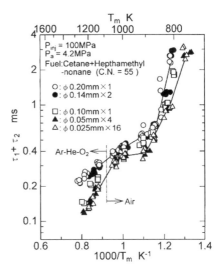

Although the data is not shown here, when the injection pressure dropped from 100 to 30 MPa for a nozzle with 0.2 mm diameter orifice, the ignition delay showed a significant increase, suggesting the dominant role of physical delay under low mixing rate conditions.

Diesel engines equipped with turbo-chargers provide maximum cylinder pressures as high as 15 MPa, and the effect of ambient pressure on ignition delay was investigated at ambient pressures up to 12 MPa for a nozzle with four 0.15 mm orifices. Figure 6.9 shows the Arrhenius plots of ignition delay data at different ambient pressures. T_a in the abscissa is the gas law temperature calculated from measured cylinder pressure. As seen, the gas law temperature, T_a, is 100–200 K lower than T_m because of the uneven temperature distribution in the combustion chamber. This suggests the need for clear definition of gas temperature in Arrhenius plots. Figure 6.9 shows that ignition delay shortens in proportion to reciprocal ambient pressure at relatively low air temperatures, because chemical reactions progresses proportionally to oxygen partial pressure. When $T_m = 1000$ K and $P_a = 12$ MPa, ignition delay is as short as 0.2 ms at injection pressure of 100 MPa.

Diesel engines for automobiles and off-road application employ generally exhaust gas recirculation (EGR) to reduce NO formation, and the resultant reduced oxygen concentration in the intake charge prolongs ignition delay. The effect of EGR on ignition delay was investigated using a constant volume combustion vessel [18]. Diesel fuel was injected from a 0.2 mm diameter orifice at a pressure of 70 MPa into an environment with gas density of 16.2 kg/m^3. The results show that the oxygen concentration has strong effect on ignition delay in a high temperature range between 800 and 1200 K where the chemical delay is dominant. For instance, at an air temperature of 1000 K the ignition delay of 0.25 ms at 21% O_2 volume concentration was elongated to 0.31 ms at 17% O_2 concentration and 0.4 ms at 13% O_2 concentration.

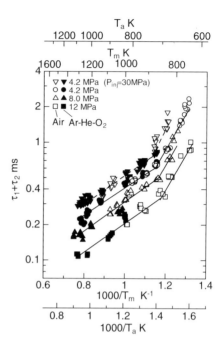

Fig. 6.9 Effect of ambient pressure on ignition delay [23] "Permission is granted by the Council of IMechE"

6.3.2 Effect of Pilot Injection

Pilot injection, an effective way to shorten ignition delay, is commonly used to improve the NO_x – particulate trade-off, because pilot injection permits engine operation at much retarded injection timings than without pilot injection. The pilot injection is also indispensable for reducing engine noise, particularly in small high speed diesel engines which has higher heat loss fractions than larger engines.

Regarding the effect of pilot injection, a question arises about which is the major cause for shortening ignition delay, increased ambient temperature due to pilot fuel combustion or hot gases, which are formed by pilot combustion, drifting near the nozzle tip. In order to investigate this, an experiment was conducted using the small rapid compression machine mentioned in the preceding section. A square combustion chamber with a dimension of 60 mm × 60 mm was used in this experiment.

The parameters investigated were the ratio of pilot fuel mass to the total fuel mass and the injection dwell. The injection dwell is the interval between the end of pilot injection and the start of main injection. Diesel fuel was injected from an orifice of 0.2 mm diameter at 30 MPa. The compressed air density was kept constant at 17.8 kg/m³ by changing the charge pressure and the piston stroke at all conditions investigated, allowing the spray characteristics being unchanged.

Figure 6.10 shows the effect of injection dwell on ignition delay for main injected fuel, τ_{main}, in an Arrhenius plot at a condition of 5.7% pilot injection amount. Temperature, T_a, on the abscissa is calculated from the cylinder pressure at the start of main fuel injection. The main fuel is injected into an environment with pressure

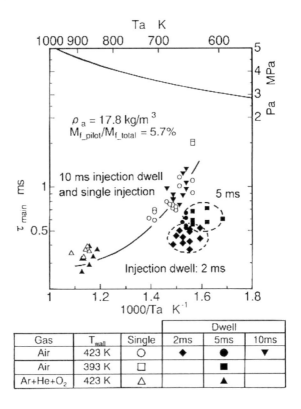

Fig. 6.10 Effect of injection dwell on ignition delay of main injection fuel, τ_{main}, at a fixed pilot injection fraction of 5.7%, showing shorter ignition delay times for shorter injection dwell times

			Dwell		
Gas	T_{wall}	Single	2ms	5ms	10ms
Air	423 K	O	◆	●	▼
Air	393 K	□		■	
Ar+He+O$_2$	423 K	△		▲	

and temperature increased by pilot fuel combustion. As the data plots show, the ignition delay, τ_{main}, with 10 ms injection dwell coincides approximately with that of no pilot injection obtained at the same temperature level. This shows a simple temperature effect of pilot injection. On the contrary, when the injection dwell becomes shorter, 5–2 ms, the ignition delay, τ_{main}, is obviously shorter as compared to that of no pilot injection case at the same ambient temperature. The result suggests that the increased temperature and pressure by pilot combustion shorten naturally the ignition delay of main injection, and in addition, if hot gas clouds formed by pilot fuel combustion are drifting near the nozzle, they will help the main fuel sprays ignite earlier. A high speed imaging study on the effect of pilot injection under swirling conditions reveals that the effect of pilot injection is most pronounced when the pilot flames being conveyed by air swirl contact the main sprays from neighboring orifices. A flamelet combustion modeling provides detailed explanation on the roles of pilot injection in ignition event of main fuel [9].

6.3.3 Modeling of Auto-Ignition Focusing on Turbulent Mixing

There are two possible ways to model what takes place during the ignition delay; one is a precise and detailed description of the entire events taking place, taking into consideration every physic-chemical aspect. Physical factors may include motions of fuel droplets, turbulent eddy motions, heat and mass transfer between droplets and surrounding gases, whereas chemical factors might be thermal decomposition, pre-flame reactions and oxidation reactions. Theoretically speaking, this approach is possible but not practical even at present for the long computational time required.

The other is a simplified manner in which only some significant factors are dealt with, all remaining factors being either simplified or ignored. Most models rely on this latter approach. One of the most familiar model is Wolfer's one in which ignition delay data gained from his bomb tests are indicated on a simple Arrhenius curve [37]. Wolfer's model relies on an implication that a chemical reaction is rate-determining step due to the fact that there are many fine fuel droplets that may form combustible mixtures, which are prepared even at the very beginning of injection. This may be valid in the case when the pressure and temperature are low enough, but it fails to hold in the case of shorter ignition delays as in the case of modern diesel engines. This is due to the fact that both physical and chemical factors of mixture formation and heat-liberation and -absorption might affect the length of ignition delay appreciably unlike Wolfer's model.

The model to be presented below focuses on turbulent mixing and places emphasis on the role of mixture heterogeneity prevailing within the spray [16]. Heterogeneity refers to place-to-place fluctuations in temperature and fuel–air ratio, or fuel concentration, both caused due to random nature of mixing of fuel and gas in the spray. As time goes on, the heterogeneity tends to decrease by strong turbulent mixing. This process may be described in a stochastic manner based on a stochastic model that employs essentially zero-dimensional approach [12]. For the overall scope and the approach of the stochastic model of diesel combustion, refer Sect. 6.6.2. Here, explanations will be limited only to the ignition delay period. In the present stochastic spray model, chemical reactions during ignition delay are expressed by a simple one-step irreversible mechanism of Arrhenius type, in which ignition is assumed to be reached when the concentration of certain active species exceeds a threshold for each fluid element whose state may change from time to time according to the details of mixing. Basically, this assumption is equivalent to assume the Livengood-Wu integral, which has been pretty often applied to predict the ignition delay under time-dependent pressure and temperature from the delay data gained at a fixed pressure and temperature.

Calculations have been performed on a computer in which fifty thousand Monte-Carlo fluid elements are assumed to simulate mixing and chemical reactions undergoing within and around the spray. Some elements originate from gas fuel and the remainder from entrained air, both being subject to random binary collision and re-dispersion uniformly in a closed spray volume without considering any structure of spray, consistently with that described by Curl's model. In this model, collision frequency ω and turbulence energy k, both being essential to turbulent mixing, are assumed to be given by $\omega = 0.3266 \, k/L$, where L is the integral scale of turbulence and is given by $L = d_N \, (\rho_f / \rho_a)^{1/2}$ where d_N is the nozzle orifice diameter and ρ_f and ρ_a are the densities of fuel and air, respectively. k-eqation takes into account the following mateers; turbulence production due to shear force in the spray and to gas expansion by chemical reaction, turbulent eddy dissipation, and increase in spray volume. Even during the ignition delay, heat liberation and endothermic processes are weak but no longer negligible once the delay length is desired to determine. For this reason, heat absorption per amount of fuel injected is taken into account to describe the effect of heat for fuel gasification and thermal decomposition. In determining model constants including chemical reaction, data have been selected with great care.

Figure 6.11 shows the comparison of predicted pressure-time trace reproduced by the present model with those of experiments made on a free-piston-type rapid compression machine using a 0.25-mm-diameter single orifice nozzle. In this figure, p_f-p_i stands for the difference between fired and non-fired pressures and p_n for the pressure of nitrogen cycle where a fuel is injected into the combustion chamber filled with nitrogen. It is noted that the predicted pressure trace shown in dotted line well reproduces the experimental one. In Fig. 6.12 Arrhenius diagrams are given for the ignition delays of τ_1 (onset of heat release) and $\tau_1+\tau_2$ (recovery to non-fired pressure) against reciprocal air-temperature $1/T_i$ for a few different pressures. Degree of reproduction looks fairly well, suggesting the soundness of the proposed model and empirical constants employed. It is true to say that there is a room for improvement in the case when air temperature is lower. The discrepancies of predicted results from measured ones might be attributed to too simple chemical kinetics adopted. Nevertheless, it may safely be stated that turbulence induced in the spray plays a dominant role in determining heat release during ignition delay period.

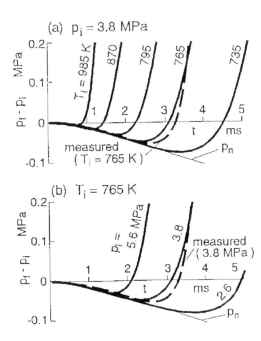

Fig. 6.11 Predicted pressure history under several conditions

Fig. 6.12 Predicted and measured ignition delayS of τ_1 and $\tau_1 + \tau_2$ against air temperature at the start of injection T_i

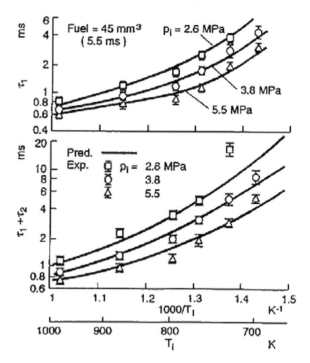

This stochastic ignition-delay model has been extended to predict the initial explosive heat release and pressure rise just following the ignition delay [34], and also to afford the description of the ignition delay and ignition process of premixed charge compression ignition (PCCI) combustion [17]. The latter analysis, chemical kinetics is based on Schreiber's five-step quasi-global mechanism.

6.3.4 Visualization of Auto-Ignition in Transient Reacting Sprays

A reacting diesel spray shows a rapid increase in its volume immediately after the start of combustion in accordance with the rapid heat release during the initial combustion phase. Later in this period, faint luminosity appears in the flame images taken by the direct photography, showing the start of soot formation, but the locations, where exothermic reactions occur first within the spray, can't be identified with the direct photography.

Formaldehyde is one of the relatively stable intermediate species formed in the low temperature oxidation of alkane fuels. Imaging formaldehyde LIF was performed to study knocking in a homogeneous charge spark ignition engine and it was demonstrated that formaldehyde formed in the end gas extinguished at multi-points where spontaneous ignition occurred prior to knocking [11]. This suggests that formaldehyde formed in the low temperature oxidation reaction is consumed rapidly upon the start of the high temperature oxidation reaction.

On the basis of this study, the same technique was applied to transient reacting sprays to image locations where spontaneous ignition occurs [26]. The third harmonic of a Nd:YAG laser with 355 nm was used to excite formaldehyde in the spray of 0-solvent (21.4% $C_{12}H_{26}$ + 22.0% $C_{13}H_{28}$ + 51.6% $C_{14}H_{30}$). The fluorescence given off from formaldehyde was imaged by a CCD camera through a band pass filter with a center wavelength of 398 nm. The test fuel was injected from a 0.15-mm diameter orifice at a pressure of 85 MPa.

The temporal change of the formaldehyde LIF in a transient reacting spray during ignition delay time and initial combustion phase was imaged at three ambient temperatures of 620, 660 and 790 K. The ignition delays for these three conditions were 0.49, 0.35 and 0.22 ms, respectively. The three series of LIF images acquired showed a common feature that formaldehyde LIF appeared across the entire cross section of the spray immediately after the start of ignition at $\tau_1 + \tau_2$, and after a short period of time dark regions indicating the consumption of formaldehyde due to high temperature reactions appeared in the spray head and grew with time. Figure 6.13 shows LIF images taken for the three ambient temperatures at 1.05, 0.78 and 0.43 ms, respectively when part of LIF images began to darken. It is seen that at ambient temperatures of 620 and 660 K the LIF intensity starts to drop mainly in the periphery of the spray head, while at a higher temperature of 790 K it decreases mainly in the central region behind the spray head. The difference between the locations of exothermic reactions for the three conditions will be discussed qualitatively in the following paragraph by considering the spatial distributions of local mixture temperature and equivalence ratio in the spray.

Figure 6.14 shows calculated ignition delay times as functions of initial gas temperature, T_{gini}, with equivalence ratio of n-decane/air mixture as a parameter [27, 8]. T_e in the figure stands for a thermodynamic equilibrium temperature for equivalence ratios of 0.5 and 2.0. The initial gas temperature, T_{gini}, was calculated using the perfect gas law from measured cylinder pressure and therefore the gas temperature in the central region of the combustion chamber, where spray ignition takes place, is higher than T_{gini} at least by several ten degrees. When looking at ignition delay curves for different equivalence ratios, we notice that ignition delay

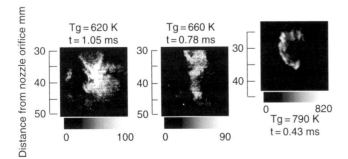

Fig. 6.13 Images of formaldehyde LIF of reacting sprays at different ambient air temperatures "Reprinted with permission from SAE Paper 2000-01-0236 © 2000 SAE International"

Fig. 6.14 Effects of initial
temperature and equivalence
ratio of homogeneous
n-decane/air mixtures on
ignition delay [27]
"Permission is granted by the
Council of IMechE"

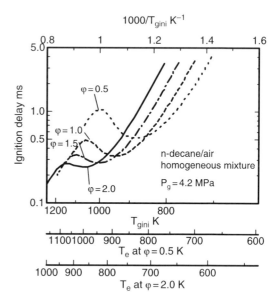

is shorter for lean mixtures when T_{gini} is below around 850 K, while it is shorter with
rich mixtures when T_{gini} exceeds 950 K. The trend of these four curves suggests that
at two ambient gas temperatures at 620 and 660 K, fuel lean mixtures in the spray
periphery are most likely to ignite first. At an ambient gas temperature of 790 K, the
temperature in the center of the combustion chamber is estimated to exceed 900 K
(see Fig. 6.9), and this may cause early ignition in the fuel rich mixtures in the
central part of the spray head.

6.4 Soot Formation and Oxidation in Transient Reacting Sprays

6.4.1 Formation of Soot Precursors

Acetylene, an unsaturated hydrocarbon, which is formed by thermal decomposition of fuel during ignition delay period reacts each other under high temperature
conditions to form poly-aromatic hydrocarbons (PAH) which are known as the soot
precursor [22].

The soot precursor and young soot particles in a transient reacting spray achieved
in the RCM were visualized simultaneously using LIF and the laser induced incandescence (LII) techniques, respectively [2]. The LIF emission excited by a 355 nm
laser sheet and detected at 400 nm wavelength represents the broad-band emission
from three- to six-ring PAHs. 0 solvent was injected from a 0.15 mm diameter
orifice at an injection pressure of 55 MPa into compressed air at 2.9 MPa and 760 K.
LIF and LII images obtained simultaneously at 1.3 ms after the start of injection
(at around 0.5 ms after the start of ignition) are given in Fig. 6.15 (a) and (b),

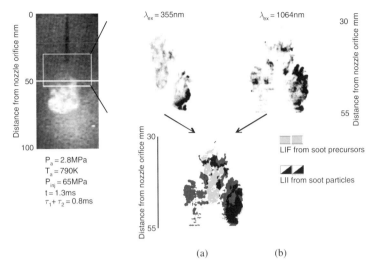

Fig. 6.15 Simultaneous images of PAHs LIF at (a) 355 nm and soot LII at (b) 1,064 nm in a transient reacting spray obtained at around 0.5 ms after the start of ignition

respectively. A comparison of the two images tells that the soot precursor exists mainly in the central and upstream region of the spray, while young soot particles are thinly distributed in the spray periphery.

6.4.2 Soot Formation

Figure 6.16 shows a series of high speed direct photographs of a typical transient reacting spray achieved in the RCM. At 1.0 ms after the start of injection, clusters with weak luminosity appear in the periphery of the spray head and the luminous flame evolves with time, increasing luminosity and volume. The luminosity, i.e., thermal radiation in visible wavelength range, includes line-of-sight information of soot temperature and soot mass concentration, but the information is not enough to

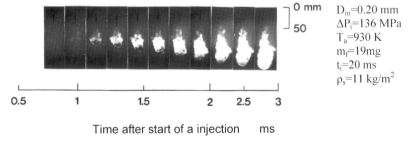

Fig. 6.16 Direct photographs of a transient reacting spray in RCM (orifice diameter = 0.2 mm, injection pressure = 136 MPa, air temperature = 930 K, air density = 11 kg/m^3)

study the soot formation process within the flame. For this reason, the planner soot scattering imaging technique was applied to reacting sprays [38].

Figure 6.17 shows three series of two-dimensional soot scattering images taken at injection pressures of 55, 85 and 134 MPa. It is seen that the scattering intensity decreases obviously with the increase in injection pressure, showing less soot formation at higher injection pressures. It is also noticed that soot is distributed rather in the periphery of the spray head, particularly at the highest injection pressure of 134 MPa. At injection pressures of 55 and 85 MPa, wavy or snaky shape clusters exist in the central region and connect the soot layers in the spray head.

In order to investigate the relationship between the soot distribution patterns observed in the figure and the flow pattern inside the flame, two-dimensional local velocity vectors in a vertical laser sheet plane in the spray was measured by the cross-correlation method focusing on a set of two scattering images taken at a small time interval [39]. Figure 6.18 shows the two two-dimensional local velocity vector distributions, where image A on the left is the velocity vector distribution based on the stationary axis, while image B presents the velocity vectors relative to the flame tip velocity. Image B clearly shows velocity vectors pointing to the spray head along the spray axis and those going upward along the flame surfaces. It is worthy of note that the soot particles formed in the central region are conveyed downstream to the spray head and swept upward along the flame surfaces by a large scale vortex

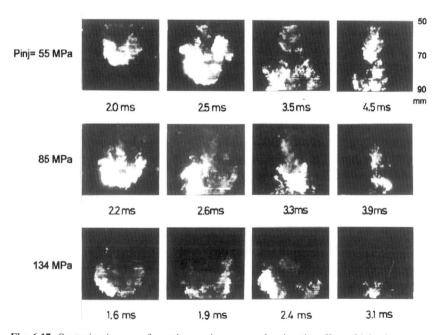

Fig. 6.17 Scattering images of soot in reacting sprays showing the effect of injection pressure "Reprinted with permission from SAE Paper number 910223 © 1991 SAE International"

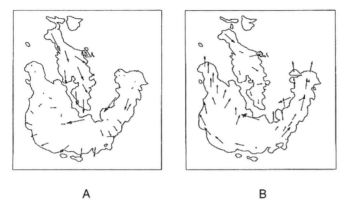

A   B

Fig. 6.18 Velocity vector mapping showing a flow field induced by a spray head vortex (A: velocity vectors in the static field, B: velocity vectors relative to the velocity vector at a location in the spray head) "Reprinted with permission from SAE Paper number 920114 © 1992 SAE International"

formed in that region. This observation gives a reasonable explanation to the soot distribution patterns observed in Fig. 6.17.

In addition to the soot scattering images and velocity vector distributions, the images of soot mass concentration, particle size and particle number density were obtained by analyzing pairs of LII and scattering images assuming the Rayleigh approximation to the soot particles in the reacting spray [24]. The qualitative two-dimensional distributions of these values were acquired at a condition when fuel was injected from a 0.15-mm diameter orifice at 100 MPa into compressed air at 2.9 MPa and 760 K.

Figure 6.19 shows the temporal series of three images, mass concentration on the top, particle size in the middle and particle number density on the bottom. The soot concentration images obtained by the LII technique are similar to scattering images discussed previously. The particle diameter shows the smallest value near the nozzle and increases with distance from the nozzle as the soot clusters in the central region go downstream toward the spray head because the coalescence between particles and coagulation progress while they travel downstream. As expected the particle number density shows an oposite trend to that of particle diameter.

6.4.3 Soot Oxidation

The soot oxidation process during the expansion stroke is as important as the formation process since it plays a decisive role in determining the soot emissions levels as explained in the gas sampling analysis in Fig. 6.1. OH molecule, one of the intermediate combustion products, is highly reactive in such reactions as soot oxidation and NO formation. The soot oxidation process in transient reacting sprays was studied using soot scattering and OH LIF imaging techniques [25]. In the experiment, polypropylene glycol, an oxygenated fuel, was added by 10% to the base 0

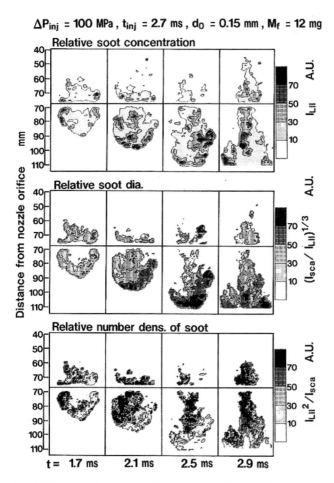

Fig. 6.19 Temporal sequences of soot concentration, soot diameter and soot number density distributions in a reacting spray "Reprinted with permission from SAE Paper number 952451 © 1995 SAE International"

solvent to control soot concentrations in the flame for the easy access of the laser diagnostics employed. Figure 6.20 shows a series of reacting spray images that include both soot scattering (red) and OH LIF (blue) distributions. It is seen that OH is distributed in the most outer regions of the reacting spray until the end of injection at 3.7 ms, and becomes dominant in the entire regions at 5.0 ms after the end of injection.

The information given above allows us to draw a picture on the soot formation and oxidation processes in a transient reacting spray as shown in Fig. 6.21. The fuel emerging from the orifice atomizes and evaporates, and the vapor undergoes immediately thermal cracking, generating a number of saturated and unsaturated split hydrocarbons. Some of the decomposed unsaturated hydrocarbons such as

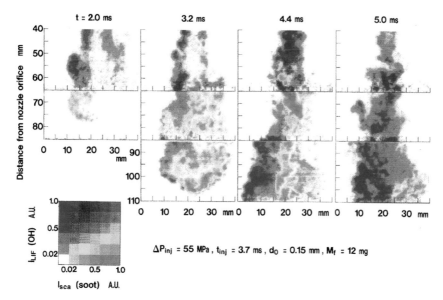

Fig. 6.20 Temporal sequences of soot and OH images in a reacting spray. "Reprinted with permission from SAE Paper 960834 © 1996 SAE International"

acetylene form PAHs, soot precursor. Heavy PAHs form soot nuclei and young soot particles, showing a low mass concentration but the highest number density in upstream regions. They agglomerate each other by collision and grow in size while traveling downstream, and the soot concentration increases simultaneously by condensation and dehydrogenation. The soot shows the highest concentration. in the spray head and soot clouds are conveyed upward along the spray surface by a large vortex generated in the spray head. The soot clusters flowing upstream along the spray periphery are re-entrained into the spray by vortex motions in the spray periphery and are oxidized by OH radicals.

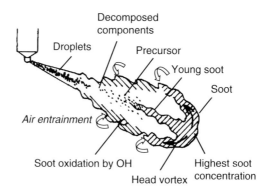

Fig. 6.21 Mechanism of soot formation and oxidation in a transient reacting spray

6.5 Effects of Flow and Turbulence on Combustion and Emissions

6.5.1 In-Cylinder Gas Motions and Their Effects on Combustion

In cylinder gas motions may significantly affect the entire processes of diesel combustion in various ways; e.g., speeding-up burning, enhancing air utilization, and promoting formation and destruction of pollutants. It is true to say that modern diesels look more relying on high-pressure fuel injection than those used to be, nevertheless gas motions and their effects on combustion are still of great concern as one of the central problems of the combustion chamber design.

In-cylinder gas motions may basically be classified into three categories for the case of direct injection engines. One is air motion induced until the start of injection, such as swirl and squish, the second is combustion-driven gas motion like reversed squish, and the other is turbulence generated with various bulk flows including the fuel jet flow. Current understanding of these gas motions is still limited because of the difficulties in both flow measurement and theoretical approach. This is especially the case for combustion-driven gas motions and turbulent motions during combustion. Yet some diagnosing methods and computational fluid dynamic simulation have been applied successfully to a certain extent to elucidate such gas motions and their effects on combustion. As a result, knowledge has been accumulated in depth. Among these findings as such, some will be briefly described below.

In high-speed direct-injection diesel engines having a deep-bowl in the piston head, various kinds of air motions are created during the compression stroke; they consist of swirl and squish, the latter being induced when the air charge is forced to flow into the bowl. Swirl is the air rotation around the cylinder axis and is generated during the suction stroke by giving an angular momentum using either helical or tangential suction port. The intensity of the swirl is usually expressed by the swirl ratio, defined as being the ratio of rotational speed of the swirl to the engine speed. Swirl itself may bring about favorable effects in fuel distribution, air utilization and speeding-up of combustion. However, too high swirl- ratio may not only decrease volumetric efficiency but also cause poor combustion as will be shown later.

The rotational speed of the swirl increases as compression proceeds since the radius of rotation of air is reduced during the compression stroke when the air is forced into the bowl that has smaller diameter than the cylinder diameter. The degree of the speeding-up depends on the ratio of bowl diameter d to the piston diameter D. For example, in the case of a cylindrical cavity having d/D = 0.5 with zero top-clearance, the swirl speed at the compression end is four times that before compression, if the swirl is a simple forced vortex and if its angular momentum is conserved throughout compression without any friction loss. In actual situations, neither frictional loss nor effects of top clearance is negligible, so that the speed ratio of before and after compression would be 3 or less in the case of d/D = 0.5.

Figure 6.22 shows a result of CFD simulation of swirl and squish induced in a cylindrical deep-bowl chamber near at the end of compression (predicted by

Fig. 6.22 Flow induced in a
deep-bowl at motored run
(swirl ratio is 2.1 before
compression and
compression ratio 16:1; C_m is
the mean piston speed, and
relative velocity the ratio of
tangential velocity to that at
cylinder radius before
compression)

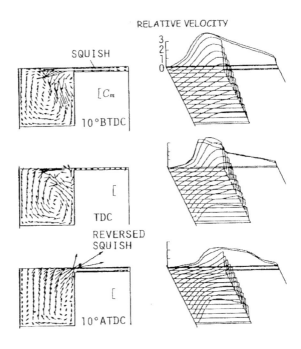

FREC-AXSY code developed by M. Ikegami and his colleagues). In this case,
swirl is assumed to be a simple forced vortex like a solid rotation until the start of
compression. As might be seen from the figure, air swirl motion inside the bowl is
not a simple forced vortex but is that having such a non-linear profile that the tangential velocity increases with radius faster than that of solid body rotation. Radial flow
from the clearance space to the bowl is sharply bent at the corner, thereby creating
a flow towards deeper into the bottom. This flow is unlike a toroidal motion such as
that might be conceived in the past. This is attributed to an interaction of two flows
of swirl and squish. Generally speaking, such secondary flows may be induced in
different ways depending on the conditions including the geometry of the bowl.

Again in the present case, if injection is made before top dead center, the fuel
jet may hit the sidewall of the bowl and, as a result, the fuel splash may spread
along the swirl motion and in the direction towards the bottom. From the figure, it
may also be noted that flow reverses its direction from the bowl to the clearance
after top dead center during the expansion stroke. Such a flow is called the reversed
squish. Once viewed from the bowl, the clearance volume may act as a sink that
carries an air and even the prepared mixture out of the bowl into the clearance. In
the case when combustion takes place within the bowl, the reversed squish is especially significant, since a fuel-rich mixture that may only partially be burnt might
outflow into the clearance and be quenched there by the contact with walls when
passing the narrow clearance at a pretty high speed. If the walls of sufficiently lower
temperatures quench partially burnt rich mixtures, a lot of soot might be formed.
This would be explained by a simple chemical equilibrium of a system consisting of

N_2, CO, H_2, free carbon and other minor species, in which the Boudourd reaction and the water–gas reaction may hold. The conclusion of the analysis would tell us that a carbon-free rich mixture might form free carbon in the mixture once the temperature is changed from well higher than 1000 K to temperatures low enough.

To avoid this situation and to achieve more satisfactory combustion, it is necessary to find out the optimum combination between number and diameter of injection nozzle hole, injection pressure, direction of spray, swirl intensity and geometry of combustion chamber. For the case of a given injection nozzle and chamber geometry, the best condition may be met somewhere at a swirl ratio. Figure 6.23 compares high-speed flame photos at the beginning of outflow for different swirl ratios [14]. In the case without swirl at swirl ratio r_s being zero, the fuel spray directly hits the bowl sidewall, around which a luminous flame is formed semi-spherically. At the same time, a part of flame bursts into the clearance, showing a bright front. Behind

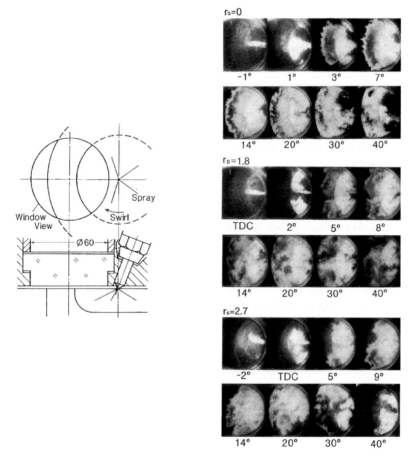

Fig. 6.23 High-speed direct photographs of diesel flame for different swirl ratios in a deep-bowl chamber (a five 0.22 mm-diameter hole nozzle; 140° spray angle

the front is a dark zone that might have been formed as the result of wall quenching. Clearly, the reversed squish and the flame expansion may play a significant role in the flame development. Under this quiescent condition, much fuel–air mixture accumulates in the vicinity of the bowl, and as a result an early outflow might have taken place. In the swirled case at r_s being 1.8, initial spill-off into the clearance looks significantly suppressed. This is due to the fact that the swirl makes the spray jet bent with an action like crosswind against the spray, thereby permitting longer spray path. At the same time, the secondary flow induced in the bowl may afford to distribute the fuel in deeper positions, both keeping the fuel from out-flowing in earlier stages. In a highly swirled case at r_s being 2.7, flame in the bowl exhibits shrunken, no longer reaching to the outer periphery inside the bowl. This results from greater deflection of the spray jet due to stronger cross-wind than in the cases of lower swirl ratios. Thus the air in the periphery remains unused for a longer time, thereby causing poorer combustion with a lot of soot being formed than at a correct swirl ratio.

The situation as such is called over-swirling, which should be avoided by all means. Over-swirling may take place if the spray penetration is not enough for the intensity of swirl prepared in the combustion chamber. Since the spray penetration increases with the nozzle-hole diameter, the number of the spray should be reduced to keep the penetration enough at higher swirl ratios.

Over-swirling may also take place in the case of a reduced spray angle, where the spray angle denotes doubled angle of the geometrical spray path with the cylinder axis. This is attributed to the localized flaming zone that is confined in the middle lower part in the cavity. To avoid this, the spray angle ranging from 140° to 160° should be employed normally.

In summary, at correct intensity of swirl motion, the swirl may improve spatial distribution of fuel within the piston bowl and simultaneously avoid unfavorable excessive outflow of the charge into the top-clearance. However, over-swirling that takes place at a swirl having an excessive intensity may weakens the penetration of fuel in radial direction, bringing about a localized flaming zone that may lead to smoky combustion and deterioration of air utilization in the outer peripheral zones.

The geometry of the piston cavity is diverse among a variety of high-speed direct-injection diesel engines; there are traditional design of shallow-dish type and deep-bowl type, either with or without central projection in the bottom. In Fig. 6.24 some typical ones are shown. Some cavities are served by either squish lip, or re-entrant lip, or toroidal ring cavity. The precise function of the reduced diameter of the cavity entrance is not well known, but it may safely be stated that the narrowed entrance may enhance the velocity of squish flow that might be helpful in ensuring better mixing on the one hand, and may keep the fuel within the cavity from excessive spill-off into the clearance space on the other hand. This might especially be important at a retarded injection, because the reversed squish that is capable of drawing out the charge inside the cavity becomes stronger at later timings due to higher piston velocity. According to the computer analysis and experiments made by AVL, a toroidal ring cavity may keep the swirl motion in the cavity stronger even until

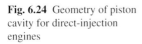

Fig. 6.24 Geometry of piston cavity for direct-injection engines

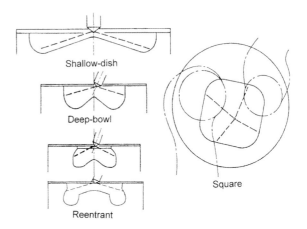

Shallow-dish

Deep-bowl

Reentrant

Square

later crank angles than otherwise, thereby achieving favorable combustion. For this reason, modern high-speed diesels are often served by this design concept.

6.5.2 Role of Turbulent Mixing of Fuel–Air in Combustion and in Formation of Nitrogen Oxides

While the bulk flows mentioned above basically governs spatial distribution of fuel inside the cylinder and determines the degree of air utilization in the diesel combustion chamber, turbulence plays most essential role in mixing of fuel and air at microscopic molecular level not only during the spray development but also during combustion. Such a microscopic point of view will be helpful in giving a better insight into the rate of heat release during diffusive combustion, which follows initial explosive combustion, as well as into the formation of pollutants, such as nitrogen oxides and particulate matter.

As might easily be expected, each fluid element consisting of the injected fuel and the entrained air within the spray has a fuel concentration, which may differ from spot to spot at a certain moment. This state, which may be called "heterogeneous state," or simply "heterogeneity," prevails over the entire process. This heterogeneity is subject to turbulent mixing, forming a mixture of less heterogeneous states as time goes on. Even during combustion following the fuel injection, such heterogeneity may still remain and undergo turbulent mixing as well. Unlike the homogeneous charge in spark-ignition engines, instantaneous concentration of the heterogeneous charge as such may vary from zero, which corresponds to pure air, to even that of pure fuel vapor during the course of turbulent mixing. In between there will be fluid elements that have a fuel–air ratio at and around the stoichiometric ratio, which may arrive at so high temperature that a lot of oxide of nitrogen may be yielded. This may well explain why the concentration of tail-pipe nitrogen oxides is much higher than that estimated from overall fuel–air ratio on a simple assumption of a homogeneous charge.

Description of the heterogeneity of fuel concentration and its decay with time towards a uniform mixture by turbulent mixing in diesel combustion has success-fully been made relying on a stochastic approach, based on Curl's binary collision-redispersion model for turbulent mixing [12]. The Curl model basically assumes a stirrer in which the heterogeneity of concentration prevails uniformly under the presence of turbulence. Random collision of two arbitrary-chosen fluid particles having different concentration is assumed to take place at an equal probability in terms of time, thereby producing two identical particles having the same state of arithmetic mean concentration as the colliding pair. The number of collision of a single fluid particle during a unit of time, or simply "collision frequency," can be determined so as to reflect the characteristics of turbulence field.

The model of diesel combustion relying on this concept is basically zero-dimensional and assumes that a uniform turbulent field holds and that the hetero-geneity in fuel concentration is uniformly distributed either in the spray zone or over the entire space once combustion is initiated, without taking into account the details of the bulk flows induced. Instead, the heterogeneity and its temporal change are described in a stochastic manner, using a Monte Carlo method to avoid solving a complicated integro-differential equation.

Figure 6.25 gives a result of calculation for a simple case at a constant-volume reaction vessel without chemical reaction, showing how the probability density function (PDF) of the fuel mass fraction, Y, progresses with time by turbulent mixing. As the initial condition, it is assumed that there is 20% wt. of pure fuel ($Y = 1$), the remainder being pure air without any mixture that has an interme-diate fuel mass fraction. In the figure, η denotes the non-dimensional time from the start of mixing, defined as being the integral of collision frequency. This value

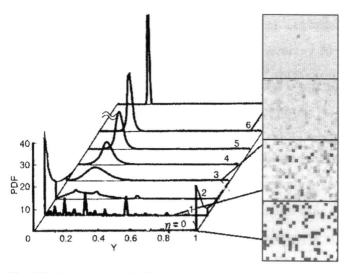

Fig. 6.25 Heterogeneous distribution of concentration and its temporal change due to turbulent mixing

inherently expresses the average cumulative collision frequency in terms of a fluid particle. Sketch in each circle shows how the spatial distribution looks like at the corresponding time. PDF against Y shows a discrete distribution until $\eta = 2$, after which the profile becomes smoother and narrower with time, approaches normal distribution at $\eta = 6$, and exhibits almost a single peak at average mass fraction of 0.2 at $\eta = 12$. Incidentally, the collision frequency ω may be found by a relation $\omega = 0.4\ u'/L$, where u' denotes the root-mean-square intensity of turbulence and L the integral length scale.

In actual cases when combustion takes place in the engine cylinder, heat may be exchanged and liberated at each collision due to chemical reaction. For this reason, the specific enthalpy should be included as one of the dependent variables that may describe the state of each fluid particle. Heat liberation is taken into consideration on the assumption of chemical equilibrium, which holds for the instantaneous state of each fluid particle. The change in enthalpy by expansion due to heat release is also considered in the employed mathematical framework. For the further details consult the reference.

In Fig. 6.26 some predicted relations of cylinder pressure and the rate of heat release versus crank angle by the present model are shown for different injection timings, together with the measured ones. The results shown are for the case of a small high-speed engine having a cylinder diameter of 90 mm, a stroke of 105 mm, a compression ratio of 16:1 and a deep-bowl chamber on piston head. In initial burning stages a higher rate of heat release is seen due to rapid burning of combustible mixture that has been accumulated. During the period of diffusive

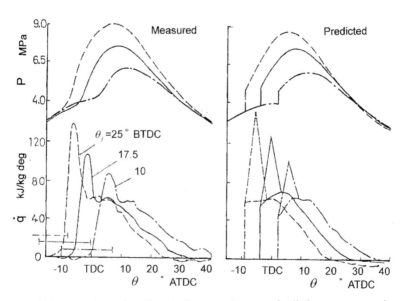

Fig. 6.26 Comparison of predicted and measured course of cylinder pressure p and rate of heat release for different injection timings θ_j (overall equivalence ratio is 0.62, engine speed 1,800 rpm and injection made at 45 MPa by a three 0.30 mm-diameter hole nozzle)

burning, the rate of heat liberation looks primarily determined by turbulent mixing of a rich mixture with either leaner mixture or that containing excess oxygen. Judged from a reasonable degree of reproduction for different injection timings, a proper description of the heterogeneity in fuel concentration and temperature in the prevailing field of turbulence may successfully be achieved. In the case presented here, the turbulence intensity and the scale have been determined by using the concept of ordinary k-εmodel, taking into consideration the spray process, gas motions such as swirl, and flame expansion. Also, the effects of ignition delay and heat transfer through walls are included.

Figure 6.27 shows the changes of PDF of temperature and of nitric-oxide concentration (NO) with crank angle, for two cases shown in the previous figure showing pressure versus crank angle relations. Since most nitrogen oxides are formed as thermal NO, the Zeldovitch mechanism is assumed to hold for each fluid particle during the mixing process. In the earlier stages of burning, temperature exhibits a very wide spread due to a greater heterogeneity. The highest end of the temperature PDF may be thought to be responsible for rapid production of nitric oxide. Partly because the mixture tends towards less heterogeneous state from crank angle to crank angle, and partly because the average temperature decreases with an increase in the cylinder volume by the piston motion, the NO formation is slowed down and finally ends up by $20°$ after the start of combustion. In the case of injection timing at $17.5°$ BTDC, the temperature PDF has a wing spreading up to only slightly higher temperature than in the case of injection at $10°$ BTDC. Even though the difference

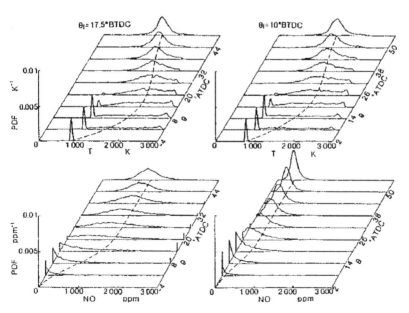

Fig. 6.27 Probability density functions of NO and temperature T against crank angle$θ$ for two injection timings$θ_j$

is quite small, this may lead to a substantial difference in the NO formation, because the sensibility of rate of the NO forming reaction with temperature is very high. For this reason, an earlier injection will finally give much higher concentration of nitric oxide than otherwise.

Figure 6.28 gives the predicted and the measured results of the peak cylinder pressure p_{max} and the concentration of tail-pipe nitric oxide against overall equivalence ratio ϕ_0, that is, the reciprocal of the air ratio, for different injection timings at a fixed swirl ratio, r_s, of 2.7 and for different swirl ratio at an injection timing θ_j of 17.5° BTDC. The degree of reproduction is pretty well at least qualitatively although it is far from being perfect as regards from quantitative aspects. The discrepancy may primarily be attributed to the assumption of zero-dimensional modeling approach. Nevertheless it might safely be stated that the present model rather successfully describe the heterogeneity prevailing in the mixture and its decay by turbulent mixing that will essential in the diesel combustion process. Thus we may conclude that what might be drawn from the results of this stochastic model would be that the NO formation essentially lies in hot spots formed in the heterogeneous nature of combustion.

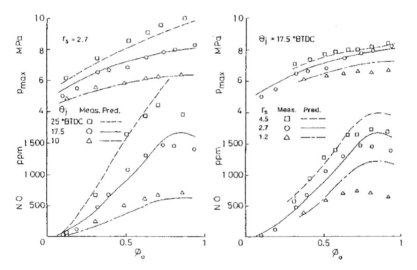

Fig. 6.28 Predicted and measured exhaust concentrations of nitrogen oxides NO_x and peak cylinder pressure p_{max} against overall equivalence ratio φ_0, showing effects of injection timing (*left*) and of swirl ratio (*right*)

6.5.3 Soot Formation and Oxidation as Viewed from Basic Aspects

Soot formation has been dealt with already from the aspects of fuel spray and gas motions within the direct-injection system. What has been emphasized has been that the fuel injection should entrain the surrounding air as much as possible to prepare a well fuel–leaner mixture and that gas motions and the sprays should be selected so as to maximize the degree of air utilization. In addition to those macroscopic viewpoints, there are some arguments arising from more fundamental levels. They will be touched on briefly below.

First, more recent study has shown the possibility of predicting the formation of free carbon, or soot, by the stochastic approach explained already. The heterogeneity of fuel concentration and specific enthalpy may remain at later crank angles and even until exhaust stroke. In such later burning stages oxidation may no longer proceed due partly to decreases in the amount of oxidizer and in gas temperature, and partly to the weakened gas motions and turbulence. As a result, part of free carbon can be frozen and emitted into the exhaust valve as soot. This prediction has been applied with a certain degree of success to analyze the effect of exhaust gas re-circulation on the NO and soot formation [15].

There are some experimental evidences that support the view that the enhancement of the turbulence intensity may enable one to promote combustion and thereby suppressing soot formation. High-pressure injection will reinforce not only atomization but also entrainment of air into the spray core making the mixture leaner. At the same time, the increased velocity may make turbulence stronger and the associated reduction of the nozzle-orifice diameter may decrease the turbulence scale, every contributing to attain favorable burning. Another example is the use of auxiliary cell called "combustion chamber for disturbance (CCD)" into which a fuel is injected in the later burning stages in a direct-injection engine [28]. Experiments made have shown that over a range of auxiliary injection timing the smoke level is much lower than without injection. This is clearly attributed to the elevated level of turbulence.

As is summarized in Fig. 6.29, there are several factors that govern the in-cylinder turbulence; some factors enhance the intensity of turbulence and the other weakens turbulence. However, kinetics of in-cylinder turbulence is not well established at present unfortunately. This is especially true for the combusting case in which some sort of thermal effects might exert an influence on turbulence; laminarization is one of such effects and may be caused by a decrease in kinematic viscosity at elevated temperatures, whereas flame-generated turbulence may be effective in reinforcing turbulence. Probably, laminarization would play a significant role in dissipation of smaller turbulent eddies in hotter flaming zones. This would reduce mixing between

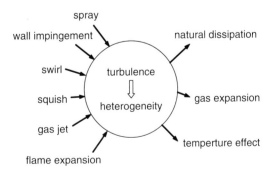

Fig. 6.29 Factors affecting turbulence during diesel combustion

Fig. 6.30 Three routes of soot emissions from direct injection diesel engines

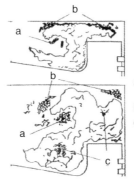

a: separated flamelets due to turbulent eddies
b: quenched by contract with wall
c: dropped reburning speed during expansion stroke

fluid, thus hindering fuel-rich spots to arrive at leaner spots and as a result bringing about a poorer combustion.

According to observations performed by laser-light sheet methods [33], many soot clouds are formed in the combustion chamber in the outer periphery of visible luminous flame. They exist as clusters having a size around a few millimeters and exhibiting a complicated outline. A greater part of these clouds might be formed in such a manner that the surrounding air having a relatively lower temperature in the tip of the spray quenches partially burned hot fuel formed by thermal decomposition. In much later stages of burning, different types of soot clouds may be observed as is sketched in Fig. 6.30. They might result from flamelets segregated into the surrounding air by turbulent eddies, those formed by quenching through the contact with the colder chamber walls, and those left unburned due to a decrease in gas temperature and to the increased turbulence scale during the expansion stroke.

Figure 6.31 shows a transmission electron microscopy (TEM) picture of a typical soot aggregate sampled from exhaust of a 3-l common-rail turbo-charged intercooled diesel engine. The diesel soot aggregate comprises several to hundreds

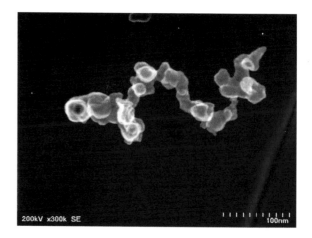

Fig. 6.31 Transmission electron microscopy picture of soot aggregate sampled from a 3 liter common-rail turbo-charged diesel engine

200kV x300k SE   100nm

carbonaceous primary particles with a diameter of about 30 nm. It was shown previously that the soot mass concentration in the cylinder drops to 1–2% of the peak concentration due to oxidation during the expansion stroke. The temporal change in soot particle size distribution during combustion and oxidation processes was investigated by a direct sampling method using a 2.2-l common-rail turbo-charged diesel engine [31]. Surprisingly, the size distribution pattern and its peak mode scarcely vary during the expansion stroke despite the drastic reduction in particle number density. The size distribution of soot particle in exhaust was also found quite similar to those in the cylinder. As Heywood points out in his textbook, no primary particle could survive during the expansion stroke if all the particles are exposed equally to hot and oxygen rich environment in the cylinder [10]. This allows a hypothesis that the inhomogeneous mixture distribution in the cylinder makes a small amount of soot to remain unburned during the expansion stroke without encountering oxidizers such as OH, oxygen and others.

The effect of symmetrical nature of the in-cylinder air motion on soot and CO emissions were investigated by means of LII flame visualization and a multidimensional CFD simulation involving reduced kinetics for soot formation and oxidation [1]. The CFD calculation of the in-cylinder air motion, which was performed with PIV data as an initial flow field, showed non-symmetric flow patterns at 40 CA BTDC. Figure 6.32 shows the calculated soot distribution in and above the piston bowl in direct comparison to the flame light and LII measurements. It is clear

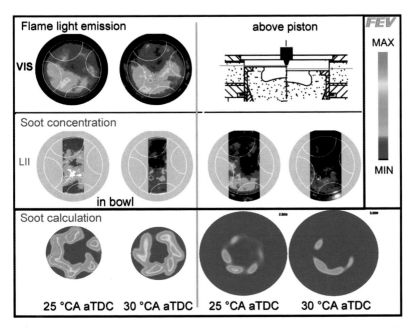

Fig. 6.32 Measured soot distribution (Flame light and LII) versus calculated soot distribution "Reprinted with permission from SAE Paper 2006-01-1417© 2006 SAE International"

that high soot concentration occurs in regions where flow velocity is low due to the non-symmetric swirl motion. This result suggests that symmetric swirl in the cylinder may bring about further reduction in soot emissions.

6.6 Combustion Control by EGR, High Pressure Charging and Multiple-Injection

6.6.1 EGR, High Pressure Injection and High Pressure Charging

6.6.1.1 EGR and High Pressure Injection

As shown in Fig. 6.33, increasing EGR leads to decreased oxygen fraction in intake charge, allowing significant reduction in NO formation through lowered adiabatic flame temperature. For instance, 40% EGR rate reduces oxygen concentration in intake charge from 21% to 15% in the case when the excess air ratio,λ, is 1.5. The resultant 300 K drop in adiabatic flame temperature yields at least one order less NO formation. On the other hand, high EGR rate inevitably brings about decreased volumetric efficiency and increased soot emissions. In order to compensate these disadvantages high boost charging and high pressure injection are employed.

The effect of initial injection velocity, orifice diameter and ambient air density on fuel air mixing can be qualitatively calculated by the quasi-steady spray theory [36]. The mean equivalence ratio,$\overline{\phi}$ in a quasi-steady spray is given by

$$\overline{\phi} = \frac{3}{2} f_{st} \left(\frac{r_0 \sqrt{\rho_f / \rho_a}}{2u_0 t \tan(\theta/2)} \right)^{0.5}$$

Where f_{st}: stoichiometric fuel air ratio,r_0: orifice radius, ρ_f: fuel density, ρ_a: air density, u_0: initial injection velocity,t: time after the start of injection, θ: spray

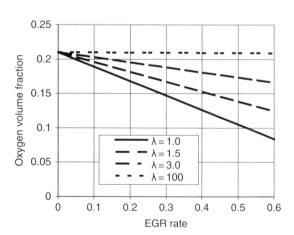

Fig. 6.33 Relationship between EGR rate and oxygen volume fraction in the intake charge with excess air ratio, λ, as a parameter

angle. When inspecting the effect of injection pressure on $\bar{\phi}$, we should fix the injection amount and injection duration, because these are two of the basic parameters governing combustion and emissions in actual engines. And hence, the orifice diameter has to be changed according to the change in injection pressure. On the basis of this assumption, the effect of injection pressure on mean equivalence ratio, $\bar{\phi}$, initial spray momentum per unit fuel mass and initial spray kinematic energy per unit fuel mass were calculated varying orifice diameter d_0. The initial spray momentum governs the spray tip penetration and air entrainment, while the initial kinematic energy represents the total turbulent energy involving in the micro-scale mixing through the eddy cascading process.

Figure 6.34 shows the result when the injection amount and duration were fixed at 38.5 mg and 3 ms, respectively. The mean equivalence ratio, $\bar{\phi}$, decreases with increasing injection pressure owing to the increased initial spray momentum per unit fuel mass and the decreased orifice diameter. The changes in both $\bar{\phi}$ and the spray momentum become moderate when the injection pressure exceeds 200 MPa, but the kinematic energy keeps increasing in proportion to the pressure drop between the injection pressure and ambient pressure. This suggests that increase in injection pressure is meaningful in terms of mixing enhancement. As stated in the preceding chapter, increasing injection pressure leads to increased lift-off length, resulting in leaner combustion with less soot formation.

The injection pressure affects also droplet size that relates to evaporation and liquid core length. The mean droplet size decreases steeply with injection pressure in a pressure range between 50 and 150 MPa and approaches asymptotically to a value with the increase in injection pressure [41]. In other words, much higher injection pressures will be of minor benefit as far as atomization is concerned.

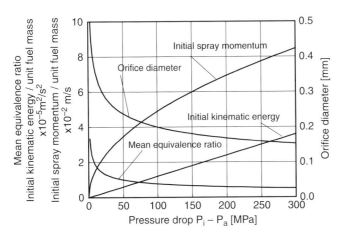

Fig. 6.34 Effect of injection pressure and orifice diameter on mean equivalence ratio in the spray, spray momentum per unit fuel mass and spray kinematic energy per fuel mass at a condition when injected fuel mass = 38.5 mg and injection duration = 3 ms (fuel density/air density = 40)

6.6.1.2 EGR and High Pressure Charging

An experimental study was conducted which focused on the effect of EGR rate on engine performance and emissions under high boost pressure conditions [4]. A single cylinder test engine which had a bore × stroke = 135 mm × 140 mm and a compression ratio of 15 was used. The pressurized intake air was supplied from an external supercharger. An injection nozzle with six 0.17 mm diameter orifices was used and the injection timing was controlled so that the rate of heat release starts at TDC at all conditions tested. The injection pressure was 150 MPa. Figure 6.35 compares the changes in cylinder pressure and the rate of heat release at full load when EGR rate was changed from 0% to 40% at a high boost pressure of 351 kPa. 40% EGR rate corresponds to 16% oxygen concentration at full load when the excess air ratio was nearly 2. The rate of heat release curves reveal that the diffusion combustion starts immediately after the start of injection with nearly zero ignition delay and takes place very fast, showing its completion before 50° ATDC even at 40% EGR rate. The fast combustion is obviously due to the high air density and high injection pressure. The maximum cylinder pressure was as high as 20 MPa, while the maximum gas law temperature was nearly 100 K lower in the case of 40% EGR rate as compared to that in the case without EGR.

Figure 6.36 shows the result of emissions. NO_x emissions decreased with the increase in EGR rate and attained very low levels below 1.0 g/kWh at high EGR

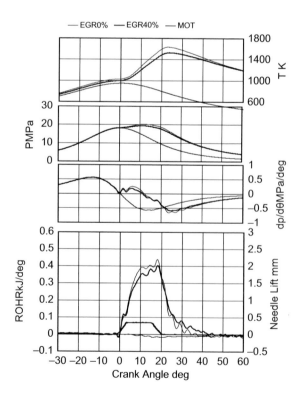

Fig. 6.35 Effect of EGR on cylinder pressure and the rate of heat release at full load (boost pressure = 351 kPa, engine speed = 1,200 rpm, injection pressure = 150 MPa, injection amount = 250 mg/stroke, λ = 2.5–1.5 when EGR rate was changed from 15 % to 50%) [4]

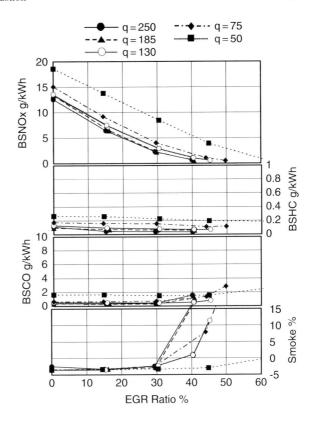

Fig. 6.36 Effect of EGR on emissions (boost pressure = 351 kPa, engine speed = 1200 rpm, injection pressure = 150 MPa, parameter, q, stands for injection amount mg/cycle)

rates over 40%. The smoke emissions was very low at levels near 0.01 g/kWh in low EGR range and rose steeply when EGR rate exceeded 30%. This study showed that high pressure charging allowed simultaneous reduction of NO_x and soot emissions at high EGR rates between 30% and 40%.

6.6.2 Effects of Multiple-Injection on Performance and Emissions

The effect of multiple-injection on combustion and emissions was first discussed in a pioneering paper by Tow et al. [35] and since then a number of experimental studies have been implemented worldwide. Figure 6.37 illustrates a schematic showing the role of each injection event of the multiple injection strategy. The early injection allows HCCI like combustion and its effect on combustion and emissions will be discussed in detail in Chap. 7. The pilot injection is intended to shorten ignition delay of the main injection fuel. The post injection is able to enhance the soot oxidation if implemented properly. The early late and late injections are both aiming at raising exhaust temperature to enhance the active regeneration of diesel particulate filters. The early late injection increases the exhaust gas temperature to raise the temperature of an oxidation catalyst placed upstream a diesel particulate

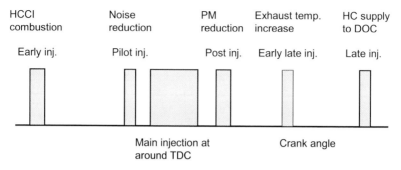

Fig. 6.37 Multiple injection strategy showing the role of each injection event

filter. When its temperature reached a certain level, fuel is supplied by the late injection to the catalyst and burns. The hot exhaust gases thus generated initiate combustion of soot accumulated in the diesel particulate filter. The late injection event, including amount, timing and frequency, is controlled so that the gas temperature supplied to the filter could be high enough for soot oxidation and below the catalyst damage temperature. Likewise, the late injection strategy is employed for elevating the temperature of the de-NO_x catalyst to its working temperatures at low loads.

6.6.2.1 Pilot Injection

The effect of pilot injection on engine performance and emissions was studied by Carlucci et al [6]. The test engine was a four-cylinder 1.9-l common-rail diesel engine equipped with an inter-cooled turbocharger but without EGR devise. The fuel injection pressure was kept at 135 MPa and the timing of the main injection was fixed at 3.4° BTDC and 4.8° BTDC at 1400 and 2000 rpm, respectively.

Figure 6.38 shows the effect of pilot injection timing on NO_x emissions with the pilot injection amount as a parameter. The notation of the injection parameters is illustrated beside the figure and the particular set of injection parameters is indicated as: Ai [deg.BTDC]/AiP [deg.BTDC]/EtP [μs]. At all three load conditions, NO_x emissions decrease with the increase in pilot injection amount, EtP, particularly when the pilot injection timing was advanced to 32.7° BTDC, as compared to the original emission level expressed by a dark circle. At pilot injection timing of 32.7° BTDC, ignition of the pilot injection fuel started later at around 15° BTDC after the low temperature reaction occurred at around 20° BTDC. The partially premixed charge compression ignition (PCCI) combustion of the pilot injection fuel observed here is beneficial for reducing NO_x emissions at all loads studied.

Another important role of pilot injection is to produce internal EGR gases to suppress NO_x formation in the main fuel sprays. The reduced NO_x emissions observed in the figure might be partly due to this effect.

Soot emissions under the same conditions remained unchanged or increased. The pilot injection fuel could produce less soot if burned in PCCI like combustion, but

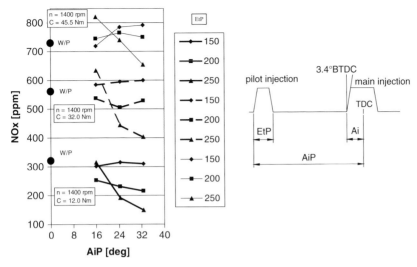

Fig. 6.38 Effect of pilot injection amount on NO_x emissions with the pilot injection timing as a parameter "Reprinted with permission from SAE Paper 2003-01-0700 © 2003 SAE International"

the resultant short ignition delay of the main injection fuel gives rise to the fraction of diffusion combustion phase, favorable for soot formation.

The effect of pilot injection on engine performance and emissions were also investigated under high engine speed and high load conditions. The trends were quite different from those observed under low speed and low load conditions. The larger the fuel amount of pilot injection was, and the earlier the pilot injection timing was advanced, NO_x emissions increased and soot emissions decreased. The ignition delay was sufficiently short under high speed and high load conditions because of the high combustion chamber wall temperatures and low heat losses, and as a result the early and large amount of pilot injection acted as if the injection timing of a single injection was advanced. Thus, the pilot injection was useless at high load and high speed conditions.

6.6.2.2 Post or After Injection

The effect of post or after injection on engine performance and emissions was investigated using a 1.91-l four-cylinder, common-rail, and turbo inter-cooled engine [5]. The authors employed the triple-injection strategy that comprises pilot, main and post injections. The double injection strategy at the reference point (2500 rpm and 8 bar b.m.e.p.) was: start of pilot injection = 38° BTDC, duration of pilot injection = 4.2° CA, start of main injection = 3° BTDC. In the triple injection strategy, post injection was added to the double injection profile, where the time interval between the end of main injection and the start of post injection, DT_{after}, and the injection amount of the post injection, were changed respectively. The main injection amount was adjusted so as to keep the b.m.e.p. constant at 8 bar. The result showed that post injection was obviously effective in reducing soot emissions (up to 40%) in a DT_{after} range between 8° and 20° CA. Late post injection produced a substantial increase in soot emissions as well as in CO emissions, while post injection too close to the main injection resulted in a drastic increase in soot emissions. NO_x emissions showed a slight decrease by less than 10%.

Figure 6.39 illustrates an example of how the multiple injection strategy is applied to a 2-l production engine [29]. The single injection mode is used in high speed and high load range, while the double or triple injection mode is applied in medium speed and medium load range. In low speed and low load range, a quartople injection mode is applied, where the first two injections correspond to the pilot and main injection and the latter two are late injections to raise exhaust temperature. The late injection is divided into two injections to prevent the injected fuel from impinging against the liner walls.

Fig. 6.39 An example of multiple injection strategy in engine speed and load map [29]

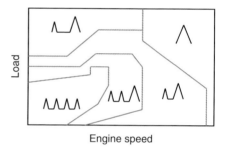

Engine speed

Summary

As described in this chapter the conventional diesel combustion comprises many physical and chemical processes. Various types of hardware to control these processes, including injection, turbo-charging and variable valve actuation systems, have showed notable progress in the past decade. Parameters involved in combustion event have to be carefully varied in the engine tuning to achieve lower emissions while seeking higher engine performance in a wide range of operating condition. This requires enormous time and cost because of the large number of parameters concerned.

In order to do this effectively in the course of engine development, computer flow dynamics (CFD) codes incorporating combustion models are being used to simulate the effects of these parameters on combustion. The model calculation predicts the process of mixture formation in the combustion chamber by considering the interaction between the spray, air motion and piston bowl and intends to determine the nozzle specification, the air management system specification and the piston bowl geometry. The calculated cylinder pressure time record and NOx emissions show reasonably good agreement with the corresponding experimental ones if adjustable coefficients are carefully selected. However, the calculated soot emissions hardly provide satisfactory result quantitatively and sometimes even qualitatively. This is mainly due to the lack of the understanding of the soot formation and oxidation process including chemical reactions on the surface of soot aggregates with complex structure and poly-disperse size distribution. The soot oxidation models presently incorporated in the simulation codes are based on classical empirical equations. In order to raise the reliability of predicting soot emissions the soot oxidation models need to be largely improved.

Regarding sprays the effect of cavitation bubbles generated inside the nozzle on atomization has been intensively studied. However, the current technology on nozzle hole shaping made it possible to suppress cavitation to occur, and this resulted in the increase in the effective orifice flow coefficient that benefits the spray tip penetration. There are several reports on the marked effect of nozzle hole shaping on reducing soot emissions from engines equipped with this type of nozzle. The CFD simulation of the internal nozzle flow has made a significant progress in the past and the precise velocity distribution at the orifice exit provided by the simulation will help improve the accuracy of the spray modeling.

Acknowledgments The contents of this chapter were mostly cited from work conducted at Kyoto University and Tokyo Institute of Technology. The authors express sincere appreciation to those who cooperated in the work. They also thank authors of referenced papers who positively provided materials to complete this chapter.

References

1. Adomeit P et al. (2006) Potential soot and CO reduction for HSDI diesel combustion systems. SAE paper 2006-01-1417.
2. Aizawa T, Kosaka H and Matsui Y (2002) 2-D imaging of soot formation process in atransient spray flame by laser-induced fluorescence and incandescence techniques. SAE paper 2002-01-2669.
3. Aoyagi Y, Kamimoto T, Matsui Y and Matsuoka S (1980) A gas sampling study on the formation processes of soot and NO in a DI diesel engine. SAE Trans., vol. 89, no. 2, pp 1175–1189 (SAE Paper 800254).
4. Aoyagi Y et al. (2004) Diesel combustion and emission study by using high boost and high injection pressure in single cylinder engine. Proceedings of International Symposium, COMODIA2004, JSME, Yokohama, pp 119–126.
5. Badami M, et al. (2003) Experimental investigation on the effect of multiple injection strategies on emissions, noise and brake specific fuel consumption of an automotive direct injection common-rail diesel engine. Int. J. Engine Res., vol. 4, no. 4, pp 299–314.
6. Carlucci P et al. (2003) Effects of pilot injection parameters on combustion for common-rail diesel engines. SAE Paper 2003-01-070.
7. Eisfeld F (1993) Atomization and spray formation. Advanced combustion science. ed. by T. Someya, Springer-Verlag, Germany, pp 98–104.
8. Halstead M P et al. (1977) The auto ignition of hydrocarbon fuels at high temperatures and pressures-fitting of a mathematical mode. Combust. Flame, 30, pp 45–60.

9. Hasse C and Peters N (2005) Modeling of ignition mechanism and pollutant formation in direct-injection diesel engines with multiple injection. Int. J. Engine Res., vol. 6, no. 3, pp 231–246

10. Heywood J B (1988) Internal combustion engine fundamentals. McGraw-Hill, Inc., p 645

11. Hoffman F et al, (1994) 2D investigation of hot spots in the unburnt gas end of I.C. engine using formaldehyde as tracer. Proceedings of International Symposium. (COMODIA 94), JSME, Yokohama, pp 517–522.

12. Ikegami M et al. (1984) A stochastic approach to model the combustion process in direct-injection diesel engines. Proceedings of 20th Symposium (International) on Combustion Institute, pp 217–224.

13. Ikegami M et al. (1988) Diesel combustion and the pollutant formation as viewed from turbulent mixing control. SAE Paper 880425.

14. Ikegami M (1990) Roles of flows and turbulent mixing in combustion and pollutant formation in diesel engines. Proceedings of International Symposium COMODIA 90, pp 49–58.

15. Ikegami M, et al (1998) Effect of exhaust gas recirculation and injection pressure on exhaust emissions from a diesel engine. Proceedings of International Symposium COMODIA 98, pp 87–92.

16. Ikegami M et al. (1999) Prediction of ignition process in diesel sprays by stochastic model (in Japanese). Trans., JSME B, vol. 65, no. 638, pp 3489–3496

17. Ishiyama T et al. (2003) Analysis of ignition processes in a fuel spray using an ignition model including turbulent mixing and reduced chemical kinetics. Int. J. Engine Res., vol. 4, no. 3, pp 155–162

18. Ito M et al. (2003) Effects of ambient gas conditions on ignition and combustion process of oxygenated fuel sprays. SAE Paper 2003-01-1790.

19. Kamimoto T and Bae M-H (1988) High combustion temperature for the reduction of particulate in diesel engines. SAE Paper 880423.

20. Kamimoto T and Matsuoka S (1977) Prediction of spray evaporation in reciprocating engines. SAE Trans., vol. 86, no. 3, pp 1792–1802 (SAE Paper 770413).

21. Kamimoto T, Yokota H and Kobayashi H (1989) A new technique for the measurement of sauter mean diameter of droplets in unsteady dense sprays. SAE Trans., vol. 93, no. 3, pp 387–408 (SAE Paper 890316).

22. Kitamura T et al. (2002) Mechanism of smokeless diesel combustion with oxygenated fuels based on the dependence of the equivalence ratio and temperature on soot particle formation. Int. J. Engine Res., vol. 3, no. 4, pp 223–248

23. Kobori S, Kamimoto T and Aradi A A (2000) A study of ignition delay of diesel fuel sprays. Int. J. Engine Res., vol. 1, no. 1, pp 29–39.

24. Kosaka H et al. (1995) A study of soot formation and oxidation in an unsteady spray flame via laser induced incandescence and scattering techniques. SAE Paper 952451.

25. Kosaka H et al. (1996) Simultaneous 2-D imaging of OH radicals and soot in a diesel flame by laser sheet techniques. SAE Paper 960834.

26. Kosaka H et al. (2000) Two-dimensional imaging of formaldehyde formed during the ignition processes of a diesel fuel spray. SAE Paper 2000-01-0236.

27. Kosaka H, Aizawa T and Kamimoto T (2005) Two-dimensional imaging of ignition and soot formation processes in a diesel flame. Int. J. Engine Res., vol. 6, no. 1, pp 21–42.

28. Murayama T et al. (1989) Reduction of smoke and NOx emission by active turbulence generated in the late combustion stage in DI diesel engines. Proceedings of 18th International Congress on Internal Combustion Engines, Diesel Engines, vol. 2, pp 1129–1142.

29. Nakai E et al. (2005) Introduction of new MZR-CD. Mazda Technical Review, No. 23, pp 98–103.

30. Pfahl U and Adomeit G (1997) Self-ignition of diesel-engine model fuels at high pressures. SAE Paper 970892.

31. Plungs A et al. (2000) Analysis of the particle size distribution in the cylinder of a common rail diesel engine during combustion and expansion stroke. SAE Paper 2000-01-1999.

32. Roth H, Gavaises M and Arcoumanis C (2002) Cavitation initiation, its development and link with flow turbulence in diesel injector nozzles. SAE Paper 2002-01-0214.

33. Shioji M et al. (1990) Characterization soot clouds and turbulent mixing in diesel flames by image analysis. Proceedings of International Symposium COMODIA 90, JSME, pp 613–618.

34. Shoji M et al. (1999) Stochastic approach for describing mixture formation and initial combustion in diesel sprays. Proceedings of The 15th Internal Combustion Engine Symposium, Seoul, July 13–16, Joint symposium by Society of Automotive Engineers of Japan and Korean Society of Automotive Engineers, pp 471–476.

35. Tow T, Pierpoint A and Reitz R D (1994) Reducing particulates and NOx emissions by using multiple injections in a heavy duty DI diesel engines. SAE Paper 940897.

36. Wakuri Y et al. (1957) Studies on the penetration of fuel spray of diesel engine (in Japanese). Trans. JSME, vol. 25, no. 156, pp 820–826.

37. Wolfer H H (1938) Der Zundverzug im Dieselmotorer. VDI Forschungsheft, vol.392, pp 15–24

38. Won Y H et al. (1991) 2-D soot visualization in unsteady spray flame by means of laser sheet scattering technique. SAE Paper 910223.

39. Won Y H et al. (1992) A study on soot formation in unsteady spray flames via 2-D soot imaging. SAE Paper No. 920114.

40. Yeh C-N et al. (1994) Quantitative measurement of 2-D fuel vapor concentration in a transient spray via laser-induced fluorescence technique. SAE Paper 941953.

41. Yokota H et al. (1991) Fast burning and reduced soot formation via ultra-high pressure diesel fuel injection. SAE Paper 910225.

Chapter 7
Advanced Diesel Combustion

Katsuyuki Ohsawa and Takeyuki Kamimoto

7.1 Introduction

Conventional diesel combustion that is governed by turbulent diffusion produces inevitably high temperature zones in the flame at near stoichiometric fuel–air ratios where NO is formed in high concentrations. If one could successfully burn very lean homogeneous mixtures charged in the combustion chamber, the resultant low combustion temperature would offer a significant reduction in both NO_x and PM emissions. Homogeneous mixture combustion in spark ignition engines, however, depends on turbulent flame propagation, and its lean limit is generally lower than an equivalence ratio of 0.65 at which the temperature is not low enough to suppress NO_x formation. Accordingly, a type of combustion in which spontaneous ignition occurs simultaneously throughout the combustion chamber at leaner equivalence ratios has been pursued. This is the concept of advanced diesel combustion.

In 1979, several researchers reported operating homogeneous charge two-stroke cycle engines in a compression-ignition mode [19, 23]. In 1983, Najt and Foster analyzed the homogeneous charge compression ignition combustion using a CFR engine and found that the ignition process is controlled by the low temperature hydrocarbon oxidation kinetics and the energy release process is controlled by the high temperature hydrocarbon oxidation kinetics [17]. Oppenheim proposed the concept of homogeneous charge compression ignition (HCCI) combustion in 1984 [24]. In experiments involving two-stroke cycle engines, stable combustion was achieved by controlling the fraction of residual gases in the cylinder, while Oppenheim proposed a method to control the amount of residual or EGR gases to improve thermal efficiency and reduce NO_x emissions. In 1989 Thring demonstrated stable operation of HCCI combustion using an engine accommodating an intake-port injector for both gasoline and diesel fuel [34]. Although there were several differences in the hardware for diesel and gasoline HCCI engines due to

Katsuyuki Ohsawa
Toyota Central R&D Labs., Inc., Yokomichi, Nagakute, Aichi, Japan

Takeyuki Kamimoto
Tokai University, 1117 Kitakaname Hiratsuka-shi, Kanagawa, Japan

C. Arcoumanis, T. Kamimoto (eds.), *Flow and Combustion in Reciprocating Engines*,   353
DOI: 10.1007/978-3-540-68901-0_7, © Springer-Verlag Berlin Heidelberg 2009

the different fuel properties, HCCI combustion was found to be possible for both types of fuel.

Afterwards, a variety of advanced diesel combustion concepts, such as uniform bulky combustion system (UNIBUS) [37], premixed charge compression ignition (PCCI) [3], premixed lean diesel combustion (PREDIC) [18, 33], modulated kinetics (MK) [14, 15], low temperature rich combustion (LTRC) [29] and premixed charge ignition (PCI) [32] were proposed. All these are characterized by homogeneous, or sometimes "premixed", and quasi-homogeneous or "partially premixed" low temperature combustion. The advanced diesel combustion concepts have been studied worldwide in the last decade because of their potential to achieve ultra low emissions without after-treatment systems.

7.2 Various Types of Advanced Diesel Combustion

7.2.1 Diesel HCCI or PCCI Combustion

HCCI is the basic idea of advanced diesel combustion. In the first experiment of HCCI combustion, a gasoline fuel injector was mounted in the intake manifold to create well-premixed mixtures [34]. However, the manifold fuel injection is not appropriate for diesel fuel because of its lower volatility. Therefore, in-cylinder direct fuel injection in the intake stroke or in the early compression stroke is adopted in diesel HCCI combustion engines. The diesel HCCI combustion strategy is often referred to as PCCI concept, which is based on non-uniform or partially premixed mixtures achieved during a shorter period of crank angle between injection and combustion event than in port injection HCCI strategy. The stratified fuel mixtures present in the cylinder include fuel rich mixture elements, which are effectively utilized to expand load range by preventing instantaneous combustion specific to HCCI combustion. New diesel combustion strategies including UNIBUS and PREDIC are categorized in the PCCI combustion regime.

Figure 7.1 illustrates a comparison between diesel HCCI (or PCCI) and conventional diesel combustion, showing typical features of PCCI combustion achieved by early injection timing. The typical features of PCCI combustion are summarized in the following sections.

7.2.1.1 Early Injection and Long Induction Duration

Diesel HCCI needs a sufficient mixing time before auto-ignition occurs to produce premixed mixtures of fuel and air including residual or EGR gases. In other words, a long induction time between the start of injection and the start of heat release features diesel HCCI combustion.

7.2.1.2 Two-Stage Heat Release

As the rate of heat release curve for the HCCI combustion in Fig. 7.1 shows, the HCCI combustion features two-stage heat release, one for the low temperature

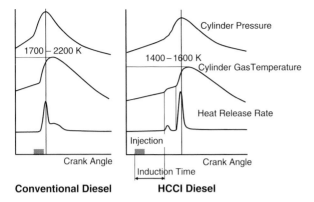

Conventional Diesel **HCCI Diesel**

Fig. 7.1 Comparison between conventional and diesel HCCI combustion, showing longer induction time for diesel HCCI combustion

oxidation reactions and the other for the high temperature oxidation reactions. It is known that the oxidation process of hydrocarbon fuels involves the negative temperature coefficient range in 700–950 K as shown in the previous chapter. Knock in gasoline engines is essentially based on auto-ignition that is dominated by the reactions in this range. In HCCI combustion, as the in-cylinder mixture temperature rises during the compression stroke, weak heat release due to the low temperature reactions occurs first, resulting in slight increases in both in-cylinder pressure and gas temperature. The low temperature reactions in HCCI combustion occurs when the mixture temperature undergoes a temporal change in the negative temperature coefficient range.

Oxidation of hydrocarbon fuels mainly takes place via two representing routes shown in Fig. 7.2 [38]. In the low temperature oxidation reactions, intermediate species including peroxides are produced through abstraction reactions of hydrogen, and auto-ignition, i.e., cool flame reaction, starts by the reaction of OH with RH produced from peroxides such as aldehydes. In a short period of crank angle after the cool flame appears, the thermal flame reaction starts, accompanied by the main heat release.

The in-cylinder gas temperatures at which the low and high temperature oxidation reactions start respectively were investigated using a single cylinder engine. The

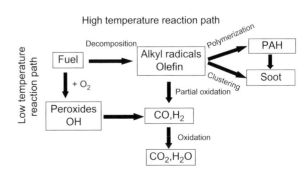

Fig. 7.2 Reaction paths of hydrocarbon fuels in low temperature and high temperature reactions [38]

Fig. 7.3 Temperatures at which LTO and HTO reactions of n-pentane air mixture start respectively with ϕ, EGR ratio and air temperature as parameters "Reprinted with permission from SAE Paper 2006-01-0028 © 2006 SAE International"

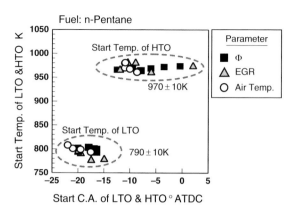

results obtained for n-pentane are shown in Fig. 7.3 [11]. As seen in the figure, the low temperature oxidation (LTO) and the high temperature oxidation (HTO) start at around 790 and 970 K, respectively, independently of equivalence ratio, EGR rate and intake air temperature. Thus, the in-cylinder gas temperature governs dominantly the ignition process in HCCI combustion.

7.2.1.3 High Heat Release Rate of Main Combustion

The main heat release is dominated by the high temperature oxidation reactions of hydrocarbon fuel through which intermediates such as olefins and alkyl radicals are rapidly consumed. These intermediate species are produced by thermal decomposition reactions including the chain breaking of C–C bonds in the fuel.

The in-cylinder mixture is nearly homogeneous when ignition starts, and once high temperature oxidation occurs locally, it triggers auto-ignition instantly throughout the combustion chamber except in the quench layers. The steep pressure rise produced by the auto-ignition generates intense combustion noise, one of the concerns of HCCI combustion.

7.2.1.4 Fuel Lean and Low Temperature Combustion

When EGR is not employed, HCCI combustion is possible only for fuel lean mixtures in a limited range of air fuel ratio (see Fig. 7.5). As shown in Fig. 7.1, the peak cylinder gas temperature in HCCI combustion ranges from 1400 to 1600 K and is significantly lower than 1700–2200 K in conventional diesel combustion. The low cylinder gas temperature suppresses both NO_x and soot production, but, on the other hand, involves the slow oxidation reactions that cause misfiring and quenching, resulting in the higher HC and CO emissions as compared with those in conventional diesel combustion.

7.2.1.5 Non-Luminous Flame

Figure 7.4 is a ϕ-T map, showing the operating regions of the advanced combustion regimes and a belt-like region for conventional diesel combustion representing the temporal changes of equivalence ratio, ϕ, and temperature, T, in the flame [2, 12]. The advanced combustion regions are located in the left bottom corner away from NO and soot formation peninsulas. In HCCI combustion, the fuel air mixtures are nearly premixed within the long induction duration, and as a result the luminous flame observed in conventional diesel combustion does not appear.

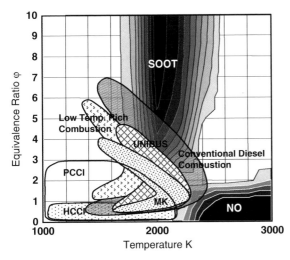

Fig. 7.4 Operating regions for conventional and advanced diesel combustion regimes in relation to NO and soot formation regions

7.2.1.6 Low NO$_x$ and Soot Emissions

Figure 7.5 illustrates a comparison of NO$_x$ emission characteristics between three combustion modes; compressed natural gas (CNG) homogeneous mixture spark ignition combustion, glow plug assisted CNG direct injection combustion, and the new diesel combustion, including PREDIC, UNIBUS and PCCI combustion [18]. NO$_x$ emissions of spark ignited homogeneous mixture combustion depend on the equivalence ratio as the NO chemistry suggests. While, NO$_x$ emissions for direct injection combustion remains at high levels from 10^2 to 10^3 ppm over a wide range of equivalence ratio, which is common in turbulent diffusion combustion. All the data for NO$_x$ emissions of the advanced diesel combustion regimes stay in a domain bounded by two extrapolated lines for CNG direct injection and CNG homogeneous combustion. This means that the advanced diesel combustion regimes are of partially homogeneous combustion. NO$_x$ emissions achieved in these combustion regimes are lower than 100 ppm although the operation is possible only in a narrow range of equivalence ratio between 0.2 and 0.4.

Fig. 7.5 Relationships
between NO$_x$ emissions and
mean equivalence ratio for
various combustion strategies
without EGR "Reprinted with
permission from SAE Paper
970898 © 1997 SAE
International"

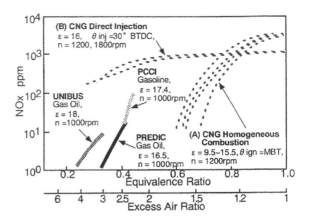

7.2.1.7 High HC and CO Emissions

The early fuel injection in diesel HCCI combustion causes spray interaction with the
cylinder liners or piston walls, resulting in the wall quenching and lubricant dilution
that may cause fuel penalty and wear problems. It is known that the fuel dilution to
lubricant oil contributes significantly to the increased fuel consumption under early
injection HCCI operation. The spray interaction with the cylinder liners as well as
the low temperature combustion is a major cause for the high HC and CO emissions
and high indicated specific fuel consumption (ISFC).

Figure 7.6 is a comparison of the curves of ISFC versus indicated mean effective pressure (IMEP) for two early injection modes and standard diesel combustion
mode of a 0.48-l single cylinder engine. In the early injection modes, EGR rate was
changed from 33% to 56% according to load [8]. In the early injection mode, fuel
was injected in a series of five short injections to minimize the spray impingement

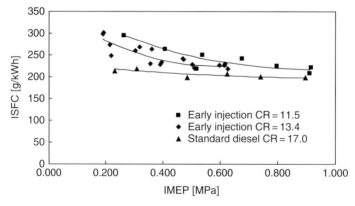

Fig. 7.6 ISFC for early injection (compression ratio CR = 11.5 and 13.4) and standard diesel
operation (CR = 17.0) "Reprinted with permission from SAE Paper 2004-01-0935 © 2004 SAE
International"

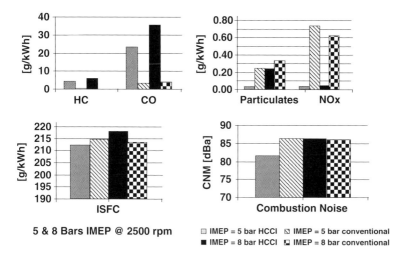

Fig. 7.7 Comparison of emissions, ISFC and combustion noise of a HCCI combustion engine to those of corresponding diesel engine "Reprinted with permission from SAE Paper 004-01-1904 © 2004 SAE International"

on the walls, and the start of the first injection was set at 90° BTDC. The ISFC values for the early injection modes are obviously higher than that for the standard diesel combustion mode, particularly at light-load. This is partly due to lower combustion efficiencies (86% at 0.3 bar and 94% at 0.9 bar) and partly due to the fuel wetting on the wall. The multiple-injection shortens the fuel penetration and is an effective way to reduce fuel wetting on the cylinder walls [9, 26]. The use of high volatile fuels is also beneficial to reduce of HC and CO emissions [28].

Figure 7.7 shows a typical result of emissions, fuel consumption, and combustion noise under HCCI combustion mode, comparing it to that of corresponding conventional diesel combustion [6]. It is seen that HC and CO emissions are quite high for HCCI combustion despite very low NO_x and particulate emissions. The fuel consumption of HCCI combustion is slightly deteriorated because the narrow margin obtained by the fast combustion was canceled out by the low combustion efficiency. The thermal efficiency of HCCI combustion is practically restricted by the high noise emission due to the high heat release rate.

Since the exhaust temperature is lower than in conventional diesel combustion, the advanced diesel combustion systems need special exhaust gas treatment systems that are active at lower temperatures to reduce HC and CO emissions.

7.2.1.8 Narrow Operating Range

Figure 7.8 shows a diesel HCCI combustion region in which equal IMEP lines are included. The data were obtained at an engine speed of 1500 rpm using a single cylinder test engine which had a ϕ 76 mm bore × 110 mm stroke and a compression ratio of 18 [25]. In this study, n-heptane (cetane number 56) was injected into the

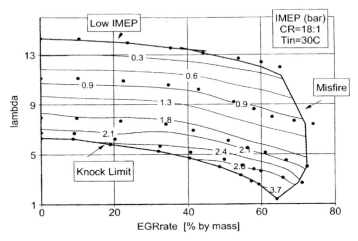

Fig. 7.8 Diesel HCCI operating region and IMEP map for a compression ratio of 18 and intake charge temperature of 30°C "Reprinted with permission from SAE Paper 2003-01-0747 © 2003 SAE International"

intake charge to eliminate the difficulty of fuel evaporation. The abscissa represents the EGR rate (% by mass), while the ordinate gives the overall excess air ratio, lambda, of the cylinder charge. It can be seen, under such operating conditions, diesel HCCI combustion can be achieved over a narrow range of excess air ratios and EGR rates. IMEP is governed mainly by excess air ratio, and 4.0 bar of maximum IMEP is achieved at a lambda of 1.5 and EGR rate of 65%. The upper limit of the HCCI region is bounded by knock, while misfire occurs at the right and top-right boundary of the region. The richest lambda attainable at zero EGR is approximately 5.0–6.0, significantly high as compared to 3.0 for gasoline HCCI combustion. However, the diesel HCCI combustion can tolerate very high EGR rates, up to 70% by mass. The operation of diesel HCCI combustion engine depends also on fuel injection timing, which affects the extent of in-homogeneity of mixtures produced during the induction time.

7.2.2 Uniform Bulky Combustion System (UNIBUS)

UNIBUS employs a centrally mounted Pintle-type fuel injector to control mixture formation and avoid fuel impingement on the cylinder walls. The test engine was a 4-l four-cylinder diesel engine with a compression ratio of 18 [37]. Figure 7.9 shows the effects of injection timing on IMEP and emissions under conditions without EGR. As injection timing is advanced from TDC to 40° BTDC, NO$_x$ increases and soot decreases as conventional combustion dominates in this range of injection timing. However, when injection timing is further advanced to 60° BTDC, both NO$_x$ and soot emissions decrease, showing a transition to premixed charge compression ignition combustion. Further advance in injection timing allows NO$_x$ emissions

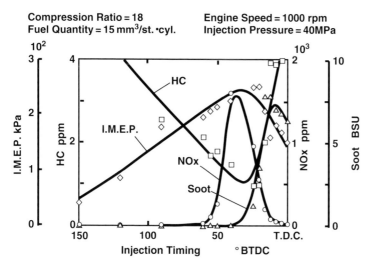

Compression Ratio = 18
Fuel Quantity = 15 mm³/st. ·cyl.

Engine Speed = 1000 rpm
Injection Pressure = 40MPa

Fig. 7.9 Effects of injection timing on emissions and IMEP in UNIBUS combustion [37]

lower than 10 ppm and no soot emissions, but is accompanied by increased HC emissions and lowered IMEP. The optimum injection timing of the UNIBUS combustion regime was found to lie between 40 and 60° BTDC. As the figure indicates, the major concerns with UNIBUS are high HC emissions and reduced IMEP. High HC emissions can be attributed to locally distributed mixtures with too lean to burn equivalence ratios and the cylinder wall wetting. Preventing the injected fuel from impinging on the cylinder walls is an important challenge in UNIBUS combustion regime. [13] The low IMEP, i.e., deteriorated thermal efficiency, is due to premature ignition, and accordingly, the key to improve thermal efficiency is how to postpone the spontaneous ignition timing. Many trials have been made to prolong the ignition delay by focusing on the compression ratio, EGR rate, injection control [7], CO_2 addition, and water injection, all aiming to lower the compressed cylinder gas temperature.

Figure 7.10 shows two series of shadow-images, comparing the non-luminous UNIBUS combustion achieved at an injection timing of 60° BTDC with the conventional luminous flame combustion at an injection timing of 20° BTDC.

7.2.3 Premixed Lean Diesel Combustion (PREDIC)

PREDIC, which employs early injection strategy, is a diesel HCCI combustion concept [18, 33]. A single cylinder test engine with a displacement of 2004 cc and a compression ratio of 16.5 was operated using a fuel with a cetane number of 19 at an engine speed of 1000 rpm. Two injectors were mounted on the side-walls of the combustion chamber so that the sprays could evolve across the combustion bowl, minimizing the spray impingement on the walls.

Fig. 7.10 Comparison of shadow images of UNIBUS combustion (*top*) with those of conventional combustion (*bottom*) [37]

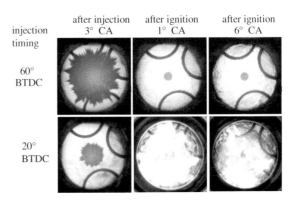

Figure 7.11 shows the effect of injection timing on NO_x emissions for various excess air ratios. For each excess air ratio, a set of three data plots forms a short curve indicating a range of injection timing in which heat release occurs moderately near TDC with substantially low NO_x emissions. The combustion mode observed here was denoted as PREDIC. If the injection timing is retarded beyond the end of the curve, excessive and harsh combustion-knock critical to the operating limit occurs because the shorter induction duration produces much heterogeneous mixture distribution, causing fuel rich mixtures to ignite earlier. Conversely, when the injection timing is advanced earlier beyond the opposite end, the engine suffers the unstable ignition and eventual misfire because the long induction period generates more uniform mixtures that are too lean to burn. This suggests that non-homogeneous mixtures are desirable for PREDIC to assure stable spontaneous ignition. Thus the extent of the homogeneity of the mixture is an important parameter to determine ignition timing together with the overall equivalence ratio. When the overall excess air ratio is decreased to 2.5, the injection timing has to be advanced to $-120°$ ATDC to achieve PREDIC. The injection timing appropriate for PREDIC is limited in a very narrow crank angle window of less than $10°$ CA for each excess air ratio.

Fig. 7.11 NOx emission characteristics of PREDIC (without EGR) "Reprinted with permission from SAE Paper 970898 © 1997 SAE International"

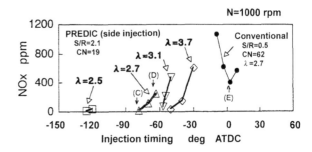

7.2.4 Modulated Kinetics (MK) Combustion

MK combustion strategy is based on significantly retarded fuel injection timing and high EGR rate, and requires auto-ignition to occur after the end of fuel injection, allowing lean and partially premixed mixtures to be formed before auto-ignition [15].

The characteristics of MK combustion were investigated using a four-cylinder DI engine, which has a bore of 85 mm and an injection nozzle with $5 \times \phi\, 0.22$ mm [14]. The swirl ratio was set relatively high at 4 to enhance mixing during the short induction duration. Shown in Fig. 7.12 is a comparison between the rates of heat release at three different injection timings and EGR rates. IT in the figure stands for the injection timing. When the fuel is injected at TDC in compressed charge containing 45% cooled EGR gas, combustion starts after the end of injection, showing a single peak heat release rate in the expansion stroke that is specific to MK combustion.

Figure 7.13 shows the effect of injection timing on intake gas temperature, excess air ratio, emissions and BSFC with EGR rate as a parameter. Compression ratio and swirl ratio employed in this experiment were 18 and 4, respectively. In the case of cooled EGR, the EGR rate was increased to 45% to maintain the excess air ratio at the same value as in the case of 37% EGR. It can be seen that at an injection timing of $3°\text{ATDC}$, NO_x is dramatically reduced by 95% with a cooled EGR rate of 45%, whereas smoke and particulates remain at the same levels as in the case w/o EGR, respectively. Despite the late combustion phasing after TDC, the thermal efficiency penalty is relatively small. The authors claim that both the low combustion temperature due to the fuel lean combustion and the reduced flame interaction with the piston surfaces due to high swirl air motion act effectively to reduce heat losses to the combustion chamber walls. HC emissions are lower than

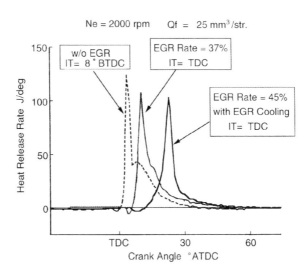

Fig. 7.12 Effect of EGR rate and injection timing (IT) on the rate of heat release in MK combustion [14]

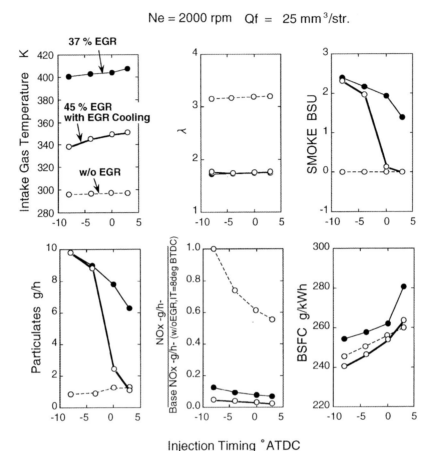

Fig. 7.13 Effect of injection timing on intake gas temperature, excess air ratio, PM and NO_x emissions and BSFC with EGR rate as a parameter in MK combustion [14]

in PCCI combustion because of the reduced wall wetting and reduced quench layers near the cylinder walls.

The operation of MK combustion is limited to a low-medium engine speed and load range because, at high load conditions, ignition tends to occur before the end of fuel injection due to high cylinder gas temperature.

7.2.5 Low Temperature Fuel Rich Combustion (LTRC)

The injection timing for LTRC is the same as or slightly earlier than that for the conventional diesel combustion. The LTRC features rich or near stoichiometric mixture combustion with highly cooled turbocharged intake air and high EGR rates near 50%. The LTRC was developed with the aim of producing fuel rich spikes in

Displacement	1995 cc
Cyl.No.-Bore x Stroke	L4-82.2 x 94
Combustion System	Direct Injection
Compression Ratio	18.6
Intake System	Turbocharged with Intercooler
Valve Train	4-Valve DOHC
Fuel Injection System	Common Rail System

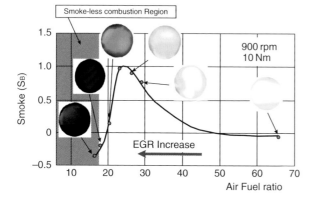

Fig. 7.14 Effect of EGR rate on changes in air to fuel ratio, smoke emission and flame color in LTRC [29]. Flame images were taken by direct photography

engines for the NO_x storage and reduction catalyst. Sasaki et al. performed experiments on LTRC combustion using a 2-l, four-cylinder inter-cooled turbocharged engine with a compression ratio of 18.6 [29]. Figure 7.14 shows the changes in smoke emission and flame images in LTRC with the increase in EGR rate, which reduces the air fuel ratio simultaneously. The smoke level increases with the increase in EGR rate, exhibiting luminous colors due to the conventional high combustion temperature, and takes a maximum at an air–fuel ratio of 24. Further increase in EGR rate establishes smoke-less combustion, exhibiting non-luminous flame at air–fuel ratios lower than 17.5.

7.3 Classification of Advanced Diesel Combustion

A diagram illustrating the relationships between injection timing and combustion event for advanced diesel combustion strategies is given in Fig. 7.15. The advanced diesel combustion regimes can be categorized into two groups, according to the method to achieve low temperature combustion. The low temperature combustion in the first group including HCCI, UNIBUS, and PREDIC is realized by early fuel injection, and as a result the rate of heat release in this category offers two peaks

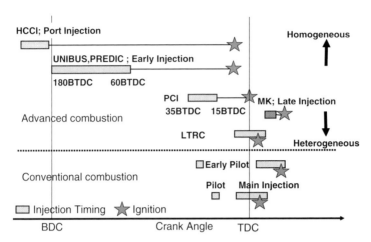

Fig. 7.15 Relationships between injection timing and combustion event in advanced and conventional diesel combustion

representing the low temperature and the high temperature reactions, respectively as shown in Fig. 7.1.

MK combustion, PCI combustion and LTRC are in the second group and employ near TDC fuel injection timing and high EGR rate. MK combustion involves late fuel injection in high EGR rate environment and the low temperature combustion takes place in the expansion stroke. LTRC strategy features intake-air cooling and high EGR rate to prolong the ignition delay time. As the duration between the end of injection and the start of combustion is shorter than those in the first group, MK and LTRC strategies make use of high swirl and high injection pressures to complete the formation of partially premixed mixtures within this narrow duration. The differences that characterize combustion events in the first group from those in the second are longer induction duration and the two-stage heat release, both of which are typical in HCCI combustion.

7.4 Fundamental Aspects of Diesel HCCI Combustion

7.4.1 Factors Affecting HCCI Ignition

The effects of engine parameters on HCCI combustion were investigated using a port injection single cylinder diesel engine [11]. The test engine had a bore × stroke = 94 mm × 100 mm, a displacement of 694 cc and a compression ratio of 16. The effects of three factors on HCCI combustion including fuel amount, EGR rate, and air temperature were systematically examined using n-pentane with a cetane number of 29.2 under conditions listed in Fig. 7.16.

The effect of equivalence ratio is shown in Fig. 7.17, in which LTO and HTO stand for low temperature oxidation and high temperature oxidation, respectively.

Fig. 7.16 Engine operating conditions in HCCI combustion experiment

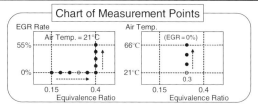

Engine Speed	1400rpm
Fuel	n-Pentane (CN:29.2, T_b:36°C) / iso-Pentane
Injection System	Type used in Gasoline engine
Amount of Injected Fuel (Equivalence Ratio)	7 to 19 mg/stroke (Equivalence Ratio:0.15 to 0.4)
EGR Rate	0 to 55%
Air Temperature	21 to 66°C

As seen in the figure, the increase in equivalence ratio enhances low temperature oxidation, giving rise to the in-cylinder temperature. Consequently, this accelerates the start of high temperature oxidation, causing a higher rate of heat release at higher equivalence ratios. The resultant steep pressure rise tends to produce harsh combustion noise.

Figure 7.18 shows the effect of EGR rate at an equivalence ratio of 0.4. The increase in EGR rate suppresses low temperature oxidation, leading to longer delay to the start of high temperature oxidation. At the same time, a high EGR rate yields longer combustion duration due to the higher specific heat of gases and reduced oxygen concentration. The resultant mild combustion benefits noise reduction.

The effect of intake air temperature at an equivalence ratio of 0.3 is presented in Fig. 7.19. It is obvious that high intake air temperature allows earlier initiation

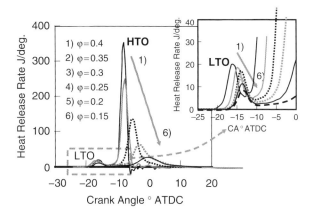

Fig. 7.17 Effect of equivalence ratio on heat release rate without EGR

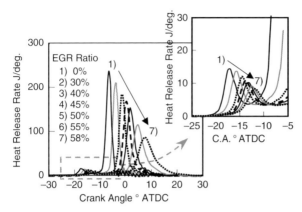

Fig. 7.18 Effect of EGR rate on heat release rate at a constant fuel injection amount "Reprinted with permission from SAE Paper 2006-01-0028 © 2006 SAE International"

of both low temperature and high temperature oxidation reactions. It also delivers a higher rate of heat release during the high temperature oxidation reactions.

The effect of blending ratio of n-pentane with a cetane number of 29 and iso-pentane with 10.5 is given in Fig. 7.20. The increase in the blending ratio of iso-pentane causes a decrease in the cetane number, resulting in delays in both low temperature and high temperature oxidation reactions. It should be noted that the peak value of the heat release rate is less sensitive to the change in fuel blending ratio as compared to the change in EGR rate.

The cylinder wall temperature also affects the HCCI combustion because it changes thermal condition in the cylinder gases through the heat transfer during the induction and compression stroke [4].

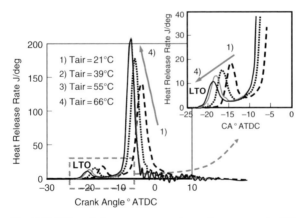

Fig. 7.19 Effect of air temperature on heat release rate at $\phi = 0.3$ and without EGR

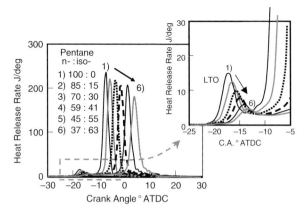

Fig. 7.20 Effect of fuel blend ratio on heat release rate (total $\phi = 0.35$, EGR = 0) "Reprinted with permission from SAE Paper 2006-01-0028 © 2006 SAE International"

7.4.2 Role of In-Homogeneities in Mixture

If the in-cylinder mixture is completely homogeneous and compressed progressively with time, auto-ignition will take place instantly throughout the combustion chamber, causing infinitely steep pressure rise. However, the in-homogeneities in temperature and equivalence ratio produced during the induction and compression stroke fortunately moderates the auto-ignition event.

Planar laser induced fluorescence (PLIF) measurements of formaldehyde and OH were made using an optical engine to investigate the effects of in-homogeneities in the mixture on HCCI combustion [5, 10]. The magnitude of in-homogeneity was evaluated from the spatial distribution of formaldehyde LIF intensity in the combustion chamber. The fluctuation of ignition timing in HCCI combustion achieved by port fuel injection or in-cylinder direct injection before 50° BTDC was unexpectedly large as compared to that for the case of late injection timing after 50° BTDC. The experiments made it clear that the large fluctuation of the ignition timing in HCCI case is caused by the in-homogeneous temperature distribution in gases due to the heat transfer from the walls, while the small fluctuation in the latter case is due to the wider distribution of equivalence ratio in the in-homogeneous mixtures. The mixture elements near stoichiometric equivalence ratio tend to ignite faster than other elements with rich or lean fuel–air mixtures, minimizing the fluctuation of ignition timing. The successive ignition of rich and lean mixture elements generates simultaneously a moderate heat release rate. Thus, in-homogeneities both in the equivalence ratio and temperature play an important role to achieve stable ignition and moderate heat release in HCCI combustion.

It was observed in this experiment that formaldehyde appeared first in the cool flame and then disappeared upon the appearance of OH. The results were consistent with the behaviors of OH and formaldehyde in the cool flame reactions simulated by the chemical kinetics.

A large eddy simulation (LES) model in a CFD code was performed to simulate temporal and spatial fluctuations in temperature and equivalence ratio in a turbulent flow-field in the engine cylinder [27]. The fluctuating turbulent flow-field was given as the initial condition for the calculation using an initial turbulent kinetic energy of 0.74 C_m^2 (C_m: mean piston speed) and the integral length scale of 10 mm (Large scale) or 5 mm (small scale). The reaction kinetics for iso-octane is based on the five-step global model proposed by Schreiber et al. [30]. The calculation was made on the PCCI combustion in an engine with bore × stroke = 82.3 mm × 114.3 mm and a compression ratio of 10. The engine was assumed to operate at an engine speed of 600 rpm, overall equivalence ratio of 0.4 with initial fluctuation ratio ϕ' of 0.2 or 0.1, and a uniform mixture temperature at 450 K.

Figure 7.21 presents the cylinder pressure and the rate of heat release calculated for the four initial conditions with different equivalence ratio fluctuations and different length scales. The zero-dimensional calculation, which is also included in the figure, shows the fastest heat release with the highest peak value among the five cases calculated. The case of homogeneous mixture with $\phi' = 0$ shows a slower rate of heat release than the one-dimensional one because the heterogeneous local temperature distribution that is generated during the compression stroke by the turbulent wall heat transfer gives different local reaction timings and rates. The other three cases all show further slower rate of heat release according to the different initial conditions. It is seen that the combination of $\phi' = 0.2$ and large length scale gives the most moderate heat release rate among the three different heterogeneous conditions.

Figure 7.22 (a) and (b) represent the local heat release rate fields during the hot flame period for the two cases of $\phi' = 0$ and $\phi' = 0.2$ with "large length scale" assumption. In Fig. 7.22 (a), uneven spatial distributions in the heat release rate are observed at 10.5° BTDC (0.2° before the peak heat release rate) in three cross sections right to the cylinder axis. The turbulent wall heat transfer during the

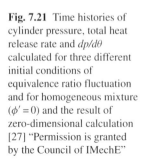

Fig. 7.21 Time histories of cylinder pressure, total heat release rate and $dp/d\theta$ calculated for three different initial conditions of equivalence ratio fluctuation and for homogeneous mixture ($\phi' = 0$) and the result of zero-dimensional calculation [27] "Permission is granted by the Council of IMechE"

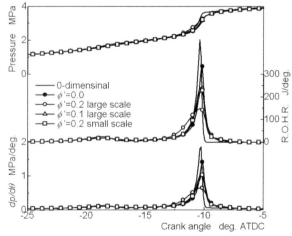

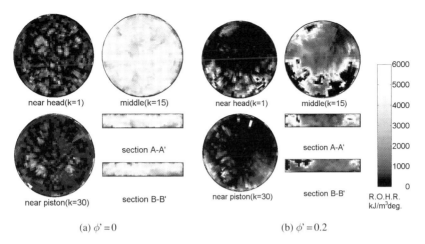

(a) $\phi' = 0$ (b) $\phi' = 0.2$

Fig. 7.22 Effects of homogeneities of equivalence ratio and temperature scale on local distribution of heat release rate. (**a**) homogeneous charge $\phi' = 0$, at 10.5° BTDC (0.2° before the peak heat release), (**b**) heterogeneous charge $\phi' = 0.2$, at 11° BTDC (0.7° before the peak heat release rate) [27] "Permission is granted by the Council of IMechE"

compression stroke caused the unevenness as stated before. Figure 7.22 (b) shows more pronounced heterogeneous heat re lease rate distributions at 11° BTDC (0.7° before the peak heat release rate) in three horizontal cross sections. The heterogeneous structure dominating in the entire cylinder is obviously the cause for the most moderate heat release rate in this case.

The results show that the HCCI combustion is not homogeneous but has a strong turbulent structure even when the fuel and air are fully premixed prior to intake. The results also indicate that the in-homogeneity are caused primarily by the thermal stratification due to heat transfer during compression, combined with turbulent transport as the above LES predicted.

7.5 Control of Combustion

Despite the low NO_x and PM emission characteristics of HCCI combustion, none of the early-injection diesel HCCI combustion concepts has been employed as yet in production engines because of the narrow load range of HCCI operation and the difficulty in combustion control under transient modes.

As shown previously, the start of combustion is determined by the in-cylinder conditions, and there is no direct control of the combustion phasing. The in-cylinder conditions can be varied by changing the excess air ratio, compression ratio, EGR rate and intake air charge temperature. When gasoline which has a high ignition resistance is used, higher mixture temperature is required during the compression stroke to ensure auto-ignition. On the contrary, the low ignition resistance of diesel fuel demands lower mixture temperatures during the compression stroke than in gasoline HCCI operation and this makes it more difficult to control the combustion phasing and rate in diesel HCCI combustion.

7.5.1 Expansion of Operation Range

Due to the difficulty of running advanced diesel combustion engines under high load conditions, switching between advanced and conventional combustion modes is required. This approach is called as dual-mode or mixed-mode operation. It is desirable to expand the operation range of the advanced combustion mode because a narrow operating range demands frequent transition of the two combustion modes under a transient mode operation.

The upper load boundary of the diesel HCCI operating region is constrained by the premature start of heat release in the compression stroke, and this demands naturally applying high cooled EGR to postpone the ignition timing to crank angles near TDC. The high cooled EGR is necessary to expand the diesel HCCI operating range as shown in Fig. 7.8. The HCCI operating range including iso-IMEP contours in the figure indicates that 65% EGR rate increases the IMEP at the knock boundary to 4.0 bar from 2.3 bar without EGR condition.

Another effective way to push the upper boundary of diesel HCCI operation to higher loads is to add chemicals in the mixture to suppress high temperature reactions [20, 36]. An experiment was performed using a 1.86-l single cylinder diesel engine with a compression ratio of 18.3 [20]. Naphthalene was used as fuel that features a low distillation temperature ($60°C$ at 50% distillation) and a low octane number of 67. With naphthalene injection into the intake port, the HCCI operation showed smokeless and NO_x-less emissions and an indicated thermal efficiency of 39% at the higher load boundary of 0.37 MPa. To expand the upper boundary by delaying the auto-ignition, an amount of methanol, a reaction inhibitor, was injected into the cylinder. A maximum BMEP of 0.9 MPa was successfully obtained in HCCI operation with smokeless and 200 ppm NO_x emissions and 45% indicated thermal efficiency at the condition of 50% fraction of methanol injection. Although the 50% methanol injection is practically unacceptable, addition of a reaction inhibiter was found to be effective to expand the operation range of HCCI combustion.

Boosting the intake pressure is an approach to produce higher specific power. An investigation was conducted using an 11.7-l, six-cylinder turbo-charged diesel engine fueled with a combination of n-heptane and ethanol [22]. A theoretical analysis supported by experiments suggests that although boosting allows high load HCCI operation, the turbocharger causes high pumping losses even if it is optimized to fit to the HCCI operation, and the peak cylinder pressure reaching 20 MPa will be probably the load limiting.

7.5.2 Functional Control of Combustion

The control of diesel HCCI combustion under transient modes and the switch from and to conventional diesel combustion are both essential to bring the diesel HCCI operation into practice in production engines. The characteristics of diesel HCCI combustion have been described in the preceding sections on the basis of engine data acquired under steady state operating conditions. However, most of the parameters such as EGR rate, compression ratio, and intake air temperature are difficult to tightly control under transient modes, and this is the major cause that prevents the diesel HCCI combustion from being introduced in practical engines.

In this section, strategies to control gasoline HCCI combustion, which is much easier, will be described first, and then the measures to control diesel HCCI combustion will be discussed.

7.5.2.1 Cycle-to-Cycle Control of Gasoline HCCI by VVA Strategies

In gasoline HCCI combustion, the combustion phasing needs to be advanced due to gasoline fuel's high resistance to auto-ignition. The variable valve actuation (VVA) technology looks promising under transient modes because it can increase the hot residual gas fraction in the cylinder and increase the compression ratio to achieve higher temperatures during the compression stroke to ensure auto-ignition. It is reported that an electro-hydraulic VVA system that can change the inducted gas composition and effective compression ratio using a closed loop control of intake- and exhaust-valve timings has successfully achieved simultaneous control of the peak cylinder pressure and combustion timing of propane HCCI combustion on cycle-to-cycle basis [16, 31].

Figure 7.23 is the schematic diagram, showing the control strategy for the simultaneous control of cylinder peak pressure and combustion phasing. The control system incorporates a physics based model that simulates the ignition delay time as a function of inducted gas composition and the compression ratio. The VVA system modulates the inducted gas composition to control the peak pressure on cycle-to-cycle basis while using intake valve closing (IVC) control on a slower time scale to vary the effective compression ratio and control phasing.

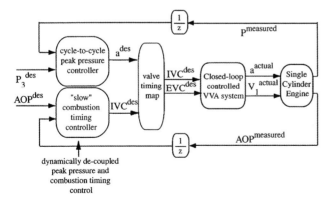

Fig. 7.23 Control strategy for simultaneous control of peak pressure and combustion timing (P_3^{des}: desired peak pressure, AOP^{des}: desired angle of peak pressure, a^{des}: desired molar ratio of inducted residual to reactant, IVC^{des}: desired intake value closing, and V_1: volume at final valve closure) [31] "Permission is granted by the Council of IMechE"

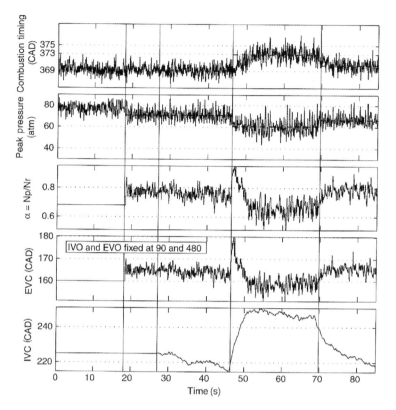

Fig. 7.24 Simultaneous control of both peak pressure and combustion timing (AOP: angle of peak pressure, a metric for combustion timing) [31] "Permission is granted by the Council of IMechE"

Figure 7.24 shows the decoupled controller performance over a range of desired pressure and combustion timings. Peak pressure and combustion tracking responses occur within 5 and 25 engine cycles, respectively. The report points out that when the engine load is changed stepwise, the phase lag in the temperature of the engine structure affects the in-cylinder pressure in the following cycles. Another work on a feedback loop control of valve timings of a gasoline HCCI engine during transients can be found in ref. [1].

7.5.2.2 Control of Diesel HCCI

The control strategy being used in diesel HCCI is different from that applied to gasoline HCCI because of the low ignition resistance of diesel fuel that demands delayed ignition and heat release timings during the compression stroke. The strategy to control the hot residual gas fraction by VVA, which works well in gasoline HCCI to ensure auto-ignition, can't be applied to diesel HCCI as it is. Thus, low ignition resistance of diesel fuels makes it more difficult to control the combustion phasing and rate in HCCI operation.

Diesel HCCI is obliged to employ high amount of cooled EGR to shift the ignition timing to crank angles near TDC. Lower compression ratio is preferable as it lowers the mixture temperature during the compression stroke. Enhancing evaporation and avoiding spray impingement on the walls are both essential in early injection HCCI when low volatile diesel fuel is used.

A study on operation of a 0.48-l single cylinder DI diesel engine with variable effective compression ratio in HCCI and conventional diesel mode is reported [8, 9]. The intake valve closing timing (IVC) was retarded effectively reducing the effective compression ratio to 9.0 and 6.5 for HCCI operation from the original compression ratio of 15 for conventional. An adjustable EGR system was fitted, which included a thermostatically controlled EGR cooler. In HCCI mode operation, diesel fuel with a cetane number of 56 was injected from a piezo injector in a series of five short injections during the compression stroke to minimize the spray penetration. Proper combustion phasing was accomplished by adjusting the EGR rate so that the point of 50% heat release coincided with TDC. The authors state that the window in which combustion without knock and misfire was achieved was only a few crank angles wide, making the EGR rate critical for successful operation.

The main operating parameters at an engine speed of 2000 rpm are summarized in Tables 7.1 and 7.2. The excess air ratio, lambda, ranges from 1.5 to 1.0, according to engine load and EGR rate. At higher loads (over 0.4 MPa), supercharging was needed to avoid lambda falling to less than 1. A comparison between the results in the two tables suggests a trend of higher EGR rate and lower lambda with HCCI operation. Both intake charge and cooling temperatures are found lower for HCCI operation to reduce compression temperature.

In HCCI mode, soot and NO_x were reduced by a factor of 10 or more, while HC and CO emissions showed significant increases against conventional combustion, although HC and CO emissions were improved a lot with a series of five short injections during the compression stroke as compared to those with double and triple injections. In order to operate HCCI mode under transients, a fast response cooled

Table 7.1 Operating conditions for the base measurements with early injection mode "Reprinted with permission from SAE Paper 2005-01-0177 © 2005 SAE International"

IMEP [MPa]	$P_{fuel\,rail}$ [MPa]	λ [-]	P_{intake} (abs) [kPa]	T_{intake} [°C]	$T_{water}\,T_{oil}$ [°C]	EGR [%]
0.3	75	1.5	102	45	70	55
0.4	90	1.2	102	45	70	58
0.5	97	1.1	145	50	70	70

Table 7.2 Operating conditions for the base measurements in conventional diesel mode "Reprinted with permission from SAE Paper 2005-01-0177 © 2005 SAE International"

IMEP [MPa]	$P_{fuel\,rail}$ [MPa]	λ [-]	P_{intake} (abs) [kPa]	T_{intake} [°C]	$T_{water}\,T_{oil}$ [°C]	EGR [%]
0.3	47	4.3	114	65	90	30
0.4	50	3.2	118	63	90	22
0.5	60	2.4	120	78	90	18

EGR devise as well as VVA devise needs to be developed to follow the quick change in load and speed.

7.5.2.3 Cycle-to-Cycle Control of HCCI by Dual-Fuel Strategies

A cycle-to-cycle control system for a dual-fuel HCCI engine was demonstrated [21, 22]. Tow different fuels, n-heptane and iso-octane, with low and high resistance to auto-ignition were injected directly into the intake-port of a six-cylinder, 11.7-l heavy-duty turbo diesel engine. The engine had a compression ratio of 18, and was equipped with an intercooler and a heater to make thermal management of the intake charge. Heating was only used at low loads to keep HC and CO emissions at low levels. No EGR devise was accommodated in the engine. The engine status such as CA50, which stands for the crank angle position at 50% heat release, and IMEP were calculated using the cylinder pressure and inlet condition data fed to a CPU in the control system. The control algorithm received the engine status data calculated as input and a closed loop control was performed to provide combustion timing and IMEP as specified by the operator. The timing control was the most difficult part and the sensitivity of timing was calculated as a function of engine speed, inlet temperature, and mixing fraction of the two fuels. When this was done, injection event such as total amount and the mixing ratio between the two fuels and intake heat were adjusted according to the controller output. The performance of the controller on the step responses to CA50, IMEP and engine speed was successfully assessed.

Another dual-fuel strategy to control PCCI combustion was proposed [11]. In this approach gasoline was supplied into the intake port, while diesel fuel was injected into the cylinder at an early timing to trigger the start of combustion of homogeneous gasoline mixtures. The ignition processes were investigated using a single cylinder engine with a displacement of 694 cc, a compression ratio of 14 and a swirl ratio of 2.3. It was found that the start of PCCI combustion was controlled by varying the mass ratio of the two fuels and that the resultant PCCI combustion took place mildly as shown in Fig. 7.25. The range of the PCCI operation was successfully extended to an IMEP of 12 bar at a high boost condition, while keeping NO_x and smoke emissions as low as 10 ppm and 0.1 FSN, respectively. The dual-fuel PCCI combustion approach introduced here seems to have potential to fit transient mode operation as it requires no slow EGR devices.

7.5.2.4 Control of Mixed Mode Combustion

Mixed or dual mode operation has the advantage that part load HCCI operation can be combined with conventional diesel operation at higher loads. As the operation range of the advanced diesel combustion is narrow, mixed mode combustion strategy, is a practical approach to implement the advanced diesel combustion in real engines as shown in Fig. 6.2. This demands a new technology to switch the combustion mode from one to the other without torque shock and emissions penalty.

There is a report on a vehicle test where the mixed combustion strategy involving conventional and PCI combustion was practically assessed [35]. The vehicle was

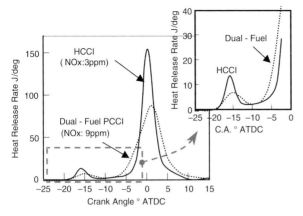

Fig. 7.25 Comparison of heat release rate of HCCI combustion to that of dual fuel PCCI combustion (total $\phi = 0.35$) "Reprinted with permission from SAE Paper 2006-01-0028 © 2006 SAE International"

powered by a six-cylinder 5-l engine, which comprises a 160 MPa common-rail injection system, variable geometry turbocharger (VGT) and a high swirl intake port. The brake mean effective pressure (BMEP) limit for PCI combustion was first investigated using EPA-47 diesel fuel at an engine speed of 1500 rpm. When the air fuel ratio (A/F) decreases with the increase in load due to the increased EGR rate, smoke emissions rose steeply when A/F was approaching unity. The limit of BMEP for the PCI combustion was determined from this experiment as 320 kPa. The engine fitted with an oxidation catalyst was mounted to a 2 WD and 4AT vehicle with an inertia weight of 5250 lbs. A vehicle test under the FTP-75 test cycle showed that over 60% of the time, the engine was operated in the PCI mode. The emission test results showed that although NO_x showed a 50% reduction, PM was increased by 40% due to an instantaneous increase in smoke emissions occurring during the switching. Other concerns are the large variations of non methane organic gases (NMOG) and CO emissions due to the unstable PCI combustion, combustion noise and torque surge during transient conditions.

7.6 Summary

The effects of chemical and physical parameters on auto-ignition during the engine cycle have been clarified quantitatively by a number of theoretical and experimental studies. Despite the understanding of the effects of parameters on auto-ignition process, its control is still hard because of its delicate sensitiveness to these parameters. The low resistance of diesel fuel to auto-ignition tends to permit auto-ignition to occur at early crank angles during the compression stroke, causing harsh knock. If the mixture temperature, which is one of the most influential parameters to control auto-ignition, is decreased to avoid knock by applying lean mixtures, high cooled EGR and low compression ratio, the engine is easily misfired if the temporal variation of the mixture temperature fails to start the low and high temperature reactions. The resultant narrow operating region in the low load range has to be extended before the advanced diesel combustion engine enters practical use.

Since the mixture temperature affects auto-ignition dominantly, the temperatures of the cylinder liner and combustion chamber walls affect eventually the auto-ignition phasing through the heat transfer between the walls and mixtures. This means that the control of auto-ignition timing under transient modes is really difficult because the temperatures of the combustion chamber walls can hardly change due to their large thermal inertia when the engine operating condition is suddenly changed. The dynamic response characteristics of engine components such as EGR valve and turbo-charger are not fast enough to follow the quick change in engine operating condition. Consequently these delays need to be compensated using a model based closed loop control of combustion.

The switch between two operation modes, advanced and conventional diesel combustion, has to be implemented under control to avoid torque shock, increased noise and smoke emissions. A sophisticated model based control system is also required to achieve smooth mode transition. Technological development in both hardware and software is demanded to make the advanced diesel combustion engines to be a reality.

Acknowledgments The authors would like to acknowledge Dr. K. Nakakita, Toyota Central R&D Labs., Inc., for his positive suggestions and support. They also acknowledge Dr. K. Akihama and Dr. K. Inagaki, Toyota Central R&D Labs., Inc., and Dr. H. Yanagihara, Toyota Motor Engineering & Manufacturing Europe, for providing valuable data and figures.

References

1. Agrell F et al. (2005) Control of HCCI during engine transients by aid of variable valve timings through the use of model based non-linear compensation. SAE Paper No. 2005-01-0131
2. Akihama K, Takatori Y and Inagaki K (2001) Mechanism of the smokeless rich diesel combustion by reducing temperature. SAE Paper No. 2001-01-0655
3. Aoyama T, Hattori Y et al. (1996) An experimental study on premixed-charge compression ignition gasoline engine. SAE Paper No. 960081
4. Chang J et al. (2005) Characterizing the thermal sensitivity of a gasoline homogeneous charge compression ignition engine with measurements of instantaneous wall temperature and hest flux. Int. J. Engine Res., Vol. 6, No 4, pp. 289–309
5. Dec J E, Hwang W and Sjoberg M (2006) An investigation of thermal stratification in HCCI engines using chemiluminescence imaging. SAE Paper No. 2006-01-1518
6. Duret P, Gatellier B et al. (2004) Progress in diesel HCCI combustion within the European SPACE LIGHT project. SAE Paper No. 2004-01-1904
7. Hasegawa R and Yanagihara H (2003) HCCI combustion in DI diesel engine. SAE Paper 2003-01-0745
8. Helmantel A and Denbratt I (2004) HCCI operation of a passenger car common rail DI diesel engine with early injection of conventional diesel fuel. SAE Paper No. 2004-01-0935

9. Helmantel A, Gustavsson J and Denbratt I (2005) Operation of a DI diesel engine with variable effective compression ratio in HCCI and conventional diesel mode. SAE Paper No. 2005-01-0177

10. Hildingsson L, Persson H and Johansson B (2004) Optical diagnostics of HCCI and low-temperature diesel using simultaneous 2-D PLIF of OH and formaldehyde. SAE Paper No. 2004-01-2949

11. Inagaki K, Fuyuto T et al. (2005) Dual-fuel PCI combustion controlled by in-cylinder stratification of ignitability. SAE Paper No. 2006-01-0028

12. Kamimoto T and Bae M (1988) High combustion temperature for the reduction of particulate in diesel engines. SAE Paper No. 880423

13. Katsumi N, Kawamura K et al. (2004) Development of variable spray pattern nozzle for diesel engine using PCCI. Proceeding of 13th symposium on atomization in Japan (in Japanese). ILASS. pp. 87–92

14. Kimura S et al. (1997) A new concept of combustion technology in small DI diesel engines. JSAE Paper No. 9732513

15. Kimura S, Ogawa Y, Matsui Y and Enomoto Y (2002) An experimental analysis of low-temperature and premixed combustion for simultaneous reduction of NOx and particulate emissions in direct injection diesel engines. Int. J. Engine Res., Vol. 3, No. 4, p. 249

16. Milovanovic N, Turner J W G et al. (2005) Active valve-train for homogeneous charge compression ignition. Int. J. Engine Res., Vol. 6, No. 4, pp. 377–397

17. Najt P M and Foster D E (1983) Compression-ignited homogeneous charge combustion. SAE Paper No. 830264

18. Nakagome K, Shimazaki N et al. (1997) Combustion and emission characteristics of premixed lean diesel combustion engine. SAE Paper No. 970898

19. Noguchi M and Tanaka Y (1979) A study on gasoline engine combustion by observation of intermediate reactive products during combustion. SAE Paper No. 790840

20. Ogawa H, Miyamoto N et al. (2005) Combustion control and operating range expansion in an homogeneous charge compression ignition engine with direct in-cylinder injection of reaction inhibitors. Int. J. Engine Res., Vol. 6, No. 4, pp. 341–360

21. Olsson J-O, Tunestal P and Johansson B (2001) Closed-loop control of an HCCI engine. SAE Paper No. 2001-01-1031

22. Olsson J-O, Tunestal P and Johansson B (2004) Boosting for high load HCCI. SAE Paper No. 2004-01-0940

23. Onishi S and Hong S (1979) Active thermal-atmosphere combustion (ATAC) – a new combustion process for internal combustion engines. SAE Paper No. 790501

24. Oppenheim A K (1984) The knock syndrome-its cures and its victims. SAE Paper No. 841339

25. Peng Z, Zhao H and Ladommatos N (2003) Effect of air/fuel ratios and EGR rates on HCCI combustion of n-heptane, a diesel type fuel. SAE Paper No. 2003-01-0747

26. Pottker S, Delebinski T et al. (2004) Investigations of HCCI combustion using multi-stage direct-injection with synthetic fuels. SAE Paper No. 2004-01-2946

27. Saijo K, Kojima T and Nishiwaki K (2005) Computational fluid dynamics analysis of the effect of mixture heterogeneity on combustion process in a premixed charge compression ignition engine. Int. J. Engine Res., Vol. 6, No. 5, pp. 487–494

28. Sakai A et al. (2005) Improvements in premixed charge compression ignition combustion and emissions with lower distillation temperature fuels. Int. J. Engine Res., Vol. 6, No. 5, pp. 433–442

29. Sasaki S, Ito T and Iguchi S (2000) Smoke-less rich combustion by low temperature oxidation in diesel engines. Aachen colloquium automobile and engine technology, Aachen, 767

30. Schreiber M et al. (1994) A reduced thermokinetic model for the autoignition of fuels with variable octane ratings. 25th Symposium (International) on Combustion Institute. pp. 933–940

31. Shaver G M, Roelle M J et al.(2005) A physics-based approach to the control of homogeneous charge compression ignition engines with variable valve actuation. Int. J. Engine Res., Vol. 6, No. 4, pp 361–375

32. Shimazaki N et al. (2003) Dual mode combustion concept with premixed diesel combustion by direct injection near top dead center. SAE Paper No. 2003-01-0742
33. Takeda Y et al. (1996) Emission characteristics of premixed lean combustion with extremely early staged fuel injection. SAE Paper No. 961163
34. Thring RH (1989) Homogeneous-charge compression-ignition (HCCI) engines. SAE Paper 892068
35. Uekusa T et al. (2005) Emission reduction study for meeting new requirements with advanced diesel engine technology. SAE Paper No. 2005-01-2143
36. Yamada Y, Furutani M et al. (2004) Advanced/retarded criterion on ignition of fuel/air mixtures with formaldehyde doping. The sixth international symposium on diagnostics and modeling of combustion in internal combustion engines, JSME, Yokohama, pp. 199–206
37. Yanagihara H (1996) A simultaneous reduction of NOx and soot in diesel engines under a new combustion system (Uniform bulky combustion system UNIBUS), 17th international Vienna motor symposium, Combustion Institute of Japan
38. Yoshihara Y and Nishiwaki K (2004) A numerical analysis of knocking phenomena in an SI engine. J. Combust. Soc. Jpn., Vol. 46, No. 136, pp. 71–81

Chapter 8
Fuel Effects on Engine Combustion and Emissions

Thomas Ryan and Rudolf R. Maly

8.1 Historical Background

The Energy Information Administration (EIA) estimates indicate that 90% of the energy consumed worldwide is fossil fuel based, and a large fraction of this is derived from petroleum. Figure 8.1 shows the world energy fuel consumption historically, and projected through 2030. It is clear that past, present, and most current future projections indicate that the world energy needs will in large part be based on petroleum.

Petroleum composition is widely variable, depending on the source, but in general the composition includes several hundred hydrocarbons, with boiling points ranging from –68°C to 900°C, and molecular weights ranging from 44 to well over 1000. Historically, petroleum processing consisted of a relatively simple separation process using distillation to take advantage of the range of boiling points and molecular weights.

As a general rule, the lighter hydrocarbon components have a higher resistance to autoignition, i.e. a higher autoignition temperature. This makes them ideally suited for flame propagation engines, where the boiling point distribution is also nearly ideal. Over time, the ignition characteristics and the boiling point distributions have been additionally tailored through refinery processing for use in premixed flame propagation engines. The first flame propagation engines were fueled using straight run gasoline, or basically the first fraction of hydrocarbons distilled from the early petroleum sources.

The heavier fractions, on the other hand, generally have lower autoignition temperatures making them more ideal for use in compression ignition engines. In addition, the boiling point distribution and viscosity characteristics of the heavier

Thomas Ryan
Southwest Research Institute, San Antonio, TeXas, USA

Rudolf R. Maly
Consultant Engines and Future Fuels (retired director fuels research at DaimlerChrysler AG),
D-71065 Sindelfingen, Germany

C. Arcoumanis, T. Kamimoto (eds.), *Flow and Combustion in Reciprocating Engines*,
DOI: 10.1007/978-3-540-68901-0_8, © Springer-Verlag Berlin Heidelberg 2009

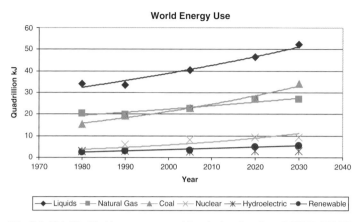

Fig. 8.1 EIA Worldwide energy usage, historical and projected (US DoE [17])

materials make them more ideal for use in direct injection engines compression ignition engines.

Early in the evolution of the petroleum refining business a very clear reality became apparent, and this is that every bit of the petroleum that enters the refinery must be converted to product. It also became apparent that there are components in petroleum that are not suited for use in either flame propagation or compression ignition engines. The incompatibilities generally arise with petroleum components that have combinations of boiling point distributions and ignition characteristics that will not work in either engine type.

This chapter includes a brief description of petroleum and petroleum refining, including the processes that have been developed to handle the troublesome components and to arrive at the appropriate product distribution. The importance of the key fuel properties are discussed in the context of refining processes and resulting compositional changes. Advanced combustion processes are discussed and future fuel requirements are suggested. The chapter concludes with brief description of the role of some of the evolving alternative feedstocks and fuels.

8.1.1 Petroleum Composition and Characteristics

As the name implies, fossil fuels are truly the processed fossil remains of sea and land plants and animals that have become buried deeper and deeper. The compression and heating of these remains resulted in the conversion to hydrocarbons. The composition of the various deposits depends on the type of the original remains and the history, in terms of the depth and age. Figure 8.2 is a simple representation of the life cycle of the fossil fuels.

PETROLEUM & NATURAL GAS FORMATION

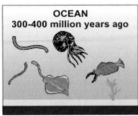

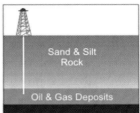

OCEAN 300-400 million years ago	OCEAN 50-100 million years ago	
	Sand & Silt	Sand & Silt Rock
	Plant & Animal Remains	Oil & Gas Deposits
Tiny sea plants and animals died and were buried on the ocean floor. Over time, they were covered by layers of silt and sand.	Over millions of years, the remains were buried deeper and deeper. The enormous heat and pressure turned them into oil and gas.	Today, we drill down through layers of sand, silt, and rock to reach the rock formations that contain oil and gas deposits.

Fig. 8.2 Life cycle of fossil fuels (EIA webpage [15])

The composition of typical petroleum is presented in Figs. 8.3 and 8.4. As can be seen, petroleum composition is very complex, with several different hydrocarbon types and many different hydrocarbon species across the entire boiling range. In addition, the composition, as indicated above, is highly variable and dependent on the location of the deposit.

Figure 8.4 demonstrates that the materials with the lower boiling points tend to be composed of the more saturated and paraffinic materials, while the heavier materials tend to be unsaturated and aromatic in nature. The figure also make reference to the heteroatoms (N, O, S) and the metals, all troublesome contaminants that are the subject of significant processing effort in the typical refinery.

8.1.2 Refinery Development

The very first refineries consisted of simple distillation systems designed to separate the lighter materials from the crude for use in the first generation of flame propagation gasoline engines. As the gasoline engines evolved and the demand increased for the lighter components, it became apparent that "straight run" gasoline (the fraction that is obtained from simple distillation of petroleum) was neither abundant enough nor of adequate quality to meet the more stringent demands imposed by the evolving gasoline engine. As will be described in the next section, the earliest quality issues related to the propensity of early gasolines to autoignite and cause severe pressure oscillations in the evolving gasoline engines. The symptoms of this autoignition phenomena were audible "knocking" sound and ultimate engine damage if allowed to persist, or if too severe.

Refineries had to be modified to both meet the expanding demand for materials in the gasoline boiling range and at the same time meet the evolving requirements for improved resistance to autoignition. Expanding the pool of material in the gasoline

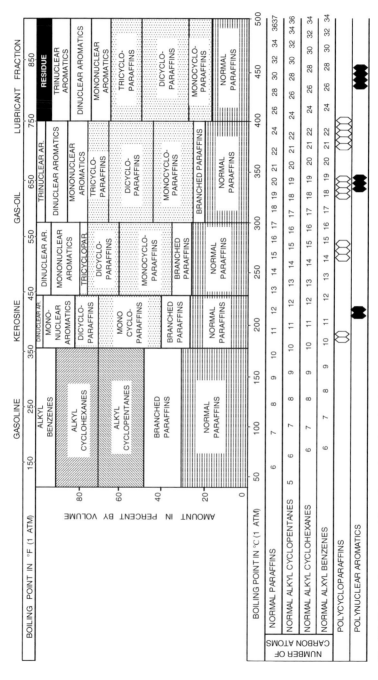

Fig. 8.3 Hydrocarbon composition of petroleum based on boiling point distribution, and fuel fraction (web locations) [14]

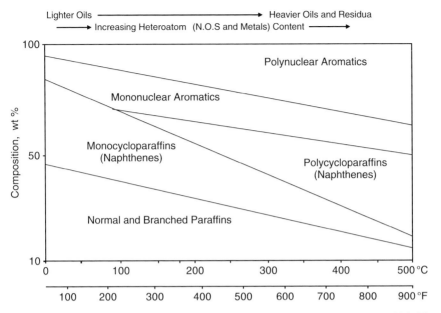

Fig. 8.4 Fractionation of petroleum showing the increase in condensed ring systems with boiling point [14]

boiling range was accomplished by adding chemical process units to the refinery to modify both the heavier (higher boiling point) and the lighter (lower boiling point) materials. The heavier materials are modified by thermal and catalytic processes including coking and cracking. Alkylation improves the ignition characteristics, as does reforming the lighter materials. These processes are shown schematically in Fig. 8.5. The typical product slat is shown in Fig. 8.6.

The focus on gasoline production drove most of the early refinery developments, and other products received little attention. As indicated above, it became very clear, even in these early refineries, that all of the material produced in the refinery had to have markets. The heavier materials, or the so called middle distillates, eventually found application in the newly developed gas turbine and diesel engines, and as a replacement for coal in residential and industrial heating. The evolution of the turbine and diesel engines did have some impact on refinery evolution in the United States, but much more impact on refinery design in other parts of the world, in particular in Europe where the diesel engine was evolving at a more rapid rate. In fact, the European refineries differ in basic design from those in the US. The European refineries have been designed to produce higher quality diesel than can be routinely produced in the typical US refinery. The primary difference between ideal gasoline and ideal diesel fuel is the ignition characteristics. It was clear early on that ideal gasoline resisted autoignition (high autoignition temperature) while ideal diesel fuels autoignite very easily (very low autoignition temperatures).

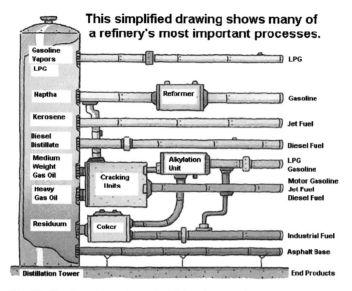

Fig. 8.5 Simple refinery schematic (EIA webpage [16])

Fig. 8.6 Simple refinery schematic (EIA webpage [16])

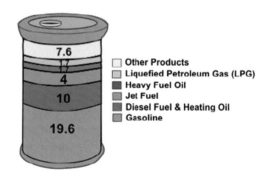

8.2 Fuel Chemistry

As indicated above, petroleum crude oils are composed of several hundred different chemical compounds. The components can be classified by hydrocarbon type. The convention for naming these compounds is described in Table 8.1 from ref. [20]. Included in the table are comments regarding various aspects of the effects of the various compounds on fuel characteristics and human health effects.

Table 8.1 Hydrocarbon type naming convention

Hydrocarbon
    Saturated alkanes (n-alkane and i-alkane)
        General formula C_nH_{2n+2}
        Boiling point and density increase with increasing the number of C atoms.
        Branched alkanes (iso-alkanes) is very small in quantity
        Boiling point of straight chains > iso-alkanes with the same no. of C
    Naphthenes or Cycloparaffins (saturated cyclic hydrocarbons, i.e. cyclohexane)
        General formula C_nH_{2n} for one ring compounds
    Alkenes or Olefins – unsaturated aliphatic hydrocarbon (i.e. ethylene or propylene)
        Very small amounts in crude oil – produced during refining
    Aromatics hydrocarbon (cyclic and polyunsaturated hydrocarbons containing conjugated
        double bonds)
    Alkylaromatics have very high Octane Number - content in gasoline is limited by
    environmental regulations – health effects due to high toxicity.
    Polyaromatic Hydrocarbons – aromatics containing more than 1 ring
        Naphthalene – 2 rings
        Anthracene – 3 rings
        Pyrene – 4 rings (very toxic)
    Hydroaromatics or naphthenoaromatics – partially saturated PAHs
Heteroatom compounds
    Sulfur compounds might be present in inorganic and organic forms. In crude oils sulfur
    concentration can range from 0.1 to more than 8 weight percent. Such as dibenzylthiophene
    (2 benzene rings separated by 1 S atom) – is most difficult to release the Sulfur
        Oxygen compounds are responsible for petroleum acidity in particular.
        Carboxylic (OH−C=O bonded to a benzene ring)
        Phenolic (OH bonded to a benzene ring)
        Nitrogen compounds
carbazole (2 benzene rings separated by 1 N atom) – neutral Quinoline (2 benzene rings with
    1 N atom on 1 ring) – basic

8.3 Engine Development Impacts

The earliest internal combustion engines were fueled with straight run gasoline. These engines had very low compression ratios due to mechanical limitations of the engine production processes and also due to the relatively poor ignition quality of the available fuels. The fundamental thermodynamic relationship between engine efficiency and compression ratio was known, but autoignition of the available fuels dramatically limited compression ratio. In fact, the early engine designers did not have a method for measuring or rating the autoignition characteristics of gasoline. They did observe, however, that increased compression ratio hastened the autoignition process and that different gasolines demonstrated different sensitivities to compression ratio. Several researchers examined this phenomenon, which ultimately became known as "knock" due to the audible noise generated in the engine during these autoignition events.

It was determined that normal combustion in an engine involves the formation of a flame front that moves in a relatively orderly fashion through the unburned fuel air mixture. As the flame front progresses, the unburned mixture ahead of the flame front is heated by heat transfer and compression. It is commonly accepted that knock is an end gas phenomenon, where the last of the unburned mixture reaches the autoignition conditions of the fuel and an uncontrolled autoignition process occurs, leading to very rapid pressure rise rates and pressure oscillations that produce the characteristic "KNOCKING" sound.

The compression ignition engine, developed by Rudolf Diesel in 1892 [29], was originally designed to operate on cheap fuels such as coal dust and/or vegetable oils. At the World Exhibition in Paris in 1900 Diesel's first engine ran actually on peanut oil. He stated in 1912 that vegetable oils may become in the course of time as important as the petroleum and coal tar products of the present time. As the engine evolved, fuel injection and cold start became the main issues. The distillate fraction of petroleum was recognized as the ideal feedstock, with appropriate density, distillation, and lubricity characteristics for fuel handling and direct in-cylinder injection. In addition, the distillate fraction also had acceptable ignition characteristics for operation in the engines available at that time.

As engine technology matured and as the demand for gasoline increased, the composition of gasoline changed and other fuel handling and engine operating characteristics placed additional demands on fuel quality requirements. During this evolution of fuel requirements, the practice of specifying fuel properties was adopted and evolved into the current practice of fuel specification. The common practice is for the engine and fuel producers to maintain working groups through organizations such as the American Society of Testing Materials (ASTM), Deutsches Institut fur Normung, Association des Constructeurs Europeens d'Automobiles (ACEA) and Japan Industrial Standard to continually monitor and update the fuel specifications for all applications in response to the industry needs and capabilities.

8.4 Fuel Specifications

8.4.1 Evolution of Fuel Specification

Knock and knock prevention became, and in fact remains, the primary focus of gasoline quality. In the1920's an industry group was formed to develop an acceptable apparatus and test method for rating the knock resistance of gasolines. The outcome of this activity was a gasoline knock resistance rating that became known as "Octane Rating". As the gasoline engine evolved with higher compression ratios and higher load ratings for ground and airplane applications, variations of the Octane Rating procedure evolved. These activities were accomplished though industry cooperative activities managed through the ASTM.

Analogous to the gasoline engine, the performance requirements of the diesel engine changed and it became apparent that a fuel ignition rating apparatus and technique were needed. Similar to the gasoline situation, an industry group was formed to develop an appropriate procedure. The result of this activity became known as the

"Cetane Rating" procedure. Both Octane and Cetane rating will be discussed further in the next section.

As indicated previously, the engine and fuel developments were evolutionary processes that continue as the engines continue to evolve. The discussion so far has dealt only with the ignition quality. Several other fuel quality specification properties were also identified. These important properties address issues that were identified over the years, relating to performance and handling characteristics ranging from cold flow, storage stability, deposit formation, volatility, and a variety of engine compatibility issues. Currently, the ASTM D4814 gasoline specification includes 13 different fuel properties or characteristics, and the ASTM D975 diesel fuel specification includes 15 different specified properties. The current ASTM D4814 gasoline specification properties and associated ASTM test methods are listed in Table 8.2. The ASTM D975 diesel fuel specifications and test methods are listed in Table 8.3.

In a break with tradition and the historical industry practice, an international group of engine manufacturers developed their recommended specifications for fuels for use in different parts of the world. They presented these requirements in a document entitled "The World-Wide Fuel Charter" [19]. There are four different standards for both gasoline and diesel fuel, basically recognizing that there are different levels of engine and emissions control technology in different parts of the world. The Category 4 properties are listed in Tables 8.4 and 8.5 for gasoline and diesel fuel, respectively. The Category 4 standards are the most stringent.

The more important properties listed in the tables are described in the following sections. As indicated above, they include a variety of characteristics that address issues that were identified to affect the fuel handling as well as several associated with engine performance. The list of properties in Tables 8.2 and 8.4 are similar, differing only in the limits imposed by the particular specification. The method of measuring and the purpose of each of the properties are either identical or similar.

Distillation Properties measured following ASTM D 86, defines the relationship between the fraction of material evaporated and temperature. These characteristics are important for both the fuel handling and mixture preparation characteristics.

Table 8.2 ASTM D4814 gasoline fuel specification properties

Distillation Properties ASTM D 86
Vapor Pressure ASTM D 5190, D 5191, D 5482
Vapor Liquid Ratio ASTM D 2533
API Gravity ASTM D 287
Density ASTM D
Oxygenate Detection ASTM D 4815, D 5599, D 55845
Sulfur Content ASTM D 1266, D 2622, D 3120, D 5453, D 3120
Lead Content ASTM D 3341, D 5059
Phosphorus Content ASTM D 3231
Octane Number ASTM D 2699, D 2700
Copper Corrosion ASTM D 130
Water Tolerance ASTM D 6422
Intake Valve Deposits ASTM D 5500
PFI Injector Fouling ASTM D 5598

Table 8.3 ASTM D975 diesel fuel specification for 2D diesel

Property	ASTM Test Method[B]	Grade						
		No. 1-D S15	No. 1-D S500[C]	No. 1-D S5000[D]	No. 2-D S15	No. 2-D S500[C,E]	No. 2-D S5000[D,E]	No. 4-D[D]
Flash Point, °C, min.	D 93	38	38	38	52[E]	52[E]	52[E]	55
Water and Sediment, % vol, max	D 2709	0.05	0.05	0.05	0.05	0.05	0.05	...
	D 1796	...	...					0.50
Distillation: one of the following requirements shall be met:								
1. Physical Distillation	D 86							
Distillation Temperature, °C 90 %, % vol recovered								
min		...	...	...	282[E]	282[E]	282[E]	...
max		288	288	288	338	338	338	...
2. Simulated Distillation	D 2887							
Distillation Temperature, °C 90 %, % vol recovered								
min						300[E]	300[E]	
max		304	304			356	356	
Kinematic Viscosity, mm²/S at 40°C	D 445							
min		1.3	1.3	1.3	1.9[E]	1.9[E]	1.9[E]	5.5
max	...	2.4	2.4	2.4	4.1	4.1	4.1	24.0

Table 8.3 (continued)

Property	ASTM Test Method[B]	Grade						
		No. 1-D S15	No. 1-D S500[C]	No. 1-D S5000[D]	No. 2-D S15	No. 2-D S500[C,E]	No. 2-D S5000[D,E]	No. 4-D[D]
Ash % mass, max	D 482	0.01	0.01	0.01	0.01	0.01	0.01	0.10
Sulfur, ppm (μg/g)[F] max	D 5453	15	...	...	15	...	...	...
% mass, max	D 2622[G]	...	0.05	...	...	0.05	...	...
% mass, max	D 129	...	...	0.50	...	...	0.50	2.00
Copper strip corrosion rating max 3 h at 50°C	D 130	No. 3	No. 3	No. 3	No. 3	No. 3	No. 3	...
Cetane number, min[H]	D 613	40[I]	40[I]	40[I]	40[I]	40[I]	40[I]	30[I]
One of the following properties must be met:								
(1) Cetane index, min.	D 976–80[G]	40	40	...	40	40	...	...
(2) Aromaticity, % vol, max	D 1319[G]	35	35	...	35	35	...	...
Operability Requirements Cloud point, °C, max or	D 2500	J	J	J	J	J	J	...
LTFT/CFPP, °C, max	D 4539/ D 6371							
Ramsbottom carbon residue on 10% distillation residue, % mass, max	D 524	0.15	0.15	0.15	0.35	0.35	0.35	...
Lubricity, HFRR @ 60°C, micron, max	D 6079	520	520	520	520	520	520	...

Table 8.4 Worldwide fuel charter fuel specifications for category 4 gasoline [19]

Properties		Units	Limit	
			Min.	Max.
'91 RON' (1)	Research Octane Number	—	91.0	—
	Motor Octane Number	—	82.5	—
'95 RON' (1)	Research Octane Number	—	95.0	—
	Motor Octane Number	—	85.0	—
'98 RON' (1)	Research Octane Number	—	98.0	—
	Motor Octane Number	—	88.0	—
Oxidation stability		minutes	480	—
Sulfur content		% m/m	—	Sulfur Free[2]
Lead content		g/l	Non-detectable [3]	
Phosphorus content		g/l	Non-detectable [3]	
Managanese content		g/l	Non-detectable [3]	
Silicon		g/kg	Non-detectable [3]	
Oxygen content		% m/m	—	2.7(4)
Olefins content		% v/v	—	10.0
Aromatics content		% v/v	—	35.0
Benzene content		% v/v	—	1.0
Volatility			See Following Tables, page 11	
Sediment		mg/l	—	1
Unwashed gums(5)		mg/100 ml	—	30
Washed gums		mg/100 ml	—	5
Density		kg/m^3	715	770
Copper corrosion		merit	Class 1	
Appearance			Clear and Bright	
Fuel injector cleanliness		% flow loss	—	5
Intake-valve sticking		pass/fail	Pass	
Intake-valve cleanliness II				
Method 1 (CEC F-05-A-93), or		avg.mg/valve	—	30
Method 2 (ASTM D 5500), or		avg.mg/valve	—	50
Method 3 (ASTM D 6201)		avg.mg/valve	—	50
Combustion chamber deposits(5)				
Method 1 (ASTM D 6201), or		%	—	140
Method 2 (CEC-F-20-A-98)		mg/engine	—	2500

Table 8.5 Worldwide fuel charter fuel specifications for category 4 diesel fuel [19]

Properties	Units	Limit	
		Min.	Max.
Cetane Number	–	55 (1)	–
Cetane Index	–	52 (2)	–
Density @ 15°C	kg/m³	820 (3)	840
Viscosity @ 40°C	mm²/s	2.0 (4)	4.0
Sulfur content	% m/m	–	Sulfur Free (5)
Total aromatics content	% m/m	–	15
Polyaromatics content (di+tri+)	% m/m	–	2.0
T90 (6)	°C	–	320
T95 (6)	°C	–	340
Final Boiling Point	°C	–	350
Flash point	°C	55	–
Carbon residue	% m/m	–	0.20
CFPP (7) or LTFT or CP	°C	–	Maximum must be equal to or lower than the lowest expected ambient temperature
Water content	mg/kg	–	200
Oxidation stability	g/m³	–	25
Foam volume	ml	–	100
Foam vanishing time	sec.	–	15
Biological growth	–		'Zero' content
Vegetable Derived Esters	% m/m		Non-detectable
Total acid number	mg KOH/g	–	0.08
Corrosion performance	–		Class 1
Copper corrosion	merit	–	Light rusting or less
Ash content	% m/m	–	0.01
Particulates	mg/l	–	24
Injector cleanliness	% air flow loss	–	85
Lubricity (HFRR scar dia. @ 60°C)	micron	–	400

Cloud Point is measured using ASTM D 2500. It is a measure of the fuel temperature at which diesel fuel becomes cloudy in appearance due to the formation of wax crystals. It is important because it provides an indication of the cold flow characteristics of diesel fuels.

Vapor Pressure can be measured by one of the following, ASTM D 5190, D 5191, or D 5482. This is a property important for the gasoline like fuels. The historical importance of vapor pressure is due to the relationship between it and cold start and vapor lock. It has gained more interest recently because of the relationship with the evaporative emissions.

Flash Point is measured using ASTM D 93. This is a safety related property that defines the minimum fuel temperature at which the vapor above a fuel will ignite in the presence of an ignition source.

Vapor Liquid Ratio is determined following ASTM D 2533. The primary purpose of this measurement is related to safety issues associated with handling gasoline like fuels. It is basically a measure of the amount of vapor formed when a fixed quantity of liquid fuel is exposed to a specified environmental condition.

API Gravity measured following ASTM D 287, is important for all fuels because of the relationship with energy content, fuel metering and mixture preparation.

Density is measured following ASTM D 1298, and like API Gravity is important because it is related to the energy content of a stoichiometric fuel–air mixture and because of its relationship to the volumetric energy content.

Oxygenate Detection is measured following ASTM D 4815, D 5599, or D 55845. It is basically a modern gasoline property related to the regulatory requirements for minimum oxygen content in gasoline.

Sulfur Content is determined following one of the following procedures: ASTM D 1266, D 2622, D 3120, D 5453, and D 3120. Sulfur content is a regulated property in both gasoline and diesel fuel and is important because it affects the durability of exhaust aftertreatment system components and the level of particulate emissions.

Aromatic Content is measured following ASTM D 1319. The aromatic content affects the ignition quality and is thus important in both gasoline and diesel fuel. Aromatic materials typically have high Octane and low Cetane. It is also a regulated environmental property because aromatics are toxic and benzene content has been limited due to the direct health effect. In diesel fuel the aromatic content affects both the NOx and PM emissions.

Lead Content is measured using ASTM D 3341 or D 5059. This is a regulated gasoline property that is important because it poisons exhaust catalysts and because it is toxic and produces toxic exhaust emissions.

Phosphorus Content measured using ASTM D 3231 is a regulated property because of the effects of phosphorus on exhaust aftertreatment devices, where three way catalyst and diesel PM filter performance is reduced.

Octane Number (ON) is determined following either ASTM D 2699 or D 2700. This is a calibrated scale based on testing the knock resistance of unknown gasolines in a standard test engine, relative to blends of reference fuels.

Cetane Number (CN) can be determined in a standard test engine using ASTM D 613, or using a constant volume combustion bomb following ASTM D 6890 or ASTM D 7170. Like ON it is a calibrated measurement using blends of specified reference fuels. CN is a measure of the autoignition characteristics of diesel engine fuel.

Copper Corrosion is measured following ASTM D 130. It is a fuel system related property because it has been found that copper speeds fuel degradation and can cause deposits that affect the fuel system performance.

Ash Content measured using ASTM D482, is a measure of the metal oxides formed during combustion of a fuel. It is important due to the effects of these ashes on engine and aftertreatment durability.

Water Tolerance measured using ASTM D 6422, is important in both gasoline and diesel fuel because of the effects of water on fuel system durability and performance. It is also important in diesel fuel because of the corrosive effects of water on fuel injection system components.

Carbon Residue is determined following ASTM D 524. It is a measure of the tendency for a fuel to form carbon deposits and is thus related to engine combustion chamber deposits.

Lubricity is measured following ASTM D 6079. It is a measure of the ability of diesel fuel to lubricate the diesel fuel injection system. It is performed in a reciprocating wear device in which the wear scar is measured after a standard test.

Intake Valve Deposits are measure using ASTM D 5500. The test is performed in a specified gasoline PFI equipped vehicle operated over a specified drive cycle. The test measurement involves quantifying the amount of deposit that forms on the intake valves over the drive cycle. This is both a performance and an emissions related characteristic because intake valve deposits affect vehicle response and hydrocarbon emissions.

PFI Injector Fouling determined using ASTM D 5598, is a measure of the propensity of a gasoline fuel to form deposits in the injectors. Injector deposits affect fuel delivery, spray pattern and nozzle response time, all in turn affect the engine performance and emissions.

As can be seen, both the ASTM and the World Wide Fuel Charter specifications consist primarily of what are called performance properties rather than more fundamental or compositional properties. The adoption of this practice is historical, and based on the fact that the capability did not exist in early petroleum refining to make significant and controlled changes in the composition of distilled materials. In fact, current refineries are capable of only limited compositional changes and the processes must be modified if the feedstock changes significantly, or the required property limits are changed significantly. Exceptions to this practice are the limits on the heteroatoms, O, N, and S, due to the effects of these materials on engine durability, and aromatics, which is limited primarily due to the early recognized impact of these materials on smoke production, ignition quality, engine durability and, more recently, human health.

8.4.2 Worldwide Fuel Charter

The historic development of fuel specifications was accomplished in a consensus approach in which representatives from both the fuel companies and the engine manufacturers agreed on the definition of the important properties as well as the limits imposed on these properties. In more recent time an international group of engine manufacturers (ACEA, Alliance, EMA, JAMA) unilaterally provided their desired fuel specifications to the fuel producers who questioned their cost benefit of the proposed new fuel specifications. The lists of specified properties are the same as the current ASTM specifications, but the limits are, in some cases, significantly different. In this process the Original Equipment Manufacturer (Auto/Vechicle Manufacturer) (OEM) engine manufacturers have treated the fuel producers as suppliers, where the component (fuel) is specified by the OEM and the fuel suppliers are expected to provide the specified product.

The primary differences between the ASTM standards and the Worldwide Fuel Charter are in those fuel properties that have the largest impact on the engine emissions. The primary differences are in the composition (aromatic content and sulfur) and in the ignition characteristics (ON and CN).

As the emissions standards become more and more stringent it is becoming apparent that fuel composition is having a more significant impact on the engine out emissions composition and concentrations. This should not be surprising, but until relatively recently, the engine calibrations were so far from optimum for low emissions that relatively minor changes in the engine design or calibration overshadowed the fuel effects. Now, as the engines approach near optimum design and calibration for low emissions, fuel is playing a significant role. The inclusion of exhaust gas treatment systems and the adoption of advanced combustion modes are likely to place even more stringent requirements on the fuel.

8.5 Diesel Fuel Studies

In view of ever more stringent emission regulations, increasing needs for diversifying feedstock's to cope with worldwide growing demands for road fuels – at affordable costs – and the necessary modernization of fuel production technologies, make detailed knowledge of the full effects of future fuel formulations in a co-optimized system of engine and fuel a shear must. Fundamental investigations into co-optimizing fuels and engines were performed by seven major European auto manufacturers in cooperation with IFP[1] under the auspices of ACEA[2] –EUCAR[3]

---

[1] IFP – Institut Français du Pétrole, 1 & 4 Avenue de Bois-Préau, 92852 Rueil-Malmaison Cedex, France, http://www.ifp.fr/IFP/en/aa.htm

[2] ACEA – European Automobile Manufacturers Association, Avenue des Nerviens 85 | B-1040 BRUSSELS, http://www.acea.be/home_page

[3] EUCAR – European Council for Automotive R&D, Avenue des Nerviens 85, B - 1040 Bruxelles, http://www.eucar.be/

Detailed results have been published in an SAE paper [25]. Practical data of a test implementation of a co-optimization study on a modern EU4 base vehicle have been published by DaimlerChrysler [11] showing surprisingly large optimization potentials. The critical issues addressed were:

- What are feasible fuel formulations providing significant reductions of regulated emissions and what are their reduction potentials?
- What are their effects on engine performance and what are the consequences for engine technologies and after-treatment systems?
- What is the additional potential for improvements if fuels and engines were treated holistically – i.e. as one system?

Subsequently, reference will be made to these papers in order to clarify the trends for clean future fuels based on fossil sources. These results provide also a sound basis for assessing the additional potentials of renewable road fuels.

8.5.1 Fuel Components

Based on an extensive literature study on fuel related engine effects, the main impacts of modified fuel formulations were scrutinized [25]. In Table 8.6, an overview is presented on the average effect of fuel components on emissions. It can be seen, that the effects of varying fuel components depend both on fuel specification and on engine design. This reflects the strong interaction of fuel specifications with engine parameters and the consequences of the dedicated optimization for the current fuel specifications. Also, there is some ambiguity in the effect on different emission species making interpretations difficult which is attributed to the use of un-adapted engines. Nevertheless, it is obvious that low concentrations of sulfur and poly-aromatics are generally favorable for low emissions. Highest emission reductions are commonly observed for fuel formulations with lowest contents of sulfur and aromatics and for low high end boiling points.

Table 8.6 Overview of the effects of fuel components on diesel emissions

Fuel modification	HC		CO		NOx	Particulates	
	l	h	l	h		l	h
Reduction of Sulfur*	0	0	0	0	0	⇓ ⇓ᵇ	⇓ ⇓ᵇ
Increase in Cetane No	0	⇓ ⇓	0	⇓ ⇓	⇓	0	0
Reduction in Total Aromatics	0	0	0	0	⇓ᵃ	0	0
Reduction in Density	⇑ ⇑	⇑ ⇑	⇑	⇑	⇓	0	
Reduction in Polyaromatics	⇓	⇓	0	0	⇓ᵃ	0	⇓ ⇓
Reduction in T90/T95	↑	↑	↑	↑	↓	0	0

⇓ ⇓/⇑ ⇑ = large effect, ⇓/⇑ = small effect, ↓/↑ = very small effect, 0 = no effect. * = for engines without aftertreatment systems, l = low emission engine, h = high emission engine, a = poly-aromatics are expected to give a bigger reduction than mono-aromatics, b = reducing S from 0.30% to 0.05% gives relatively large benefits, reducing S from 0.05% to lower levels has smaller benefits [25]

8.5.2 Base Fuels and Engine Design

The choice of the base fuels used for blending has a significant effect, too. This is shown by Tables 8.7 and 8.8 for Light Duty Trucks (LD) and Heavy Duty Trucks (HD) engines, respectively. Although the effects are roughly similar for LD and HD engines, there are still engine design related differences regarding relative impact and type of base fuel.

Table 8.7 Effect of the choice of base fuel on emissions of LD engines

Refinery base streams LD vehicles	CO	HC	NOx	Particulate	Smoke
Light Cycle Oil	>25%	>25%	20–25%	>25%	>25%
Hydroconversion of Vacuum Distillation Residue	15–20%	15–20%	15–20%	20–25%	20–25%
Kerosene	20–25%	20–25%	15–20%	<15%	<15%
Fischer-Tropsch Diesel	<15%	<15%	15–20%	<15%	<15%

Percentages indicate contributions to specified emission components in %.

Table 8.8 Effect of the choice of base fuel on emissions of HD engines

Refinery base streams HD vehicles	CO	HC	NOx	Particulate	Smoke
Light Cycle Oil	20–25%	20–25%	20–25%	20–25%	20–25%
Hydroconversion of Vacuum Distillation Residue	20–25%	15–20%	20–25%	15–20%	>25%
Kerosene	20–25%	20–25%	15–20%	<15%	<15%
Hydro-Cracked Diesel	15–20%	<15%	15–20%	15–20%	15–20%
Fischer-Tropsch Diesel	15–20%	15–20%	<15%	<15%	<15%

Percentages indicate contributions to specified emission components in %.

The Light Cycle Oil and the Vacuum Distillation Residue Hydro Conversion base fuels are the largest contributors to pollutant emissions since they have the highest aromatic contents, a very low H/C ratio, a high density and a low CN.

8.5.3 Oxygenates

The addition of oxygenates is known to reduce soot emissions. Numerous compounds have been tested to meet important selection criteria, e.g. highest efficiency in soot reduction, high oxygen content, non-toxicity, acceptable CN and good solubility in conventional hydrocarbons. All in all, in reference [25], the best oxygenated compounds turned out to be Butyrate Glycerin and Butylal. Their addition improved the NOx-smoke trade-off in all fuel blends studied and had also beneficial effects on decreasing noise and fuel consumption. However, the required amounts of oxygenates had to exceed 10 vol. % to become effective, and hence, oxygenates can not be considered as additives but rather as substantial fuel components. Thus, in spite of being quite promising, in a thorough holistic evaluation, the additional benefits of adding an oxygenate were deemed insufficient to compensate for the increase in fuel costs associated with the dedicated production of oxygenated compounds for a broad introduction in future fuel formulations.

8.5.4 General Results of the Literature Survey

Many of the fuel characteristics are non-linear and therefore allow no simple interpretation. Only by considering all results in a holistic view was it possible to derive a matrix of 11 most promising fuel specifications.

By thoroughly re-testing these 11 fuels on a single cylinder engine under identical and well controlled conditions it was consistently verified that in summary, a Fischer–Tropsch type GTL diesel – purely paraffinic – was the most promising, and the WWFC #4 Diesel the next best future fuel specification. As a reference diesel, the Standard 2005 European Diesel fuel (pump grade) was used.

The fuel specifications are shown in Table 8.9 along with data for biomass-to-liquid (BTL) Diesel and US No. 2 Diesel. Characteristic emission results are shown in Figs. 8.7 and 8.8.

8.5.5 Results of Engine and Vehicle Tests

These general results were verified by five OEMs on modern vehicles [25] covering a wide range of LD engine designs and sizes and on an HD engine as exemplified by Figs. 8.9 and 8.10.

These test bed results were substantiated by real NEDC (New European Driving Cycle) tests as exemplified by the results presented in Fig. 8.11 showing that improved fuel formulations indeed reduced exhaust emissions very substantially for virtually all components.

It is especially noteworthy that the same general trends were observed for HD trucks and for LD passenger cars: the Synthetic Diesel (GTL) offers the largest reductions in emissions and the WWFC #4 Diesel is almost as good. The major differences between the fuels are attributed to the differences in sulfur content (ideally ≈ 0), aromatic content (ideally ≈ 0), CN, viscosity, density and final boiling point.

In addition to the fuel effects, the engine settings (e.g. SOI, rail pressure, Exhaust Gas Recirculation (EGR), timing of pilot and main injection etc.) as well as the engine design parameters (e.g. combustion chamber and injector design, compression ratio, swirl level etc.) have significant, mostly well known effects on exhaust emissions. The most prominent effects are reduction of the compression ratio (e.g. from 18:1 to 15:1), optimization the fuel injection system and of the in-cylinder charge motion (e.g. swirl ratio). Hence, these parameters have to be checked and optimized individually for each fuel formulation if the lowest emission levels are to be achieved, at best engine performance.

Table 8.9 Specifications of fuels treated in this chapter

Property	Units	EU EN590 2005	WWFC #4 Diesel	GTL	BTL prelim. data	US #D2 2007
Density @15	kg/l	0,8358	0,82–0,84	0,7647	755,8	0,8421
Cetane Number		53,9	≥55,0	74,8	83,5	42,9
Total Sulphur	mg/kg	6,8	≤10	<1	<1	11,9
IBP	°C	193		169	150,0	189
5%	°C	214		175		206
10%	°C	221		180	175,0	211
20%	°C	233		187	180,0	218
30%	°C	248		200	200,0	229
40%	°C	264		219	220,0	243
50%	°C	277		251	225,0	256
60%	°C	287		267	240,0	268
70%	°C	299		283	250,0	277
80%	°C	313	≤320	297	275,0	289
90%	°C	332	≤340	312	300,0	314
95%	°C	354	≤350	321		349
FBP	°C	360		329	319,0	360
Flash Point	°C	82	≥55	59		78
Viscosity @40 Kin	cSt	2,95	2,0–4,0	1,97	1,40	2,47
CFPP	°C	−17	≤l. a. Temp	−19		<−30
Cloud Point	°C	−14,3		−18		−21,9
Total Aromatics	%m/m	26,78	≤15	0,14	<0,1	26,33
Polycyclic aromatics	%m/m	1,33	≤2	0	0	6,86
Hydrogen Content	%m/m	13,2		14,98		12,98
H/C ratio	m/m	1/6,58		1/5,68		1/6,70
H/C ratio	mol/mol	1,81		2,1		1,78
Lower Heating Value	kJ/kg	43 033		43 836		42 630
HFRR Wear Scar Diameter	μm	394	≤400	370	337	558

US No. 2 diesel is shown for comparison

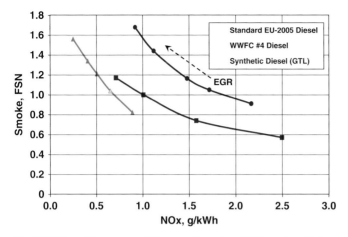

Fig. 8.7 Effect of fuel type on NOx – smoke trade-off. LD single cylinder engine, speed: 1500 rpm, BMEP 3 bar [25, IFP]

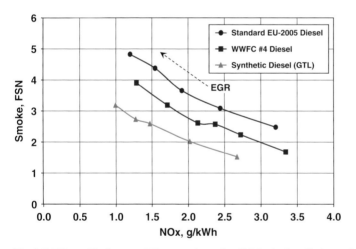

Fig. 8.8 Effect of fuel type on NOx – smoke trade-off. LD single cylinder engine, speed: 2500 rpm, BMEP 9 bar [25, IFP]

8.5.6 Results for a GTL-Optimized Vehicle

Results for a GTL co-optimized LD vehicle (see Table 8.10) were recently presented by DaimlerChrysler [11]. In the interest of minimizing adaptation costs, hardware modifications were limited to lowering the CR to 15:1 and optimizing the injection system only. In addition, the control software was adapted to cope with the different fuel properties also during transients. These results, obtained without any

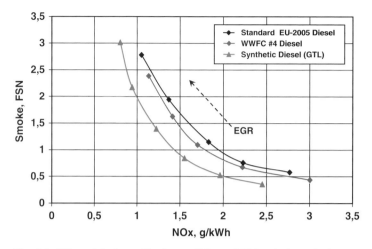

Fig. 8.9 Effect of fuel specification on NOx and PM emissions. Stationary test bench data of a modern 2.2 L Euro 4 LD engine with diesel particulate filter removed. Operating conditions: 2000 rpm, BMEP 7 bar (representative for the NEDC), variable: exhaust gas recirculation rate [25, DaimlerChrysler]

NOx after-treatment, corroborate the findings in [25] and support the need for co-optimizing future fuels and future engines in a system approach.

In conclusion, it has been demonstrated conclusively that high Cetane, purely paraffinic, sulfur-free GTL type, clean fuels improve combustion and reduce emissions drastically in modern Diesel engines without impairing engine performance (noise, fuel consumption and power). Hence, there is a very large, still unexploited

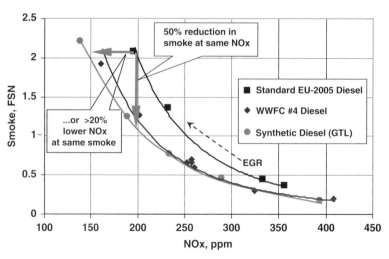

Fig. 8.10 Smoke (FSN) – NOx trade-off of a HD diesel engine. ESC test stage A100 (full load, 1200 rpm, SOI 3°BTDC) stationary test bench data. [25, Volvo]

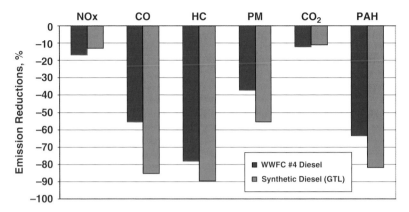

Fig. 8.11 Measured emission reductions for an un-adapted Euro 3 diesel vehicle during NEDC tests using WWFC #4 diesel and synthetic diesel (GTL). Reference is standard EU-2005 diesel fuel [25, Ford]

Table 8.10 Emission results of a GTL-optimized MB 320 CDI vehicle (originally EU4 specification) in comparison to the proposed EU5 limits. Even with only minor hardware/software adaptations and without a DeNOx catalyst, the emissions of the optimized vehicle are well below the EU5 limits

	NOx	CO	THC	PM*)
	g/km	g/km	g/km	g/km
EU5	0.08	0.5	0.05	0.005
GTL-opt. vehicle	0.075	0.2	0.041	0.0028

*) vehicle with DPF

potential for improvements in road fuels which will provide major reductions in pollutant emissions both in vehicles already on the road as well as in future dedicated co-optimized vehicles. The combination of the ideal fuel and optimized engines offers a promising near-term option. The necessary production technologies are available and proven.

8.6 Fundamental Considerations

This section is a discussion of the fundamental relationships between the fuel properties and the ignition and combustion characteristics of the fuels in engines. At the most fundamental level, composition controls all of the performance and emissions characteristics. In this sense, fuel specifications could be devised based entirely upon composition. There are, however, at least two significant problems with this approach. The first, which is very practical, is that current refineries are not designed to control composition to the level required to have purely composition based fuel specifications. The second problem is more fundamental, and that is that current kinetic models are not sophisticated enough and computers are not fast enough to handle all of the kinetic interactions that are possible in multi-component fuels. As a result, it is more convenient to consider some of the more global fuel properties, such as the autoignition temperature, global reaction rate (laminar burning velocity), and adiabatic flame temperature. These characteristics can be readily measured and correlated to engine performance and emissions.

8.6.1 Adiabatic Flame Temperature

The adiabatic flame temperature is directly related to the fuel composition. Flame temperature decreases as the fuel hydrogen content increases. Combustion in diesel engines is largely accomplished in diffusion flames that burn at the stoichiometric adiabatic flame temperature. It is the relationship between hydrogen content and flame temperature that explains the decrease in NOx emissions observed with higher hydrogen content fuels such as the GTL fuels. The relationships between hydrogen content and the flame temperature and NOx emissions are presented in Fig. 8.12. The range of hydrogen to carbon ratio covers diesel fuel through methane.

It can be seen that the adiabatic flame temperature and NOx emissions are directly related to hydrogen to carbon ratio. It is also important to recognize that soot emissions are also directly related to the fuel hydrogen content. This is demonstrated in Fig. 8.13, where soot formation rates in a well stirred reactor are plotted versus the fuel hydrogen carbon ratio.

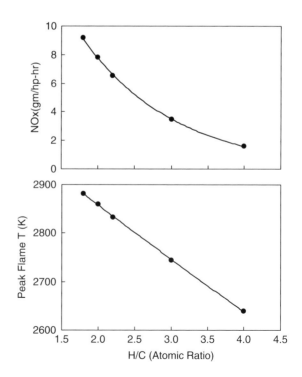

Fig. 8.12 NOx and adiabatic flame temperature versus hydrogen to carbon ratio (SwRI)

Fig. 8.13 Soot emissions versus the fuel hydrogen to carbon ratio

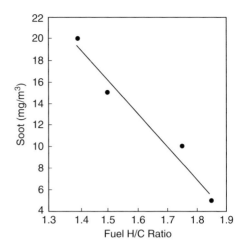

8.6.2 Autoignition Temperature

There are a number of formal definitions of autoignition temperature. Most methods, however, are performed at atmospheric pressure and are related to fuel handling safety related issues. Ryan et al. [1, 9, 31] developed a device for measuring the high pressure autoignition characteristics of fuels. The device consists of a pressure vessel that can be charged with air, or any oxidizer mixture, to high pressure and temperature. The test fuel is injected into this atmosphere and the ignition delay time is measured, based on the measurement of the pressure history in the vessel. In this context, the ignition delay time is defined as the residence time of the fuel in the vessel before pressure rise due to combustion.

The ignition delay time has been determined to be an exponential function of the oxidizer temperature, and thus it is possible to define the autoignition temperature as the temperature required to produce a selected ignition delay time. In this context, the measurement has been related to CN, ON, and ignition in Homogeneous Charge Compression Ignition (HCCI) engines (described below).

The relationship between ignition delay and CN is rather straightforward. The CN engine based rating technique (ASTM D613), as described above, involves rating the test fuel against reference fuel blends, all tested at fixed injection timing by varying compression ratio to achieve combustion at TDC. In this method, therefore, all fuels are tested at the same ignition delay time, but with different compression ratios, or with different temperatures. So that the engine tests method is based on a calibration of CN as a function of the required compression ratio. Newer rating techniques, ASTM D6890 or ASTM D 7170, are based on a direct measure of the ignition delay time in a constant volume combustion device, call the Ignition Quality Tester, or IQT.

Ryan et al. [32, 33] demonstrated that this device can be used to measure other, more detailed, ignition and combustion characteristics, useful in rating fuels for advanced combustion systems such as encountered in HCCI engines. In these

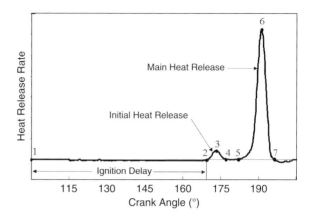

Fig. 8.14 Heat release rate diagram for HCCI reactions with cool flame

advanced combustion engines, more information is required than can be gleaned from a simple CN or ON. Fuel reactions in these engines are initiated by compression heating. There is no active ignition control, so controlling the start of reaction is a significant issue. Figure 8.14 is a heat release rate diagram for a typical HCCI engine. In this example, the fuel composition is such that there is significant low temperature reaction, as indicated by the two stage heat release event.

Research has demonstrated [34, 35] that fuel composition controls the low temperature reaction phenomenon, where normal paraffin's display significant low temperature reaction, and aromatic materials displaying little or no low temperature reactivity, and in fact the aromatics can suppress low temperature reactions [34, 35]. The other constituents, iso and cycle paraffins, olefins, etc. display some low temperature reactions, and generally fall, in terms of reactivity, in between the normal paraffins and the aromatic constituents.

Work reported by Ryan [30] indicates that there are two important ignition characteristics that control the start of the high temperature reactions in HCCI engines. The first is the temperature at which the low temperature reactions are initiated, and the second is the magnitude of the low temperature reactions, both annotated in Fig. 8.14 as the temperature at point 2 and the area under 2-3-4. Both combine to control the start of the main reaction. Figure 8.15 shows the correlation between the temperature at the start of the low temperature reactions in an HCCI engine and the ignition temperature as measured in the IQT [Elevated Pressure Autoignition Temperature (EPAIT)]. Figure 8.16 shows the correlation between the energy liberated in the cool flame and the time between the start of the low temperature reactions and start of the main reactions.

It is clear from the above data that the start of the high temperature reactions is controlled, at the practical level, by the initiation temperature and the magnitude of the low temperature reactions. It has been further demonstrated that these fuel ignition characteristics can be measured in the IQT, making these near ideal specification properties for HCCI engine fuels. Kalghatgi [22] also recognizes the shortcomings of the current ON and CN in rating fuels for HCCI engines. His approach

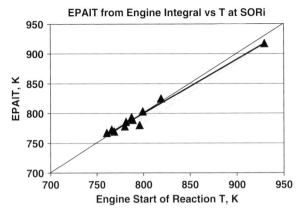

Fig. 8.15 Correlation between the temperatures at the start of the low temperature reactions in the engine and the EPAIT (SwRI)

involves the use of the difference in the RON and MON (sensitivity) as a measure of the HCCI ignition characteristics.

Another category of advanced engine combustion concepts involve accomplishment of the reaction process in the equivalence ratio – flame temperature domain in such a way as to avoid both NOx and soot formation. Kamimoto et al. [23] suggested the possibility for these Low Temperature Combustion modes, as illustrated in Fig. 8.17. A significant fuel related issue associated with this mode of combustion is related to the fact that engines calibrated to operate in these modes are very sensitive to the ignition quality of the fuel and to the propensity of the fuel to produce soot. Desirable fuel characteristics are low aromatic content and high CN, but at least consistent levels are essential. Surveys of fuel quality in the United States indicate that the gasoline properties are much more consistent than

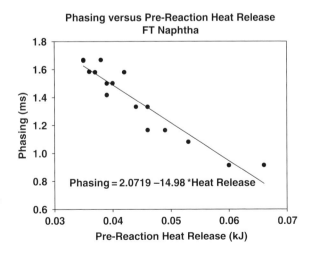

Fig. 8.16 Correlation between the magnitude of the low temperature reactions and the start of the main reaction [33]

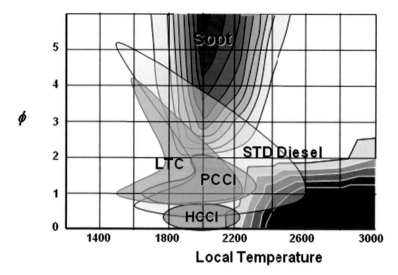

Fig. 8.17 Soot and NOx formation regions in equivalence ratio – temperature space

the diesel fuel properties. For instance, the 2005 diesel fuel survey [8] indicated that the CN varied from 38 to 58, and aromatic content varied from 13% to 50%. These variations are too large to compensate for in the control of the advanced combustion mode engines.

8.7 Alternative Feedstocks and Fuels

Since the availability of fossil oil and gas is finite, both the primary energy sources and the fuel types will have to be diversified over time [24]. Fossil based fuels will become cleaner and cleaner to comply with ever more stringent emission regulations and oil based fuels will gradually be substituted at least in part with natural gas based fuels such as CNG and preferably ultra clean liquid GTL fuels. In parallel, renewable fuels will enter the market. A vision of the temporal evolution of European road fuels, prepared by the fuels working group IG Fuels of ACEA–EUCAR [7], is presented in Fig. 8.18.

Although the fuel consumption of passenger cars will decline markedly due to the successful introduction of ever more fuel efficient technologies, the overall fuel consumption will increase because of an increase in transportation by trucks and also due to a continued increase in the total distance driven in the passenger sector. In order to meet the envisaged goals for securing energy supply and reducing CO_2 emissions, increasing amounts of gas based and renewable fuels will have to replace oil based fuels. Renewable fuels are of special interest since they contribute favorably to meeting both goals.

Petroleum exploration, recovery, transportation, refining, and fuel product distribution represent a very significant infrastructure investment, and one that must be

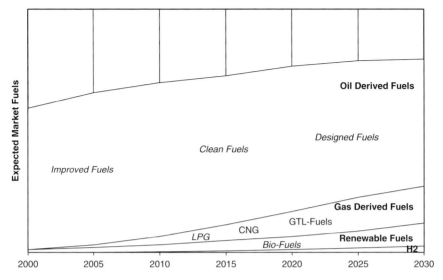

Fig. 8.18 Vision of the evolution of road fuels in Europe [7]

used as a bench mark for the alternative fuels. Basically, the world operates based on a petroleum economy. As such, the alternative fuels must be able to compete in this economic environment. The first element of the petroleum benchmark is the shear magnitude of the worldwide petroleum energy usage. The second element of the petroleum benchmark is the inclusion of petroleum and petroleum product usage in the production of the alternative fuels. In order for an alternative fuel to survive in the market it must be able to compete economically against petroleum.

The Argonne National Laboratory maintains a public domain spreadsheet based energy and greenhouse gas model that can be used to predict the wells-to-wheels analysis for any of the alternative energy pathways [21]. This model, called GREET, has been used to examine several different alternative feedstocks, processing, and transportation pathways to compare basically three different important factors: total energy balance, petroleum usage, and greenhouse gas contribution [26]. A similar detailed WTW database for European conditions and all pathways is available on the internet, produced jointly by ACEA-EUCAR, CONCAWE and EC [13]. In terms of total energy balance, the most efficient feedstock and pathway is the existing system using petroleum and petroleum refining. This pathway, while not the worst in terms of greenhouse gas contribution, does have a major impact. The worst pathways in terms of greenhouse gas impacts generally involve the use of coal and all of the other low hydrogen content fossil fuel feedstocks that do not include carbon sequestration. In addition, many of these pathways have relatively low total energy balances, especially if carbon capture requirements are included. The best pathways in terms of greenhouse gas impact include all of the nuclear and renewable sources as the primary energy source. The one important factor not included in the GREET model is the infrastructure cost associated with development of the respective pathways to a level needed to compete with petroleum.

Three alternative fuel pathways have achieved some success in the market place, these include the renewables, ethanol and biodiesel, and GTL. Ethanol and biodiesel can offer reasonable total energy balances (in some scenarios), and excellent greenhouse gas emissions benefit. GTL is of interest because it offers a significant financial return for energy companies as a means to monetize their stranded natural gas reserves.

In the EU biofuels are seen as strategic means to cope with the impending problems of energy security, economy and ecology in the transport sector. A target of 5.75% biofuels by energy in 2010 has been set by the EU Commission [2] and plans are under way for mandating 15% in 2015. About half will come from bioethanol. The bad energy and CO_2 balance and the high costs of bioethanol from cereals (1st generation biofuels) will be avoided by using ligno-cellulose as biomass source (i.e. 2nd generation biofuels) which, by the way, will be used also for producing BTL (an optimum Diesel fuel). Ligno-cellulose will come from waste and/or energy plants and will therefore not interfere with land use or crops for food production. In addition it provides high conversion efficiencies and CO_2 reduction factors.

In Germany, the pioneer of biofuels in the EU, a bill has been passed on January 1, 2007 [3], mandating in energy terms the use of at least 3.6% bioethanol in gasoline and 4.4% biogenous diesel in diesel fuel by 2010. The total biofuel market share must be 17% by 2015 under the constraint that the whole production chain provides a decarbonization of at least 10%. These figures will be reached only by resorting to 2nd generation biofuels which are promoted aggressively in Europe. The bill is enforced by fines for not complying with the minimum biofuels shares.

8.7.1 Renewables

Ethanol is of interest in the US market as a renewable source of oxygenate and Octane for gasoline. Ethanol from sugar cane has achieved some economic success in Brazil due largely to direct and indirect government subsidy. Currently US produced ethanol comes primarily from fermentation of corn starch. The energy balance for this approach has been greatly debated and it appears that it is only slightly positive. Under these conditions it is highly unlikely that ethanol will become a primary energy source. However, research at the National Renewable Energy Laboratory [26] on the production of ethanol from cellulastic material is very promising, and if proven successful, could make ethanol a primary energy resource.

Biodiesel has several different definitions, but the primary constituents are fatty acid methyl esters and petroleum diesel fuel in blended fuels. The differences in the definitions are in the concentration of the methyl ester. In the US and Europe the primary fatty acid of interest are the C18 fatty acids, consisting primarily of oleic and linoleic fatty acids (one and two double bonds), with traces of linolenic fatty acid (three double bonds). There are ASTM and ISO specifications for various biodiesel formulations, but it appears that the actual quality in the market varies greatly. The energy balance for biodiesel appears to be positive, but the key issue is the production capacity. It is estimated that 10% of the diesel demand can be met using the fatty acids before competition in the food market makes it impractical for use as a fuel.

Adding less than 10% biodiesel into diesel does not significantly alter the emission characteristics of a vehicle except that the energy content is lowered somewhat causing a lower mileage. No modifications of the engine are required. For higher biodiesel contents a slight tendency for decreasing CO, HC and Particulates but a tendency for increasing the NOx emissions are observed. Most significant, however, is the reduction of particulates which is most pronounced in older engines with calibrations EU3 and lower.

BTL diesel, produced using Fischer Tropsch processing of biomass generated producer gas, is a biofuel of the second generation with properties being superior over those of biofuels of the first generation. It can be blended with diesel in any ratio reducing emissions in proportion to the BTL fraction. Since BTL is not yet available in larger quantities [4], vehicle tests have been performed with GTL which has identical properties due to the same production processes. Only the feedstock source is different: biomass for BTL and natural gas for GTL. Due to the production process, BTL is inherently an ultra clean fuel, containing no sulfur and no aromatics. This and other favorable properties (high CN, high H/C ratio, etc.) make BTL an ideal fuel for reaching lowest emission levels. The GTL data presented previously demonstrate the potential of this biofuel fuel option which is considered to be preferable over all other alternatives.

8.7.2 Availability, Possible Market Shares and CO_2 Benefits of Biofuels

In order to be sustainable in the long term, the authors of ref. [11] use the subsequent assumptions for the assessment of the production potential of biofuels. These assumptions are believed to be very conservative and impose no undue constraints on a sustainable production. Also, there is enough room for other uses of biomass, and since the crops are grown specifically for biofuel production, no interference is expected with food production, environmental concerns or competition with power generation and home heating.

- Biofuels are produced solely from specially grown indigenous crops. Wood, waste and similar biomass (forestry and recycling) are not taken into account since it is assumed that they will be fully used in the power generation and heating sectors which compete with biofuel production about utilization of biomass resources.
- As a starting basis for crop land potentially available for biofuel production, only the land shares, actual crops and yields are considered which were recorded in

the agricultural sector in EU25 in the year 2005. Land shares are assumed to increase only slightly in the period to 2030. This in itself accounts for providing the right environmental, structural, climatic and local conditions necessary for a sustainable and efficient crop production.

- As crops for biofuel production of the 1st generation only sugar beet, cereals and rapeseed are considered. Since subsidies for sugar production are dwindling, a 20% use for Ethanol production is reasonable. It is also reasonable to assume that the lowest quality end of cereal crops amounts to 20% and is available without extra cropping efforts or negative effects on food production. These assumptions are certainly at the lower end of the biomass production potential and provide a safe margin for uncontrollable uncertainties inherent to any scenario. For the production of 2nd generation biofuels, the straw from cereal production is used as source for ligno-cellulose and requires therefore no extra cropping, and for BTL fuel production in EU25 it is assumed that a small fraction of the total crop land (up to 80% of the set aside land which is about 8% of the total crop land) will be available for growing suitable, widely different, fast growing energy crops to avoid the problem of monoculture and crop rotation. In addition, the land currently used for biodiesel production will be gradually transformed into use for energy crops.

- Since the various technologies for producing biofuels differ in their degree of maturity, different rates for increases in productivity are assumed for each crop. The same assumptions apply for the EU25 and Germany.

The summary results for the possible shares of biofuels in EU25 and Germany are presented in Figs. 8.19 and 8.20 along with the associated savings in CO_2 emissions. It is remarkable that SI biofuels will meet the political goals of 5.75%, 8% and 15% in EU25 and DE as well. Biodiesel, however, will hardly exceed the 5.75% share in Germany and will not reach this figure in EU25.

The larger targets for CI biofuels can only met by additional contributions from BTL, e.g. 2nd generation biofuel. BTL will not surpass biodiesel before 2020, due to the need for technology and infrastructure development. Since it uses the total bio-energy of the entire crop, BTL alone could provide over 16% of the German diesel demand in 2030 which is the largest contribution of all biofuel variants. Hence the high interest in quickly developing this synthetic biofuel with similar properties as the ultra clean GTL. To promote synthetic fuels, an alliance for synthetic fuels in Europe was formed in March 2006. Members are: Daimler-Chrysler, Renault, Volkswagen, SasolChevron and Shell [10].

It is important to note, that rather large quantities of biofuels >28% could be produced in DE by 2030 in spite of the very conservative assumptions if the 2nd generation biofuels are included. This indicates that biofuels constitute a realistic means for securing European energy supply, lessening markedly, the dependence on imported oil and gas. In addition, money spent on the production and use of these indigenous fuels will stay inside Europe and benefit the local population instead of benefiting regions elsewhere.

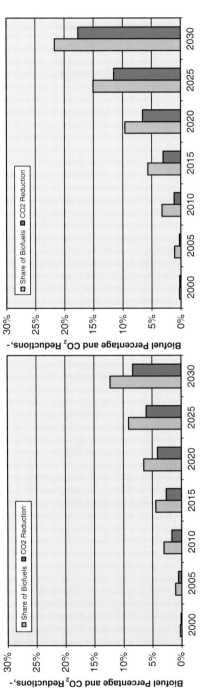

Fig. 8.19 Possible share of biofuels and related CO$_2$ reductions in EU25 [5,6]. *Left*: predominantly 1st generation biofuels. *Right*: predominantly 2nd generation biofuels

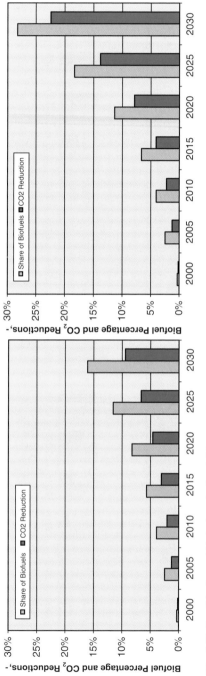

Fig. 8.20 Possible share of biofuels and related CO$_2$ reductions in DE [6, 28]. *Left:* predominantly 1st generation biofuels. *Right:* predominantly 2nd generation biofuels

8.7.3 Costs of Biofuels and Biofuel Blends

In order to provide consistent data for a cost assessment, available data for Germany are used under the assumption that these are relevant for EU25 as well. No taxes are accounted for to avoid erroneous or misleading results. Hence, production costs are all costs encountered, however without any taxes. As an example for how large possible cost changes can be, the crude oil prices for May 2005 and August 2006 and their effect on fuel production costs are shown in Table 8.11.

Undesirable as rising oil prices might be, higher oil prices will close the gap between production costs for fossil fuels and biofuels. Since biofuels most likely will not be used as neat fuels but rather in blends with fossil fuels, the effect of higher costs for biofuels shrinks further. As shown in Table 8.12, the rather high production

Table 8.11 Crude oil prices and fuel production costs [18]

		May 2004	May 2005	May 2006	May 2007	June 2008
Crude oil prices[1]	$/bl	40.24	49.84	70.92	63.40	136.74[3]
Production costs for Gasoline[2]	€/L	37.3	37.4	51.0	51.3	59.4
Production costs for Diesel[2]	€/L	35.2	43.1	51.0	49.2	75.5

[1] West Texas Intermediate (WTI), monthly average
[2] monthly averaged total net costs in Germany
[3] as of June 12, 2008

Table 8.12 Production costs of fossil fuels and biodiesel blends for different crude oil prices [6, 27]

DE Crude oil price	$/bl	50	60	70	80
Production costs of fossil fuels					
Gasoline without biofuels [1]	€/toe	538	623	707	792
Diesel without biofuels [1]	€/toe	514	597	680	762
Production costs of biofuels		2005	2010	2015	2020
Bioethanol ex Sugar [2]	€/toe	1025	1025	939	939
Bioethanol ex Cereals [2]	€/toe	939	939	854	854
Bioethanol ex Ligno Cell. [2]	€/toe	1.281	1.281	1.025	1.025
Biodiesel [2]	€/toe	811	811	811	811
BTL [2]	€/toe	1.281	1.281	683	683
Production costs of biofuel blends					
Gasoline without biofuels [1]	€/L	0,41	0,47	0,54	0,60
Diesel without biofuels [1]	€/L	0,44	0,51	0,58	0,65
2010					
Gasoline with 5.75% biofuels [1]	€/L	0,44	0,50	0,56	0,62
Diesel with 5.75% biofuels [1]	€/L	0,46	0,52	0,59	0,66
2015					
Gasoline with 8% biofuels [1]	€/L	0,44	0,50	0,56	0,61
Diesel with 8% biofuels [1]	€/L	0,45	0,52	0,59	0,65
2020					
Gasoline with 15% biofuels [1]	€/L	0,46	0,52	0,57	0,63
Diesel with 15% biofuels [1]	€/L	0,46	0,52	0,58	0,65

[1] costs calculated from WMV data for period Jan. 2004 to May 2006
[2] costs for 2005 and 2015 according to FNR 2006

costs for most biofuels of today will have an increasingly smaller effect on the costs of future fuel blends. Even for biofuel blends as high as 15%, the production costs are only 1–3 € cents above the production costs for neat fossil fuels if the oil price is at 80 $/bl.

8.7.4 Potentials for Co-Optimization

GTL, or *Gas-to-Liquids*, is a general reference to the use of Fischer–Tropsch catalyst to synthesize paraffinic hydrocarbons from producer gas. In the case of GTL, the raw material for the producer gas (mix of CO, hydrogen and water) is natural gas. There are very large known reserves of natural gas located in regions throughout the world where it is impractical to use the gas directly as a gas. GTL offers the opportunity to produce high quality liquid fuel materials that can be easily transported using conventional means. The economics for this process are highly dependent on the price of the natural gas. In cases where the gas has negative value, such as in some off-shore oil production facilities where the gas cannot be released or flared, GTL can compete at $25/bbl equivalent. For on shore remote gas, the competitive price increases depending on the contract price of the gas. GTL is a means for the energy companies to monetize their extensive remote gas resources and will, therefore, likely become a primary energy source for liquid distillate like fuels.

In Figs. 8.21–8.23 the effect of co-optimization is exemplified for a modern EU4 diesel car and a GTL fuel, representing the benefits of BTL as well. The effects of GTL diesel fuel on the exhaust emissions of a Mercedes Benz E220 CDI

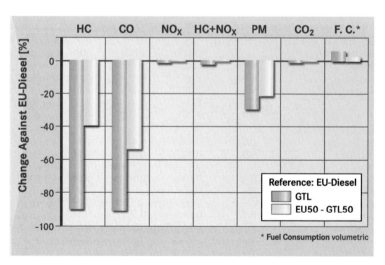

Fig. 8.21 Emission reduction potential of GTL diesel fuel in an unmodified MB E220 CDI diesel engine [11, 12]

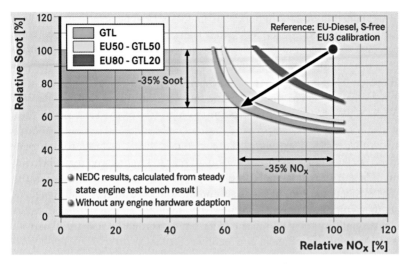

Fig. 8.22 Emission reduction potential of GTL diesel fuel in a MB E220 CDI diesel engine with software adaptation only [11]

passenger car are shown in Fig. 8.21. The tests were performed without any modifications to the engine [11, 12]. The results are of interest because the engine technology employed in this vehicle (high pressure, common rail fuel injection, cooled EGR, variable geometry turbocharger) can be considered to be representative of the majority of the current, as well as near- and medium-term future diesel passenger car fleet. Significant reductions in PM, HC and CO were found, when compared to the sulfur-free European diesel fuel that was used as the reference fuel, while NOx emissions were unchanged.

In Fig. 8.22 the results are shown for the same engine if only the engine software calibration is adapted appropriately to the fuel properties. They indicate that a

	NOx	CO	THC	HC + NOx	PM
	g/km	g/km	g/km	g/km	g/km
Euro 4 LD standard effective 01/2005	0.25	0.5	-	0.30	0.025
Euro 5 LD standard [1]) effective 09/2009	0.18	0.5	-	0.23	0.005
Euro 6 LD standard [1]) effective 09/2014	0.08	0.5	-	0.17	0.005
GTL-optimized vehicle (MB 320 CDI) [2,3])	0.075	0.2	0.041	0.12	0.0028

[1]) PM proposed to be changed to 0.003 g/km using the PMP measurement procedure
[2]) Vehicle with DPF but without DeNox aftertreatment
[3]) Base vehicle: MY 2005 MB 3.2 L CDI with Euro 4 calibration

Fig. 8.23 NEDC test results for the GTL optimized Mercedes-Benz E320 CDI vehicle [11]

simultaneous reduction of 35% in both NOx and PM could be achieved when neat GTL diesel fuel is used. Reference fuel is the clean sulfur-free European diesel fuel. Blends with GTL diesel fuel were also investigated, and it was found that 50:50 and 20:80 blends of GTL diesel and the reference fuel showed still impressive 85% and 45% of the potential of the neat GTL fuel, respectively.

In Fig. 8.23 the results are shown for hard and software changes. The compression ratio was reduced to take advantage of the high CN of the GTL diesel fuel, providing also a lower peak temperature during combustion and a more favorable volume to surface ratio. Thus NOx levels and heat losses are reduced. Nozzle modifications accounted for the lower density of the GTL diesel and for improving spray formation and air utilization. The software optimization addressed mainly the handling of transients since the steady state emission levels with GTL diesel fuel are so low that they contribute only insignificantly to the total emissions. These modifications cause only minor additional engine production costs, and are therefore considered to be fully cost-efficient and readily acceptable.

It can be seen that the very ambitious target of reducing NOx emissions below 0.08 g/km was reached, including a small safety margin, while all the other regulated emissions remained well below the proposed EU6 limits. This is an exceptional achievement in view of the total absence of any NOx aftertreatment. Due to the joint contributions by the DPF and the clean GTL diesel fuel, the PM emissions were well below the extremely low proposed EU6 limit. Since not all possible modifications were implemented, it is believed that there is still ample room for appreciable further improvements.

CO_2 emissions with the GTL diesel and the modified engine configuration were 4% lower than with the standard vehicle and EU Diesel. This translates into an increase in volumetric fuel consumption of 5%. This indicates a similar or even a slightly improved thermal efficiency of the modified GTL fuelled engine, when the 6% difference in volumetric energy density is taken into account. Therefore, the envisaged reduction in NOx has been achieved without compromising the engine efficiency in any way. Also, the rated power output was maintained at 165 kW. In summary, an excellent co-optimization result was achieved, providing a very promising outlook for the future when ultra clean fuels will be available in significant quantities.

Table 8.13 is comparison of the various fuels and energy sources in terms of physical state, energy content, and raw material

Table 8.13 Comparison of various energy sources

	Gasoline	Diesel	Biodiesel	CNG	E85
Structure	C2-C12	C10-C20	C16-C18	CH4	C2H6
CN	5–20	40–55	46–60	NA	NA
ON	86–94	8–15	~25	120+	100
Source	Petroleum	Petroleum	Soy Beans	Fossil	Corn
Energy Content (MJ/L)	30	35.6	32.5	10.5	22.2
Energy Ratio (vs Gasoline)	1.0	1.15	1.1	~3@26 MPa	~1.4
State	Liquid	Liquid	Liquid	Gas	Liquid

References

1. Allard LN et al. (1999) Analysis of the Behavior of the ASTM D613 Primary Reference Fuels and Full Boiling Range Diesel Fuels in the Ignition Quality Tester (IQT)-Part III. SAE Paper 1999-01-3591
2. An EU Strategy for Biofuels (2006) http://eur-lex.europa.eu/LexUriServ/site/en/com/2006/com2006_0034en01.pdf
3. Biokraftstoffquotengesetz – BioKraftQuG http://www.bgblportal.de/BGBL/bgbl1f/bgbl106s 3180.pdf
4. Choren Industries. http://www.choren.com/en/, http://www.shell.com/static/investor-en/downloads/presentations/2005/Goldman_Sachs_Future_Fuels_Jack_Jacometti.pdf
5. EU 25 Energy and Transport Outlook to 2030 – Part 4 http://ec.europa.eu/dgs/energy_transport/figures/trends_2030/5_chap4_en.pdf
6. FNR Biokraftstoffe – eine vergleichende Analyse (2006) http://www.fnr.de/pdf/literatur/pdf_236biokraftstoffvergleich2006.pdf
7. Fuels working group IG Fuels (2002) EUCAR/ACEA, Brussels
8. Fulton B (2006) New Feedstock's and Replacement Fuel Diesel Engine Challenge. DEER Conference, 2006
9. Gray AW and Ryan TW (1997) Homogeneous Charge Compression Ignition (HCCI) of Diesel fuel. SAE Paper 971676, May, 1997, SAE Trans.
10. GreenCar Congress (2006) Alliance for Synthetic Fuels in Europe http://www.greencar-congress.com/2006/03/alliance_for_sy.html
11. Herrmann H-O, Keppeler S, Frieß W, Botha JJ, Schaberg P, Schnell M (2006) The Potential of Synthetic Fuels to Meet Future Emission Regulations. 27th Vienna Motor Symposium, 2006
12. Herrmann H-O, Pelz N, Maly RR, Botha JJ, Schaberg PW, Schnell M (2004) Effect of GTL Diesel Fuels on Emissions and Engine Performance. 25th Motor Symposium, Vienna, 29–30 April 2004
13. http://ies.jrc.cec.eu.int/wtw.html?&no_cache=1&sword_list[]=wtw&sword_list[]=study
14. http://www.chem.tamu.edu/class/majors/chem470/Composition_and_Distillation.html
15. http://www.eia.doe.gov/kids/energyfacts/sources/non-renewable/oil.html#Howformed
16. http://www.eia.doe.gov/kids/energyfacts/sources/non-renewable/oil.html#Howused
17. http://www.eia.doe.gov/oiaf/aeo/excel/figure3_data.xls
18. http://www.mwv.de/
19. http://www.oica.net/htdocs/fuel%20quality/Final%20WWFC%204%20Sep%202006-2.pdf
20. http://www.personal.psu.edu/users/w/y/wyg100/fsc432/lecture%202.htm
21. http://www.transportation.anl.gov/software/GREET/
22. Kalghatgi G et al. (2003) A Method of Defining Ignition Quality of Fuels in HCCI Engines. SAE Paper 2003-01-1816
23. Kamimoto T, Bae M (1988) High Combustion Temperature for the Reduction of Particulate in Diesel Engines. SAE Paper 880423
24. Keppeler S, Degen W, Herrmann H-O (2007) Alternative Fuels for Internal Combustion Engines – Potential and Limitations. 6th International Colloquium Fuels, TAE Stuttgart/Ostfildern, Germany, January 10–11, 2007
25. Maly RR, Schaefer V, Hass H, Cahill GF, Rouveirolles P, Röj A, Wegener R, Montagne X, Di Pancrazio A, Kashdan J (2007) Optimum Diesel Fuel for Future Clean Diesel Engines. SAE 2007-01-35
26. McCormick R (2006) Liquid Fuels from Biomass. 2006 DEER Conference
27. MWV, 2006 MWV Foliensatz – Mineralölversorgung/Preisentwicklung/Besteuerung. http://www.mwv.de/Download/a-versorgung.pdf
28. MWV, MWV-Prognose 2020 für die Bundesrepublik Deutschland. http://www.mwv.de/Download/prognose.pdf
29. Obert EF (1970) Internal Combustion Engines, International Text Books, Scranton, PA 1970

30. Ryan TW III, Mehta D, Callahan TJ (2004) HCCI: Fuel and Engine Interaction. 2004 DEER Conference
31. Ryan TW, Callahan TJ (1996) Homogeneous Charge Compression Ignition (HCCI) of Diesel Fuel. SAE Paper 961160, May, 1996, SAE Trans.
32. Ryan TW, Callahan TJ, Mehta D (2004) HCCI in Variable Compression Ratio Engine-Effects of Engine Variables. SAE Paper 2004-01-1971
33. Ryan TW, Matheaus A (2003) Fuel Requirements for HCCI Engine Operation. SAE Paper 2003-01-1813
34. Shibata G et al. (2006) The Interaction Between Fuel Chemicals and HCCI Combustion Characteristics Under Heated Intake Air Conditions. SAE Paper 2006-01-0207
35. Shibata G et al. (2005) Correlation of Low Temperature Heat Release with Fuel Composition and HCCI Engine Combustion. SAE Paper 2005-01-0138

Printing: Krips bv, Meppel, The Netherlands
Binding: Stürtz, Würzburg, Germany